John Wood.

MODELLING THE FLOW AND SOLIDIFICATION OF METALS

This volume is dedicated to the memory of

PROFESSOR J.A. SHERCLIFF FRS

Modelling the Flow and Solidification of Metals

edited by

T.J. SMITH

GST Professional Services Ltd
Willingham, Cambridge, UK

Reprinted from Applied Scientific Research Volume 44, nos 1–2, 1987
additionally containing one review paper and an index

1987

MARTINUS NIJHOFF PUBLISHERS

A MEMBER OF THE KLUWER ACADEMIC PUBLISHERS GROUP
DORDRECHT - BOSTON - LANCASTER

Distributors

for the United States and Canada: Kluwer Academic Publishers, P.O. Box 358, Accord Station, Hingham, MA 02018-0358, USA
for the UK and Ireland: Kluwer Academic Publishers, MTP Press Limited, Falcon House, Queen Square, Lancaster LA1 1RN, UK
for all other countries: Kluwer Academic Publishers Group, Distribution Center, P.O. Box 322, 3300 AH Dordrecht, The Netherlands

Library of Congress Cataloging in Publication Data

Modelling the flow and solidification of metals.

Includes index.
1. Solidification. 2. Liquid metals. 3. Founding.
I. Smith, T. J.
TN690.M56 1987 671.2'4 87-11065

ISBN 90-247-3526-2

PRINTED IN THE NETHERLANDS

Preface

The origin of this book can be traced to a Workshop held at the University of Cambridge in December 1985 under the auspices of the Wolfson Group for Studies of Fluid Flow and Mixing in Industrial Processes. This Group was established at the University of Cambridge in January 1983 and includes members from the Departments of Applied Mathematics and Theoretical Physics, Engineering and Chemical Engineering. As its name suggests, the objective of the Group is to undertake, co-ordinate and stimulate research in various aspects of fluid flow and mixing in industrial processes. However, another equally important aim for the Group is to promote co-operation between the University and industry at all levels from collaborative research projects to joint colloquia. The Workshop in December 1985 on 'Mixing, Stirring and Solidification in Metallurgical Processes' which led to this book was one in an annual series of such meetings first held in December 1983.

The existence of the Wolfson Group is due to the enthusiasm of its original advocate, the late Professor J.A. Shercliff FRS, Head of the Department of Engineering who, together with Professor G.K. Batchelor FRS, Professor J.F. Davidson FRS, Dr J.C.R. Hunt, and Dr R.E. Britter, were responsible for the initial application to the Wolfson Foundation and for the subsequent direction of the Group's activities.

My own involvement with the Wolfson Group began in October 1983 when I was appointed as one of the three full-time staff. At that time, our objective was to further develop existing collaboration between the University and industry as well as expanding our research activities into new areas of current industrial importance. One such area was the flow and solidification of metals. At that time a number of projects on the electromagnetic stirring of metals were in progress in the Department of Engineering. Further work was instigated by Mr D.B. Welbourn, formerly Director of Industrial Co-operation at the University of Cambridge. Mr Welbourn's enthusiasm and encouragement led to a team being established in the Department of Chemical Engineering to study the interaction between flow and solidification as well as the application of such studies in computer-aided design for the metals industry.

Other phenomena currently being studied by the Wolfson Group include the dynamics of dispersed two-phase flows and the dynamics of mixing and reaction in turbulent flows. At present these studies are directed towards the problems of stirred tanks, bubble columns and flow reactors, but the results are also relevant to many metallurgical systems. Here, it must be acknowledged that many of these projects are undertaken in collaboration with particular companies or research establishments. These organisations, together with the Science and Engineering Research Council, provide a significant amount of funding for which the Group is most grateful.

Many of the papers which make up this book were presented at the Wolfson Group Workshop in December 1985. Others were presented at a meeting on 'Progress in Modelling of Solidification Processes' organised by the Institute of Metals. As the proceedings of neither meeting were published, a number of the authors were invited to contribute papers to this volume. I am grateful to all these authors for their effort and support. I am also grateful to the referees of the individual papers for their contribution to this volume.

My own direct involvement with the Wolfson Group came to an end in 1986 when I joined GST Professional Services Limited, although an association with the Group remains through a continuing affiliation with the Department of Chemical Engineering. Much of the administrative burden of actually producing this volume has been carried by my new employers. Therefore, I am extremely grateful to the company for their support, to my colleagues Dr A.B. Dunckley and Dr G.T. Armstrong for their advice and interest in the project and especially to Mrs S. Pateman, Miss P. Symonds and Miss A. Garner for their uncomplaining efforts with their word processors.

Finally, sincere thanks are due to the Wolfson Foundation without whose support this volume would not have been possible.

T.J. SMITH
Cambridge
February 1987

Contents

Applied Scientific Research 44: 1–7 (1987)

Modelling mushy regions

A.B. CROWLEY [1] & J.R. OCKENDON [2]
[1] Mathematics and Ballistics Group, Royal Military College of Science, Shrivenham, Swindon, Wilts, SN6 8LA, UK; [2] Mathematical Institute, 24–29 St Giles, Oxford, OX1 3LB, UK

Abstract. Two phase or mushy regions in which the phase boundary has a complicated morphology can exist in both pure and impure materials. In the former case, a macroscopic mathematical description can be given for the particular problem of melting by volumetric heating; no such procedure seems available for impure materials and this paper describes the reason for this. The principal mathematical difficulty is the analysis of the Stefan problem in the presence of a surface energy term.

1. The classical Stefan model

Several simple classical continuum models for solidification exhibit phase boundaries of irregular shape. The classical Stefan model of phase change, incorporating only heat conduction and latent heat release in the form of a 'free boundary' problem for a partial differential equation, is well posed at least for small times [16], and hence incapable of predicting irregular morphologies. However the introduction of even small amounts of supercooling, that is, allowing the liquid phase to have a temperature less than the melting temperature, or of small amounts of a diffusing impurity can cause a dramatic change in the mathematical structure of the solution. Examples of this are

(i) The one-dimensional 'one-phase' Stefan problem, defined as a problem in which one or other phase is precisely at the melting temperature T_M. The liquid phase is defined to be all material to one side of the phase boundary, rather than the material with temperature above T_M. Situations in which the liquid temperature $T_L < T_M$ with positive latent heat L are mathematically equivalent to those in which $T_L > T_M$ and $L < 0$. In this case it can be shown rigorously that if $T_L - T_M$ is sufficiently large and negative finite time blowup can occur, with a planar phase boundary moving at infinite speed at blowup [8,20]. The same remark applies to circular or spherical phase boundaries, even in the presence of a Gibbs-Thomson surface energy term.

(ii) The one dimensional Stefan problem with volumetric heating. This melting problem results in superheating in the solid phase, rather than supercooling in the liquid [2,21].

Paper presented at the meeting *Mixing, Stirring and Solidification in Metallurgical Processes*, University of Cambridge, 10–11 December 1985.

(iii) The two-dimensional one-phase Stefan problem with zero specific heat, which is also called the Hele-Shaw problem in the context of flow of a thin layer of viscous fluid between parallel rigid plates. Here supercooling is analogous to the suction of the fluid away from the cavity boundary (which is analogous to the phase boundary). Such suction, as opposed to blowing, produces cusp formation in the boundary in finite time. The problem is then genuinely ill-posed in the absence of a surface energy term, rather like the backward heat equation. The same probably happens with non-zero specific heat [10].

(iv) The Mullins-Sekerka instability [17] for a dilute binary alloy. This indicates unstable growth of fingers or dendrites in the free boundary when the liquid phase is constitutionally supercooled, but is only an approximate theory based on a linear analysis which neglects certain physical mechanisms [26].

In some cases it is reasonable to identify these breakdowns of simple models, which would hold in the presence of smooth phase boundaries, with the onset of a 'mushy region', where solid and liquid phases co-exist. Mushy regions thus cannot be described by classical Stefan models, and the aim of this paper is to outline some modifications which have been proposed in the mathematical and materials science literature, and to describe their mathematical status. The modifications will be presented in two sections; first modifications on a local scale, comparable to the scale of the free boundary irregularities, and then those on a large scale aimed at giving an averaged description of a mushy region. Clearly the former must be borne in mind when constructing the latter, which is the principal goal.

2. Small scale modifications to the classical Stefan problem

(i) The introduction of a Gibbs-Thomson surface energy term clearly improves the smoothness of the model. It introduces a higher derivative term in the free boundary condition, which in the general case of a binary alloy, means that the condition, usually written

$$T = T_M - m_L c_L$$

can be recast as

$$T + v = T_M + \kappa\sigma, \quad v = \begin{cases} m_S c & \text{solid} \\ m_L c & \text{liquid} \end{cases}$$

where T and the chemical activity v are continuous at the free boundary. κ denotes the boundary curvature taken as positive when the interface is concave towards the liquid, and σ is the positive surface energy. Except in the steady state [4], nothing mathematically rigorous is known about this model, but some very interesting conjectures can be made on the basis of numerical experiments [12,15]. For example [22] has conjectured that the surface energy splits the continuum of possible travelling wave fingers in a Hele-Shaw cell (Saffman-Taylor problem) into a discrete set. Also [11] has suggested that for a more general initial value problems, the boundary may break up into components, each small enough to be stabilised by surface energy.

Surface energy is clearly capable of dealing with (i), (iii) and (iv) above, although with small σ some wavelengths will still be unstable. However, the analysis even of 'travelling wave' solutions to try to model dendrites and fingers is very difficult.

(ii) The introduction of non-equilibrium effects at the free boundary in particular a kinetic undercooling term

$$T = T_M - \frac{1}{\mu} v_n,$$

where v_n is the normal velocity of the free boundary towards the liquid, is also stabilising. This term can be proved to avoid (i) in the case of one dimension [23]. However, μ is so large in typical dimensionless variables that the kinetic undercooling is only likely to be important in such rapid growth processes as pulsed laser annealing where $v_n \sim 10$ m/s.

(iii) Both the above phenomena can be regarded as limiting cases of the 'phase-field' model [3,4]. In this model which is derived from statistical mechanics arguments, the sharp phase boundary in the classical Stefan problem is replaced by a steep transition region in which the liquid fraction, f, satisfies

$$\tau \frac{\partial f}{\partial t} = \varepsilon \, \Delta f + \frac{\lambda}{2}\left(2f - 1 - (2f-1)^3\right) + T$$

$$\frac{\partial T}{\partial t} + L \frac{\partial f}{\partial t} = k \, \Delta T.$$

Here τ/λ and ε/λ are related to relaxation time and length scales respectively, k is the thermal conductivity, and we have abbreviated div grad to Δ.

In the limit as ε, $\tau \to 0$ and $\lambda \to \infty$ we can retrieve the general (and safest!) classical free boundary condition

$$T = T_M + \kappa\sigma - \frac{1}{\mu} v_n$$

for a pure material, where $\sigma \propto \varepsilon$, $\dfrac{1}{\mu} \propto \tau$.

3. Large scale modifications to the classical Stefan problem

It is not well known that some very simple devices are available to 'stabilise' ill-posed models on macroscopic scales, say those comparable to the length scale of mushy regions.

(i) Integration

It is a striking fact that the time derivative $(\partial c/\partial t)$ of the solution of the so-called oxygen diffusion/consumption problem [5]

$$\frac{\partial c}{\partial t} = \Delta c - 1, \quad c = \frac{\partial c}{\partial n} = 0 \text{ on free boundary}, \quad c \geqslant 0$$

is a function T satisfying

$$\frac{\partial T}{\partial t} = \Delta T, \quad T = 0, \quad \frac{\partial T}{\partial n} = -v_n \text{ on free boundary}, \quad T < 0$$

so if T is interpreted as a solid phase temperature, $L < 0$. However, the oxygen diffusion/consumption problem can be proved to have a unique solution for all time, at least in one space dimension! The reason for this paradox is that we insist $c > 0$ in the consumption problem; hence free boundaries constantly disappear and reappear. All the singularities in the Stefan problem are associated with regions of negative c nearing the free boundary. We thus have a very simple example of mathematical smoothing; the solution of the oxygen problem is the same as that of the Stefan problem while the latter is well behaved, but still gives us a mathematical interpretation when the Stefan solution breaks down. It has no obvious physical interpretation however.

(ii) Weak solutions

Restating the Stefan problem with volumetric heating Q

$$\frac{\partial T}{\partial t} = k\,\Delta T + Q,$$

where the melting temperature is taken as zero, in terms of the enthalpy h yields

$$\frac{\partial h}{\partial t} = k\,\Delta T + Q \text{ where } h(T) = T + LH(T)$$

and $H(u)$ is the Heaviside function such that

$$H(u) = \begin{cases} 0 & u < 0, \\ 1 & u > 0, \end{cases}$$

and whose inverse is zero for $0 < H < 1$. Here h must be interpreted as a distribution, or generalised function. Such an interpretation can be shown to be equivalent to 'smoothing' the enthalpy function $h(T)$ and then taking an appropriate limit. It may be shown that a unique solution of this formulation exists, and moreover, this is a practical proposition because there are numerical schemes which automatically converge to this solution [2]. Furthermore, in the case of instabilities caused by volumetric heating ((ii) of the introduction), it provides a model for a mushy region which is physically reasonable [13]. The idea is that an assumed distribution of a large number of nucleation sites enables an analysis to be made of the growth of small molten 'blobs' around these sites. Their existence is plausible if they are sufficiently small to be stabilised by the Gibbs-Thomson effect. An average over all these blobs can then be made to recover the weak formulation.

Attempts have been made to formulate the alloy solidification problem with a naive extension of these ideas, introducing an enthalpy which depends on both temperature and chemical potential [6,25]. The lever rule may be used to relate the enthalpy to the temperature and concentration in the metastable region between liquidus and solidus curves in the equilibrium diagram. Numerical experiments on this formulation suggest that it can predict mushy regions in certain cases where the mass diffusivity is small compared with the thermal diffusivity, corresponding to the constitutional supercooling criterion for the instability of a planar front. However, mathematically nothing is proven for this approach and in particular the idea of smoothing the enthalpy and concentration as functions of the temperature and activity does not seem to work. Nevertheless, numerical results for the one dimensional quenching problem which has an analytic solution are in good agreement.

(iii) Ad hoc averaging

In the case of alloy solidification observed morphologies can be exploited to make some progress. For example, in the case of columnar growth, the Scheil approximation can be adopted to describe the mushy region away from the tips of the dendrites [18,19]. This enables the mean liquid fraction f to be expressed in terms of the local temperature, which thus satisfies an equation of the form

$$\frac{\partial}{\partial t}(T + Lf) = k \, \Delta T.$$

A similar procedure may also be adopted for an equiaxed region [9]. The mathematical difficulties posed by the local analysis of the tip region are similar to, but worse than, those of the Hele-Shaw problem mentioned in the introduction. The proposal is an undercooling term in which $v_n \propto (T - T_E)^2$, where T_E is the equilibrium phase change temperature of the bulk melt, but this is not a non-equilibrium term in the sense of §2.(ii).

(iv) Other models

1. Irreversible thermodynamics [7,14]. Here the transport equations are written down with the fluxes defined in terms of the gradients of the inverse temperature and the chemical potential. For these models the existence of a mathematical solution may be proved. Lately classical thermodynamics arguments, assuming local equilibrium, have been applied to derive models including the Dufour and Soret effects [1].

2. Mathematical mush [24]. Suppose f again denotes the liquid fraction, and write the enthalpy as $h = T + Lf$ where $f = H(T)$ as defined above. This may be smoothed to give a rate equation which may be written as

$$\varepsilon \frac{\partial f}{\partial t} = T \quad \text{for } 0 < f < 1, \quad \text{or } f = 0 \text{ and } T > 0, \quad \text{or } f = 1 \quad \text{and } T < 0.$$

Unfortunately we cannot put a physical interpretation on ε.

It may be shown that this smoothed formulation of the problem has a unique solution, and this converges to the weak solution of the standard Stefan problem described in (ii) above. In this model a mushy region is identified with a region where $0 < f < 1$, and at phase change there will be a narrow mushy region of width $O(\varepsilon^{1/2})$. However, in problems including volumetric heating an extended mushy region is formed as in the standard weak formulation [2].

4. Conclusion

The modelling of mushy regions is, at least from the mathematical viewpoint, in a state of disarray, except for the case of volumetric heating of a pure material. Not enough is currently understood about the microstructure. The failing is at a mathematical level with problems such as that of dendrite tips, but this is itself caused by a lack of understanding of the surface energy effects.

References

1. V. Alexiades, D.G. Wilson and A.D. Solomon: Macroscopic global modelling of binary alloy solidification processes. *Quart. App. Maths.* 43 (1985) 143–158.
2. D.R. Atthey: A finite difference scheme for melting problems. *J. Inst. Maths. Applics.* 13 (1974) 353–366.
3. G. Caginalp: An analysis of a phase field model of a free boundary. *Arch. Rat. Mech. Anal.* (to appear).
4. G. Caginalp and P.C. Fife: Elliptic problems involving phase boundaries satisfying a curvature condition, U. of Arizona preprint (1985).
5. J. Crank and R.S. Gupta: A moving boundary problem arising from the diffusion of oxygen in absorbing tissue. *J. Inst. Maths. Applics.* 10 (1972) 19–33.
6. A.B. Crowley and J.R. Ockendon: On the numerical solution of an alloy solidification problem. *Int. J. Heat Mass Transfer* 22 (1979) 941–947.
7. J.D.P. Donnelly: A model for non-equilibrium thermodynamic processes involving phase changes. *J. Inst. Maths. Applics.* 24 (1979) 425–438.
8. A. Fasano, M. Primicerio and A.A. Lacey: New results on some classical parabolic free boundary problems. *Quart. Appl. Maths.* 38 (1980) 439–460.

9. S.C. Flood and J.D. Hunt: A model of a casting. *Applied Scientific Research* 44 (1987) 27–42.
10. S.D. Howison, J.R. Ockendon and A.A. Lacey: Singularity development in moving boundary problems. *Q. Jl. Mech. Appl. Math*. 38 (1985) 343–360.
11. S.D. Howison: Private communication.
12. J.D. Hunt and D.G. McCartney: Numerical finite difference model for steady state array growth. *Applied Scientific Research* 44 (1987) 9–26.
13. A.A. Lacey and A.B. Taylor: A mushy region in a Stefan problem, I.M.A. *J. Appl. Math*. 30 (1983) 303–314.
14. S. Luckhaus and A. Visintin: Phase transition in multicomponent systems. *Manuscripta Math*. 43 (1983) 261–288.
15. J.W. McLean and P.G. Saffman: The effect of surface tension on the shape of fingers in a Hele-Shaw cell. *J. Fluid Mech*. 102 (1981) 455–469.
16. A.M. Meirmanov: On a free boundary problem for a parabolic equation (in Russian). *Matem. Sb.* 115 (1981) 532–543.
17. W.W. Mullins and R.F. Sekerka: Stability of a planar interface during solidification of a binary alloy. *J. Appl. Phys*. 35 (1964) 444–451.
18. J.R. Ockendon and A.B. Taylor: A model for alloy solidification. In: Bossavit, A., Damlamian, A. and Fremond, M. (eds) *Free Boundary Problems: Applications and Theory*, Vol. 3, Pitman (1985) pp. 157–165.
19. E. Scheil: *Z. Metallkunde* 42 (1942) 70–72.
20. B. Sherman: A general one-phase Stefan problem. *Quart. Appl. Math*. 27 (1970) 427–439.
21. M. Ughi: A melting problem with a mushy region: qualitative properties. *I.M.A.J. Appl. Math.* 33 (1984) 135–152.
22. J.M. Vanden-Broeck: Fingers in a Hele-Shaw cell with surface tension. *Phys. Fluids* 26 (1983) 2033–2034.
23. A. Visintin: Stefan problem with a kinetic condition at the interface. Preprint of I.A.N. of C.N.R. Pavia (1985).
24. A. Visintin: Stefan problem with phase relaxation. *I.M.A.J. Appl. Math.* 34 (1985) 225–245.
25. R.E. White: The binary alloy problem: existence, uniqueness, and numerical approximation. *SIAM J. Numer. Anal.* 22 (1985) 205–244.
26. J.S. Langer: Instabilities and pattern formation in crystal growth. Rev. Mod. Phys. 52 (1980) 1–28.

Applied Scientific Research 44: 9–26 (1987)

Numerical finite difference model for steady state array growth

J.D. HUNT [1] & D.G. McCARTNEY [2]
[1] *Dept. of Metallurgy and Science of Materials, Oxford University, Oxford, UK.*
[2] *Dept. of Metallurgy and Materials Science, Liverpool University, Liverpool, UK*

Abstract. A control volume finite difference analysis has been used to calculate self-consistent cellular growth shapes. The results obtained are compared with the analytical models which have been proposed to describe array growth. It is concluded that the analytical models should be used with extreme caution in the cellular region. A preliminary examination of the stability of the steady state shapes has been made numerically. The results are compared with experiment and it is found that neither the minimum undercooling condition nor marginal stability analysis as applied, correctly predicts the growth conditions.

Nomenclature

List of symbols and values used in the calculations unless otherwise stated.

Symbol	Meaning	Value	Units
C_S	solid solute composition		wt%
C_L	liquid solute composition		wt%
C_∞	bulk alloy composition	0.48	wt%
D_i	thermal or solute diffusivity (D_L, D_S, α_L or α_S)		m^2s^{-1}
D_L	solute diffusivity in the liquid	3.5×10^{-9}	m^2s^{-1}
D_S	solute diffusivity in the solid	1×10^{-12}	m^2s^{-1}
e	the temperature error defined in equation (13)		K
e_j	the temperature error for interface point j		K
G	temperature gradient in bulk liquid	9000	Km^{-1}
k	solute distribution coefficient	0.18	
K_L	thermal conductivity in liquid	105	$Wm^{-1}K^{-1}$
K_S	thermal conductivity in solid	180	$Wm^{-1}K^{-1}$
L	Latent heat of fusion per unit volume	10.04×10^{8}	Jm^3
m	the slope of the liquidus	-5.5	K/wt%
n	distance normal to the interface		m
r	distance from cell centre (see fig. 2)		m
R_1, R_2	the principal radii of curvature		m
T_L	temperature in the liquid		K
T_0	liquidus temperature for alloy of composition C_∞		K
T_S	temperature in the solid		K
U_i	C_S, C_L, T_L or T_S		
V	steady state velocity in z direction		ms^{-1}
V_n	velocity normal to the interface		ms^{-1}
z	distance along axis of cell		m
α_L	thermal diffusivity in liquid	4.2×10^{-5}	m^2s^{-1}
α_S	thermal diffusivity in solid	7×10^{-5}	m^2s^{-1}
ΔT_I	tip undercooling		K
θ	curvature undercooling constant	2.8×10^{-7}	Km
δ	amplitude of perturbation		m
δe	amplitude of the temperature error		K
λ	cell spacing (see fig. 2)		m
λ_P	wavelength of the perturbation		m
μ	kinetic coefficient	1×10^{-1}	$ms^{-1}K^{-1}$

When alloys or mixtures of two components are grown directionally, as for instance by pulling a long rod-shaped specimen out of a furnace at a constant rate, arrays of cells or dendrites are produced. The cells or dendrites have spacings and tip undercoolings (e.g. ref. [1–4]) determined by the alloy composition and the imposed growth conditions (velocity and temperature gradient). For a very small range of growth conditions just after the breakdown of a planar front, two dimensional plate-like cells are sometimes found. Further from the breakdown condition (or sometimes immediately after breakdown) three dimensional finger-like cells are produced. These are smooth and are arranged in a hexagonal array. Still further from the planar front breakdown condition dendrites are formed. These have a branched tree-like structure which is not truly steady except perhaps at the tip.

Much of the theoretical work in the past few years has concentrated on isolated dendrites (e.g. ref. [5–13]). These are formed under different conditions for example, when solid nucleates and grows outwards from the centre of a bath of supercooled liquid. Under these conditions the temperature decreases infront of the dendrite tip, whereas during array growth the temperature usually increases infront of the tip and the heat is taken out through the solid.

There have been a number of approximate analytical models of array growth. Burden and Hunt [14,4] attempted to allow for the interaction of the diffusion fields between neighbouring cells using an approach suggested by Bower et al. [15], and Laxmanan [16] has extended this approach. Kurz and Fisher [17] assumed that the cell tip was elliptical and approximated the diffusion solution to that for an isolated dendrite. Trivedi [18] modified the isolated dendrite model to include a positive temperature gradient. The latter references [17] and [18] do not make an allowance for the overlapping of the diffusion fields from neighbouring cells.

The present authors believe that it is essential to consider the array to describe directional cellular or dendritic growth correctly. Clearly the array must be considered when attempts are made to predict cell or dendrite spacings.

Numerical work includes that of McCartney and Hunt [19], who describe a control volume finite difference method for calculating three dimensional cell shapes. These shapes are assumed to be growing under steady state and to have deep grooves going back to a eutectic temperature which is outside the domain being modelled. Ungar and Brown [20–22] and Ungar et al. [23] consider the early development of a two-dimensional cell-like structure from a planar front. McFadden and Coriell [24] calculate steady state two dimensional cell shapes and test them for stability. In this work and that of Ungar the composition (or velocity) is restricted so that the cell groove depth is similar in size to the cell spacing. The aim of the latter work [20–24] is to understand the initial breakdown of a planar front, while that of the former [19] is to model finger-like cells or dendrites which occur much further from the planar front breakdown condition. If C_∞ is the critical breakdown composition of a planar front McFadden and Coriell [24] consider conditions between $1.005C_\infty$ and $1.13C_\infty$ while McCartney and Hunt consider $1.39C_\infty$ to $13.6C_\infty$. As was found by McFadden and Coriell the groove depth increases rapidly with increasing composition (or velocity) above a certain depth. This is to be expected because once solute can only be rejected sideways the groove thickness will be determined by a modified Scheil type expression (this is the basis of theories of microsegregation). The depth of a deep groove is limited by a second phase being formed at the eutectic temperature or by a change in the distribution coefficient. It should be noted that the stability of two dimensional cells with very shallow grooves is unlikely to be similar to that for three dimensional cells with deep grooves.

In the present work, as in that described previously [19], the groove is assumed to go back to a eutectic temperature which is outside the domain being modelled. No attempt is made to follow the development of the cell from the planar front; instead steady state shapes are calculated and these must later be tested for stability. The present numerical model closely follows that described in reference [19]. The model has been extended to include solid state diffusion, a better method has been developed to arrange the grid points on the solid-liquid interface and a better technique is used to iterate towards self consistent solutions. The results obtained are compared with the approximate analytical models [4,14,17–18] to give an indication of their validity. An attempt will be made in later work to produce a more accurate analytical model.

In the past it has been suggested that the growth condition for cells and dendrites is determined by the minimum undercooling [4,14,16] and more recently by marginal stability of the cell or dendrite tip [11–13,17,18,25,26]. In the present work a preliminary examination of the stability of the tip has been made and the results compared with experiment.

The problem

Only a brief resume of the numerical technique will be described here as it has already been described in detail in reference [19]. A cell from an array is assumed to be a solid growing within a cylinder. The approximation involved is illustrated on Fig. 1. The initial part of the problem is to calculate the steady-state cell shape. Since the shape does not change with time, it is convenient to transform the axes to axes moving with the cell tip.

The thermal and solute diffusion equation with coordinates moving in the z direction (Fig. 2), at the steady state velocity V, must be solved for composition and temperature in both the solid and liquid phases. These equations are

$$\nabla^2 U_i + \frac{V}{D_i}\frac{\delta U_i}{\delta z} = 0 \tag{1}$$

where U_i is either composition C or temperature T, in the liquid or solid (C_L,

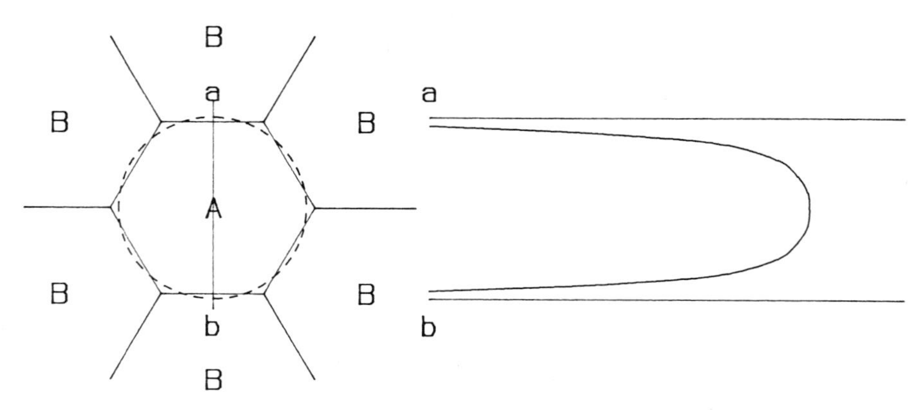

Fig. 1. Shows the approximations involved in assuming the cell has radial symmetry.

C_S, T_L or T_S see Nomenclature) and D_i is the relevant diffusivity. At the solid-liquid interface the usual interface conditions must be satisfied. The symbols used are listed in the symbol table (Nomenclature). The composition of the solid C_S, is related to that in the liquid C_L by

$$C_S = kC_L \tag{2}$$

and to conserve solute at the interface

$$V_n(1-k)C_L = D_S\frac{\delta C_S}{\delta n} - D_L\frac{\delta C_L}{\delta n}. \tag{3}$$

For heat

$$T_L = T_S \tag{4}$$

and

$$V_nL = K_S\frac{\delta T_S}{\delta n} - K_L\frac{\delta T_L}{\delta n}. \tag{5}$$

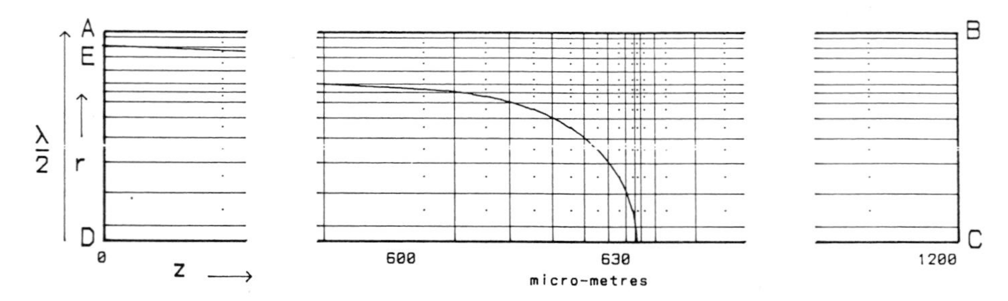

Fig. 2. A schematic view of a cell showing the mesh points, the box walls the definition of λ and the approximate dimension of the modelled region.

The temperature of the interface must be consistent with the composition curvature and kinetic mobility at each point. This may be written as an undercooling equation [27]

$$T_0 - T_I = \Delta T_I = m(C_\infty - C_L) + \theta(1/R_1 + 1/R_2) + V_n/\mu \tag{6}$$

where m is the liquidus slope, θ the curvature undercooling constant, R_1 and R_2 the principal radii of curvature and μ the kinetic coefficient.

The far field conditions are

$$C_L = C_\infty \tag{7}$$

$$T_L = T_{L\infty} \tag{8}$$

as $z \to \infty$ (in practice a gradient condition a small distance ahead of the cell was used instead of equation (8) see later).

A solution of the problem consists of solving equations (1), satisfying equations (7) and (8) for an interface shape such that equations (2)–(6) are simultaneously satisfied all along the solid liquid interface. Such a problem is difficult because the shape is not known 'a priori' but must be obtained from the analysis.

Method

Since radial symmetry has been assumed the three-dimensional problems effectively becomes two dimensional. A rectangular array of points is set up on the diametric plane r, z (Fig. 2). Box walls are erected equidistant between the points. An equation for each point is obtained by considering the heat or solute crossing four walls surrounding the point. The problem is a steady state problem so that the sum (either heat or solute) crossing the walls equals zero. For the walls perpendicular to r, only a diffusive flux is present. However, for the walls perpendicular to the z direction, a flux due to the moving coordinates must also be included. The equation of a general point for either heat or solute is

$$A_E\left(D_i\frac{\delta U_i}{\delta z} + VU_i\right)_E - A_W\left(D_i\frac{\delta U_i}{\delta z} + VU_i\right)_W + A_N\left(D_i\frac{\delta U_i}{\delta r}\right)_N - A_S\left(D_i\frac{\delta U_i}{\delta r}\right)_S = 0 \tag{9}$$

where the subscripts E, W, N, S refer to east, west, north, south of the box and A is the area of the relevant wall.

Because of the difficulty in handling curved phase boundaries with a uniform mesh, a non-uniform mesh was chosen such that mesh points lay on the solid-liquid interface (Fig. 2). Using this method the equations for the interface points can be obtained as before except that here two walls are solid and two liquid (Fig. 2). This satisfies equations (3) and (5) in a simple but accurate fashion.

In the previous work the composition and composition gradient (or temperature and temperature gradient) at the walls were assumed to be that given by a linear interpolation between adjacent points. This is a reasonably good approximation provided the box size is small compared with D_i/V where D_i is the relevant diffusivity. A better approximation for a moving box system is to use an exponential interpolation of the form

$$U_i = F + G \exp(-Vz/D_i) \tag{10}$$

where F and G are constants obtained from the composition or temperature of adjacent points. This interpolation was used in the present work so that solute diffusion in the solid could be included. The exponential interpolation has very little or not effect on heat flow and on diffusion of solute in the liquid.

The boundary conditions of the domain were treated much as before. The flux of heat or solute across the symmetry boundaries AB and CD (Fig. 2) are zero. Across BC the composition fluxes were obtained by assuming that the composition varies exponentially to a composition of C_∞ at infinity. This is the solution to equation (1) when there is no variation in the r direction (as must occur well ahead of the cell tip). A similar solution could be used for temperature. However, in practice during a directional growth experiment, a gradient is imposed near the cell tips rather than a temperature gradient was imposed along BC.

Provided the modelled groove depth is large enough compared with the spacing, the temperature and liquid compositions vary very little with r at the base of the groove. At $z=0$ (AD, Fig. 2) the temperature was assumed constant and initially put equal to zero. Later the temperature scale was adjusted to connect the temperature and composition fields.

The boundary condition for solute in the liquid at $z=0$ (AE, Fig. 2) was

$$\frac{\delta T}{\delta z} = m\frac{\delta C_L}{\delta z} \tag{11}$$

This is necessary to satisfy equation (6) when $\dfrac{\delta C_L}{\delta r}$ and $\dfrac{\delta T}{\delta r} \to 0$.

The boundary condition for solute in the solid at $z=0$ (ED, Fig. 2) was assumed to be

$$\frac{\delta C_S}{\delta z} = 0. \tag{12}$$

This is only strictly valid when $D_S \to 0$. However, in this work $D_S \simeq 10^{-3}D_L$ so that no large errors are to be expected from this approximation.

The method used to calculate the principal radii of curvature from the interface points is identical to that described in reference [19]. An analytical curve approximating the final shape was fitted to three adjacent interface points. One principal radius of curvature of the centre point could then be calculated and the slope at that point used to obtain the other principal radius of curvature [27,19].

In principle, it is possible to set up an equation for the composition and temperature at each mesh point and an additional interface equation (6), for each of the points on the solid-liquid interface. These could then be solved to give the temperature and composition at each point as well as the interface position. The equations, however, are non-linear and very large arrays and extensive computing would be needed.

In the present work and in that reported previously [19], advantage is taken of the fact that once the interface shape is fixed the equation for the composition and temperature at each mesh point are all linear. The procedure used to obtain a solution was thus to assume an interface shape. The temperature field was then simply calculated using Gaussian elimination. Knowing the temperature gradient in the groove, the composition could be calculated in a similar fashion.

This procedure satisfies all the equations with the exception of the undercooling equation (6). This equation must be simultaneously satisfied at each of the interface points by suitably modifying the cell shape.

The temperature scale was adjusted as mentioned previously so that there was no error in equation (6) for the interface point nearest $z = 0$. The temperature error at each of the other interface points could then be calculated. The temperature error e, is defined from equation (6) as

$$e = -m(C_\infty - C_L) - \theta(1/R_1 + 1/R_2) - V_n/\mu + T_0 - T_I. \tag{13}$$

In the previous work [19] an intuitive technique was used to adjust the interface shape to give a zero error at each point. The interface was moved in the z direction by an amount proportional to the error. After about one hundred iterations, errors of less than 10^{-6} K could be produced. As well as shifting the points, it is necessary to rearrange the points on the solid liquid interface after each iteration. This ensures that points do not become to close. Unfortunately, the procedure used in the earlier work led to shoulders of the cell being poorly defined by the interface points (Fig. 3). It was found that when the shoulders were well defined (Fig. 4), the iteration procedure did not converge and the undercooling errors slowly oscilated. This was because changing the position of one point effects the errors on each of the other points.

In the present work, the problem of eliminating the undercooling error was overcome by perturbing the position of the interface in a number of independent ways. In the simplest form, if there were p interface points, each interface point, other than the first, was shifted separately by a small amount in the z direction. The change on the undercooling error at that point and all the other points was obtained from the numerical calculation. This gave the terms

$$e_{ij} = \frac{\delta e_j}{\delta z_i} \tag{14}$$

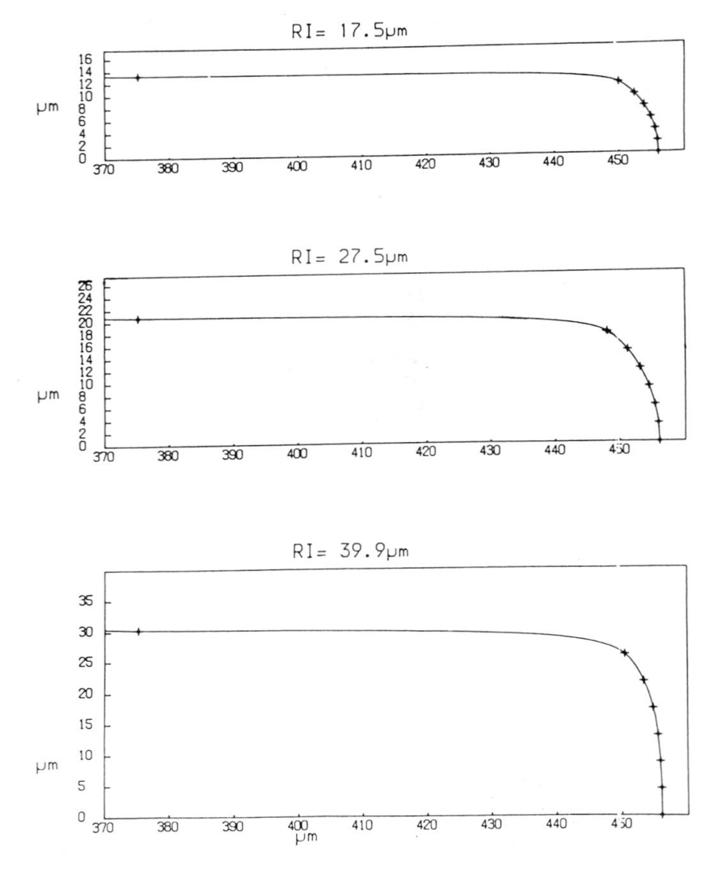

Fig. 3. Cellular shapes calculated using poorly defined shoulders [14] ($V = 8\ \mu\text{m/s}$).

(that is the change in the error of point j due to a shift δz in point i). These error gradients were then used to set up linear equations

$$\sum_{i=2}^{p} e_{ij}\delta z_i = -e_j \tag{15}$$

which could be solved to give all the $(p-1)$ δz, terms or a combined shift which should eliminate the error at all the interface points. In practice a fraction of the calculated shift was used to maintain stability. Under favourable circumstances, it was found that this procedure, gave errors of less than 10^{-6} K after only 3 or 4 iterations.

Typical results with well defined shoulders are shown in Fig. 4. The tip undercoolings at large spacings were different from those obtained in the earlier numerical work. This is because of the poorly defined shoulders of the cells which were used. It was found that provided sufficient points were

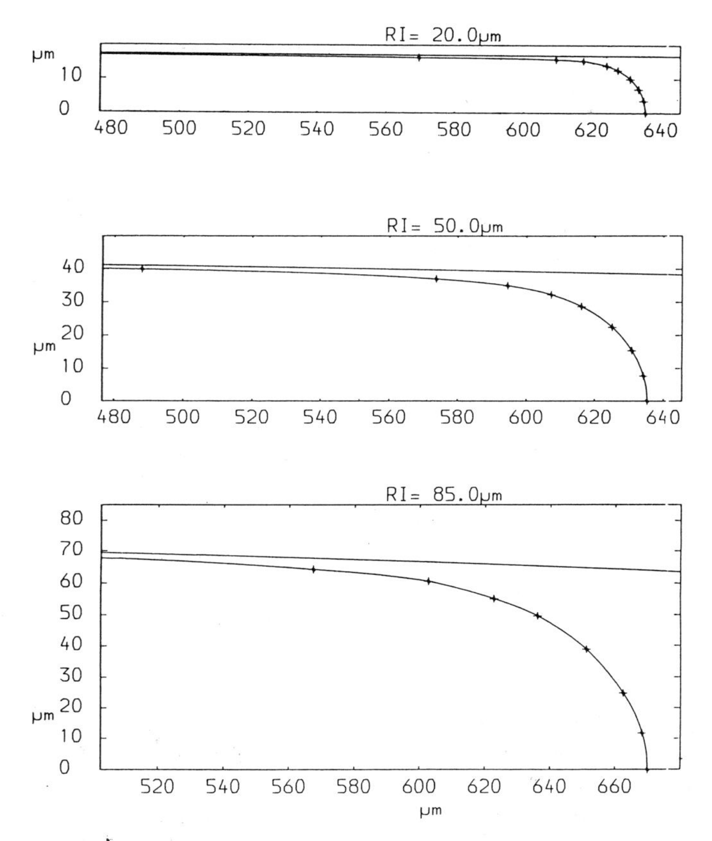

Fig. 4. Typical cell tip shapes; the modified Scheil shape is the top line ($V = 8\ \mu m/s$).

present on the shoulders, very little difference (typically 10^{-2} K) was obtained in the tip undercooling by rearranging the points in different ways.

For most of the work, twenty five columns and sixteen rows were used. A small number of calculations were carried out using double the number of rows and columns. The results were not significantly different.

The values of the constants used in the calculation are shown in Nomenclature. The alloy is based on an Al alloy investigated experimentally by McCartney [2,28]. The experimental results come from this work. The curvature undercooling constant was taken to be that measured by Gunduz and Hunt [31] in similar Al alloys.

Results

Interface shapes were calculated over a wide range of spacings, velocities, gradients and surface energies. Examples of the tip region are shown in Fig. 4.

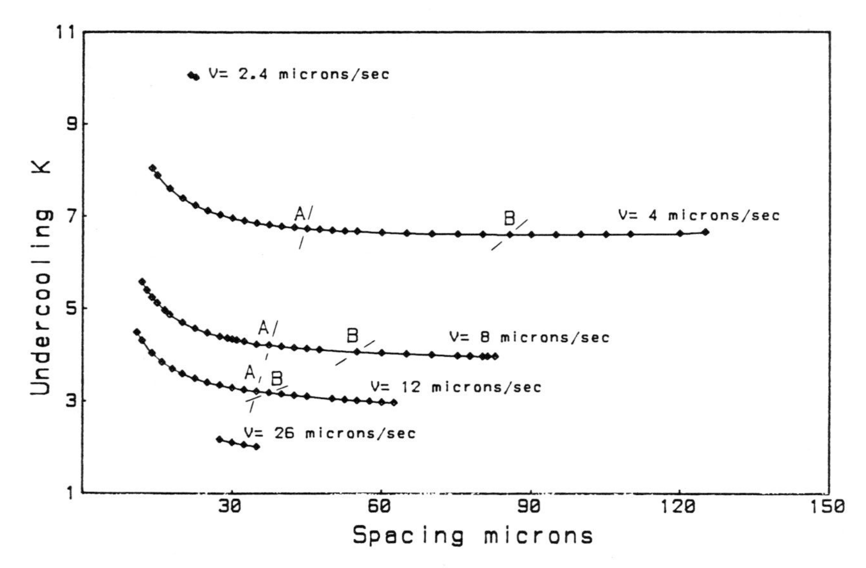

Fig. 5. Plots of tip undercooling against spacing for five different velocities. Points *A* represent the experimentally measured spacings [2,28] and points *B* those calculated using the present stability analysis (Figs. 15–17).

This figure also shows the Sheil shape modified to allow for diffusion in the z direction. As expected [4,14], the cell shapes approach the modified Scheil shape in the cell groove. The cell shapes are qualitatively in good agreement with those observed in practice, for example reference [28] and [32]. A plot of tip undercooling against spacing for five different velocities is shown in Fig. 5. The iterations procedure converged most easily at the centre of the spacings range and was suddenly more difficult towards the extremes. It is though that steady state shapes exist outside those obtained. The only limitation appears to be that as the spacing becomes wider the tips eventually became flat. Clearly this is a real limit. The tips were almost flat for the largest spacing of 4 μm/S.

The most direct comparison of the numerical work with the analytical models can be made by comparing the undercooling for a particular tip radius. The undercoolings were calculated using Burden and Hunt [14] equation 17 (ΔT_{BH}), Kurz and Fisher [17] equation 7 (ΔT_{KF}), and Trivedi [18] equation 33 (ΔT_T). The gradient in Burden and Hunt is taken to be 7.75 K/mm, as this is approximately the gradient in the liquid just behind the tip. The gradient in the Trivedi expression is taken to be $G=(K_LG_L+K_SG_S)/(K_L+K_S)$, see reference [18]. The results are tabulated in Table 1. It can be seen that Kurz and Fisher undercoolings are much too small and they have an incorrect velocity dependence. The Trivedi undercoolings are too small but have the correct velocity dependence. The Burden and Hunt results most closely follow the numerical results. The agreement however is a result of the correct

Table 1. The temperature gradient in Burden and Hunt [14] is taken to be $G = 7.75$ K/mm and in Trivedi [18] is taken to be $G = (K_L G_L + K_S G_S)/(K_L + K_S)$.

Numerical analysis				Analytical models		
Velocity (V)	Spacing (λ)	Radius (R_1)	Undercooling (ΔT)	Undercooling		
				ΔT_{EH}	ΔT_{KF}	ΔT_{T}
4 μm/s	125 μm	220 μm	6.65 K	7.42 K	0.304 K	2.93 K
4 μm/s	50 μm	40.4 μm	6.70 K	6.99 K	0.051 K	1.94 K
12 μm/s	50 μm	22.5 μm	3.06 K	2.4 K	0.086 K	0.90 K

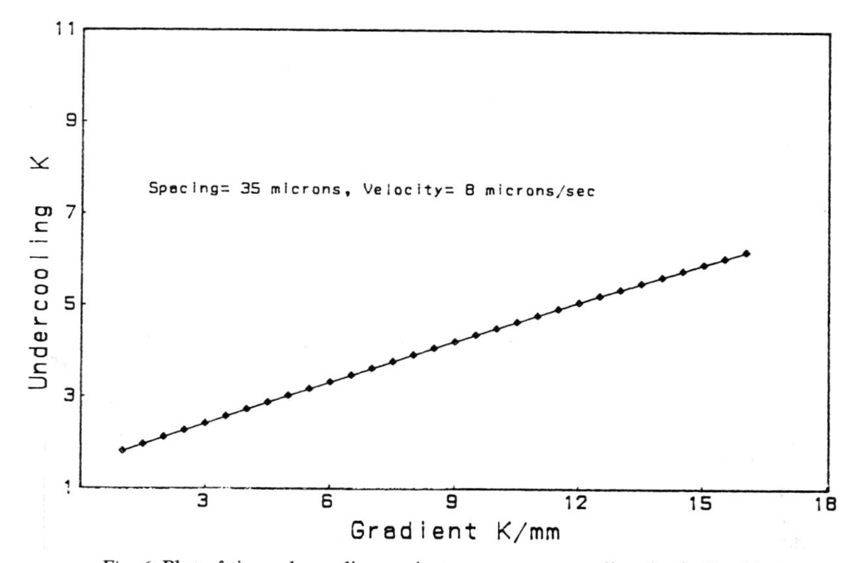

Fig. 6. Plot of tip undercooling against temperature gradient in the liquid.

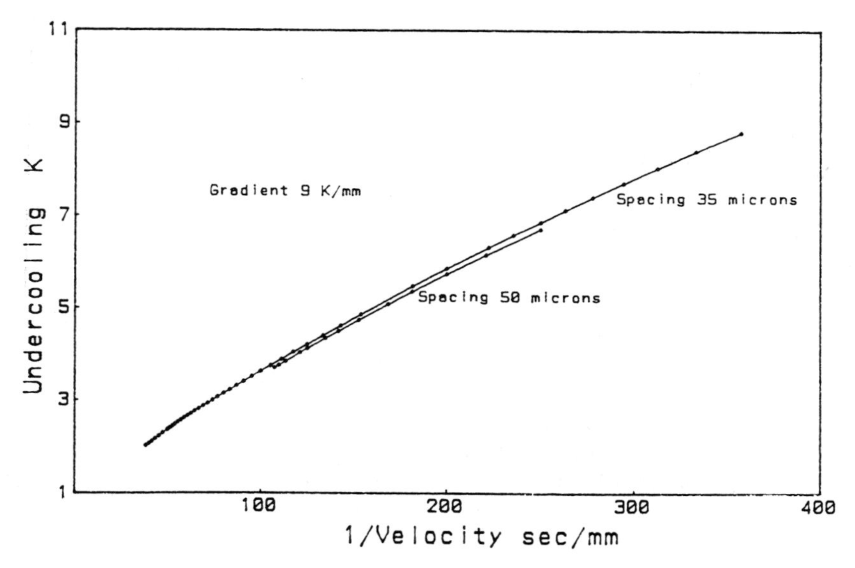

Fig. 7. Plot of tip undercooling against the inverse of the velocity for two different spacings.

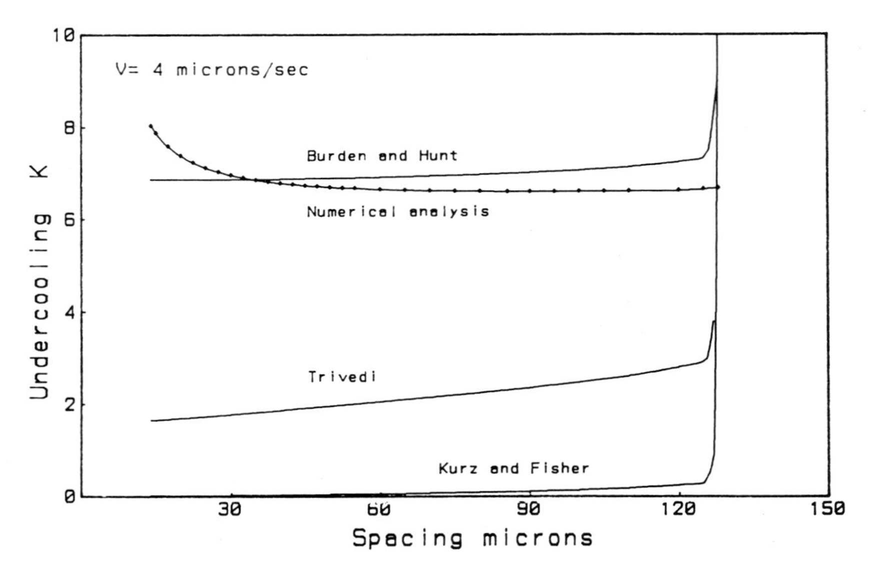

Fig. 8. Plot of tip undercooling against spacing ($V = 4\ \mu m/s$) from the numerical work and the predictions made by references [14], [17] and [18]. The tip radii used in the analytical expressions correspond to the radii found in the numerical work (note the radius $\rightarrow \infty$ as $\lambda \rightarrow 128\ \mu m$).

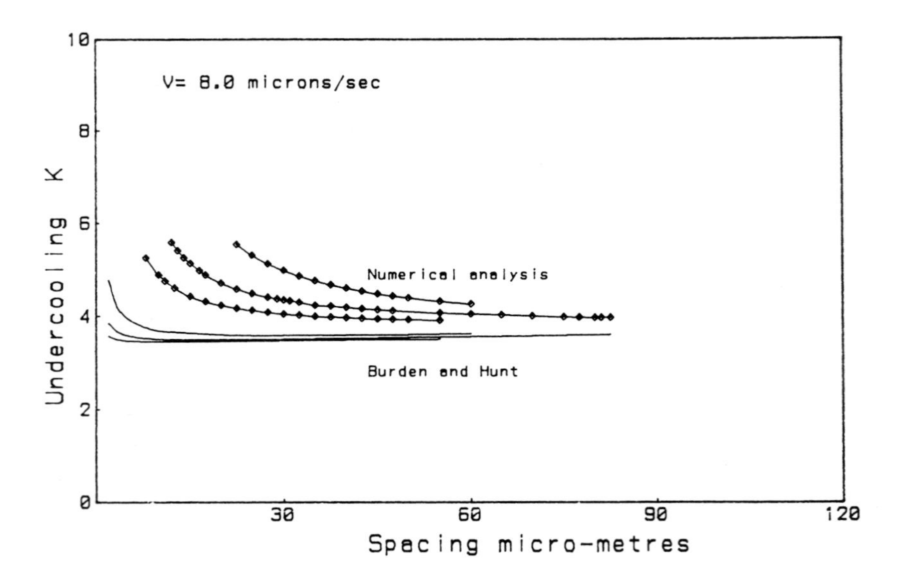

Fig. 9. Plot of tip undercooling against spacing ($V = 8\ \mu m/s$) for different Gibbs Thomson coefficients (from the top down 1.0, 0.28 and 0.1 K μm.) from the numerical work and Burden and Hunt [14] using tip radii calculated as in Fig. 8. (at small spacings the ratio is λ/R is assumed constant).

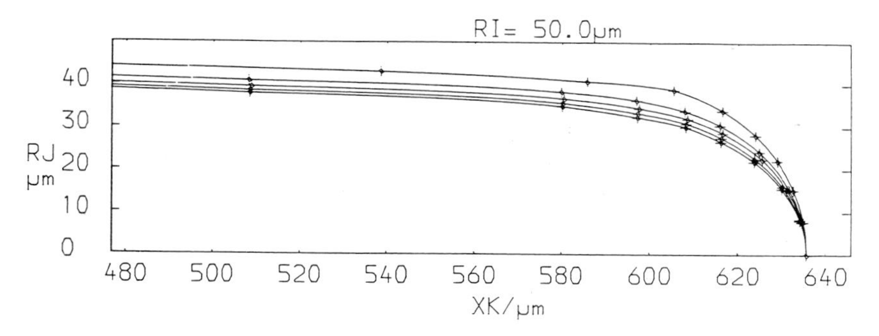

Fig. 10. Shows the effect of varying velocity from $4 \rightarrow 12$ μm on the shape of the cell tip. The tip radius of curvature decreases with increasing velocity.

prediction that the undercooling (at low velocities and high temperature gradients) depends minly on a GD_L/V term (equation 17, reference [14]). The numerical results showing a linear dependence on G and $1/V$ are shown in Fig. 6 and 7. The detailed correlation with change in spacing is not good. The Burden and Hunt undercooling increases too rapidly at large spacings and too slowly at small spacings (see Fig. 8). The latter is the result of the numerical undercoolings being much more sensitive to surface energy changes than predicted by the analytical work. Figure 9 compares the undercooling for three different surface energies.

Although McCartney did not measure the undercoolings in the alloy modelled in this work, the calculated undercoolings are consistent with those measured in similar alloys [1,3]. Figure 10 shows the variation in cell shape with change in velocity at a fixed spacing, the tips get sharper as the velocity increases.

The inclusion of solid state diffusion allows the microsegregation in a cell to be fully described. In all previous work simplifying assumptions have had to be made, particularly in the vicinity of the cell tip. Qualitatively the calculated compositions vary much as expected. Experimental work is to be carried out to check the numerical predictions.

The growth condition – stability

It is clear from Fig. 5 that solutions are present over a range of spacings for a fixed velocity. Experimentally [2,28] it is found that a unique spacing exists in practice. It has been suggested that cells or dendrites grow at the minimum undercooling [4,14,16], and more recently by marginal stability of the cell or dendrite tip [11–13,17,18,25,26]. The spacings found empirically by McCartney [2,28] are shown on Fig. 5. It is clear that these are near the minimum but in reality are at the point where the undercoolings begin to rise rapidly with decreasing spacing. The minimum is so shallow that it should not be said to define the spacing.

A fully time-dependent stability analysis is planned in future work. It has become apparent, however, that some preliminary stability information was already available in the present work.

In their classic stability analysis, Mullins and Sekerka [29] perturbed a flat interface with a sinusoidal perturbation ($\delta \sin \omega x$). They satisfy the steady state equations (1) and the undercooling equation (6). In equations (3) and (5) the velocity V is replaced by $V + \dot{\delta} \sin \omega x$. The term $\dot{\delta}$ is defined as the derivative of δ with respect to time. This term is, however, effectively a relaxation parameter because equations (1) are not time dependent and there is no truly steady state solution for an arbitrary sinusoidal shaped interface. Stability is then given when $\dot{\delta}/\delta < 0$. Clearly the approach is valid in the limit when both $\dot{\delta} \to 0$ and $\delta \to 0$. This is precisely the limiting condition which is of interest. Later work has shown the approach to be fully justified, for example reference [30].

The present author suggest that the other interface condition, the undercooling equation (6) in the form of equation (13), could have been relaxed. Instead of writing equation (13) as $e = 0$ it could be assumed that $e = \delta e \sin \omega r$. If this is done and V is kept constant the ratio $\delta e/\delta$ gives an identical numerator to that given for $\dot{\delta}/\delta$ in the Mullins and Sekerka [29] expression for a planar interface. It is the numerator which determines the sign and the most unstable wavelength of $\dot{\delta}/\delta$. Stability would then be given by $\delta e/\delta < 0$. There does not appear to be any fundamental reason that the continuity equations (3) and (5) should be relaxed in preference to the undercooling equation (6) although there is perhaps a good physical reason.

In the present numerical work the continuity equations cannot easily be relaxed. In fact, it is the undercooling equation which was relaxed to obtain self-consistent solutions. To obtain preliminary stability information it is therefore suggested that stability of a cell can be examined by perturbing a steady state shape and examining the sign of $\delta e/\delta$, the undercooling error.

Steady state shapes were perturbed by applying a perturbation to interface shape of the form

$$z = z(r)_0 + \delta \cos(2\pi r/\lambda_P) \tag{17}$$

where $z(r)_0$ is the steady state cell shape and λ_P is the wavelength of the perturbation. The amplitude δ was typically 0.1μ (for a spacing $\simeq 30\mu$). Because of the slope of the interface in the cell groove this form of perturbation effectively only perturbs the tip.

A problem arises in connecting the temperature field to the composition field. In the steady state analysis, the temperature of the first point is assumed to correspond to the temperature calculated from the composition and curvature of that point. The interface shape is then adjusted so that no error exists at all the other points. The choice of the zero error point is more difficult for the perturbed shape. It is arbitrarily suggested that the zero point should be at the base of the cell groove furthest away from the perturbed tip and that the temperature and composition fields should remain constant at this point. The

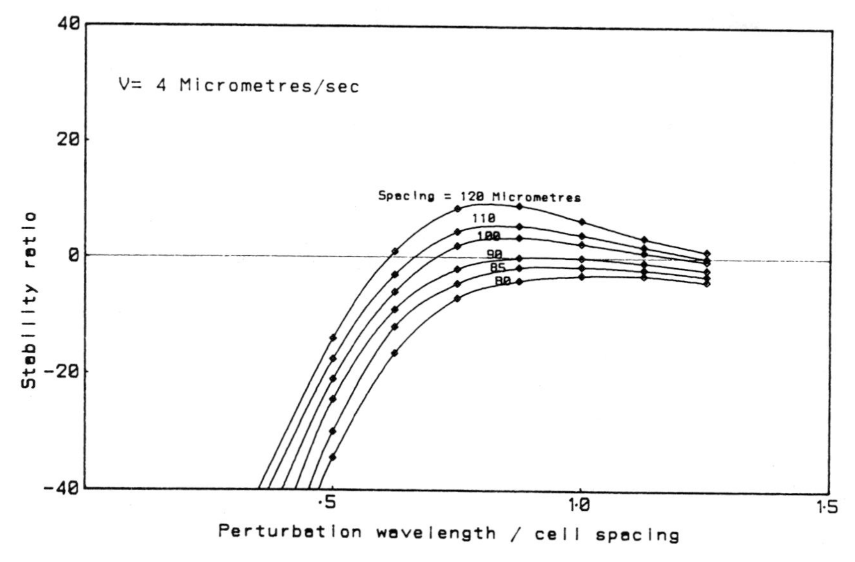

Fig. 11. Plots of the stability coefficient $\delta e/\delta$, against the normalised wavelength λ_P/λ, of the perturbation for six different spacings ($V = 4$ μm/s). The figure shows that the tip is stable ($\delta e/\delta < 0$ for all λ_P/λ) when the spacing is less than 90 μm.

composition can be made constant by replacing the gradient condition equations (11) and (12) with a composition condition (the steady state value). Alternatively the shape of the perturbation can be modified in the cell groove to give a zero error for the first few points. Both of these procedures gave tip stability curves such as those shown in Figs. 11, 12 and 13. These figures show plots of $\delta e/\delta$ against the perturbation wavelength (divided by the cell spacing) for different cell spacings. It can be seen that the results are much as would be expected. Below a particular spacing the cell tips are completely stable. It is interesting to note that the most unstable wavelength, (when just marginally stable), occurs when the ratio of wavelength to spacing is about one for each of the three velocities investigated. Initially the perturbation wavelength was divided by the tip radius since in the isolated dendrite stability analysis [13], it was predicted that the most unstable wavelength occurs at a wavelength equal to the tip radius of curvature. In the present work, however, the marginally stable ratio λ_P/R varied from about $3.7 \rightarrow 2.0$. Thus it appears in this preliminary work that it is the spacing rather than the tip radius which may be more important dimension in the array.

The limits of stability of a single cell tip are plotted on Fig. 5 and should be compared with the experimentally measured cell spacings. The marginal stability spacings vary much more rapidly with change in the velocity. Empirically the cellular structure begins to become dendritic at about 12 μm/s. Thus it appears that the marginal stability condition for a single cell, at least in this analysis, does not predict the cell spacing but may well show the breakdown from a cellular to a dendritic structure.

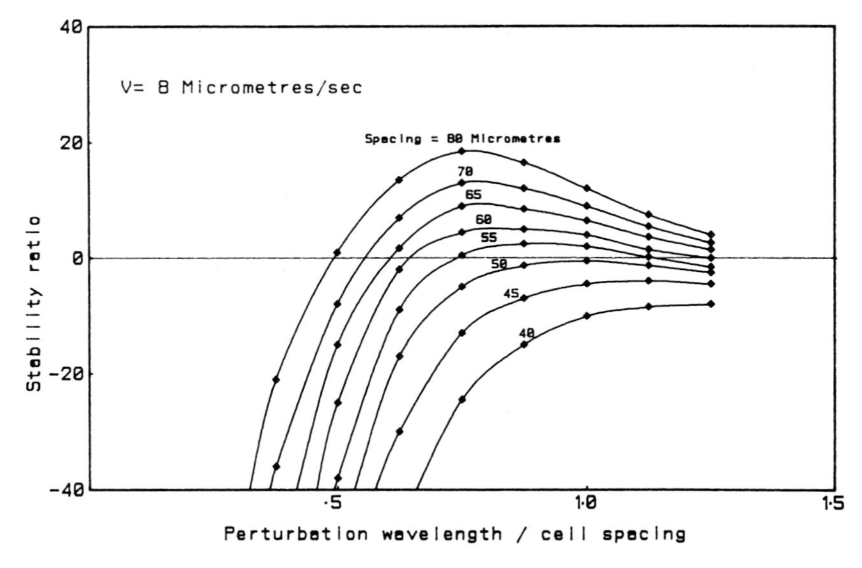

Fig. 12. Plots of the stability coefficient $\delta e/\delta$, against the normalised wavelength λ_P/λ, of the perturbation for eight different spacings ($V = 8$ μm/s). The figure shows that the tip is stable ($\delta e/\delta < 0$ for all λ_P/λ) when the spacing is less than 50 μm.

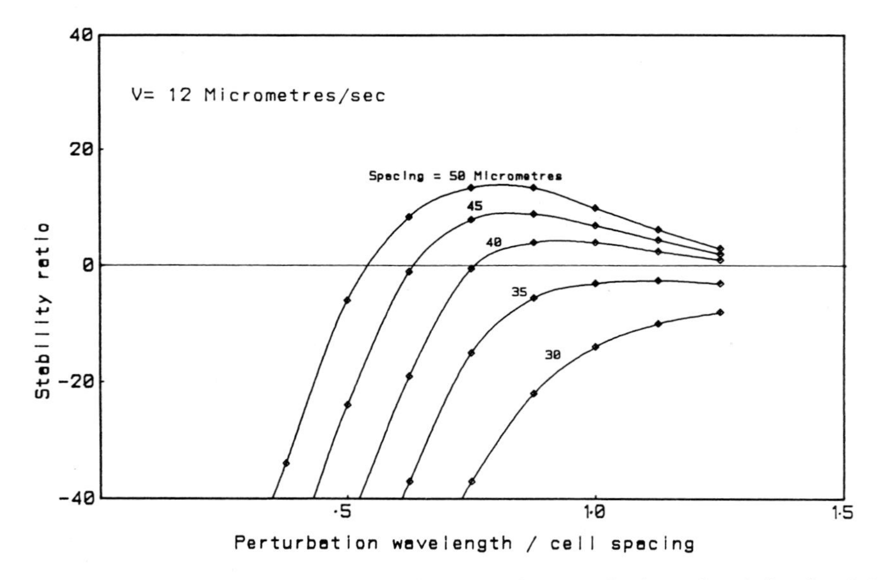

Fig. 13. Plots of the stability coefficient $\delta e/\delta$, against the normalised wavelength λ_P/λ, of the perturbation for five different spacings ($V = 12$ μm/s). The figure shows that the tip is stable ($\delta e/\delta < 0$ for all λ_P/λ) when the spacing is less than about 37 μm.

The marginal stability condition used by other authors [17,18,25,26] predicts much larger specings than found experimentally also the spacings do not agree with the values obtained in the present numerical work. A probable reason is that the earlier analyses really only consider an isolated cell or dendrite.

Discussion

Self-consistent cell shapes can be reliably calculated using the present numerical methods. The results obtained were qualitatively but not quantitatively in agreement with the Burden and Hunt model [14,4]. The agreement was more apparent than real and was mainly the result of the correct prediction of the GD_L/V term (equation 17, reference [14]). It was found that the surface energy had a much larger effect than anticipated. It is thought that the need to satisfy the undercooling equation all along the smooth cell surface leads to a much earlier truncation of the Sheil shape for small spacings or high surface energies. This result may mean that smooth cell shapes should not be used to describe dendritic shapes even approximately, as had originally been intended.

The undercooling results were not in agreement with the Kurz and Fisher [17] model probably because this model neglects the effect of the array. The good experimental agreement of this model at high temperature gradients occurs as a result of applying the marginal stability condition. The Trivedi [18] model showed the correct dependence on velocity, because the effect of the temperature gradient is incorporated, however, the undercoolings are too small. This is again probably the result of the neglect of neighbouring cells.

It appears that neither the minimum undercooling condition nor, in this preliminary work, marginal stability of a single cell tip correctly predicts the cell spacings. It should be pointed out that in the present stability work no transport of solute (or heat) was allowed between adjacent cells. By implication that means that each member of the array was perturbed in an identical way. Clearly this is too limiting a condition. What is necessary is for the macroscopic cellular interface to be perturbed in such a way that transport is allowed between the cells. It seems probable that this will give a small range of stable cell spacings. Practically, it is found that there is a range of stable spacings present on a specimen. In some areas the spacing is sufficiently narrow for a cell to be overgrown by the surrounding cells, while in other regions the spacing is decreased by a cell tip splitting mechanism. An attempt is being made to examine the stability of such a multiple cell system.

Conclusion

It is concluded that smooth steady state cellular shapes can be calculated using the present numerical method. Only one cell shape was found for a fixed spacing, velocity and gradient. Qualitatively, the calculated undercoolings vary much as anticipated [14,4] except that the undercooling depends more critically on surface energy at small spacings. It appears that the approximate diffusion solution used by [14,4,16–18] should be used with extreme caution in the cellular growth region. The inclusion of solid state diffusion means that microsegregation has been fully treated. A comparison with experimental work [2,28] indicates that neither the minimum undercooling nor the marginal stability, as applied in this work, correctly described the cellular growth condition.

References

1. M.H. Burden and J.D. Hunt: *J. Cryst. Growth* 22 (1974) 99.
2. D.G. McCartney and J.D. Hunt: *Acta Met.* 29 (1981) 1851.
3. M. Tassa and J.D. Hunt: *J. Cryst. Growth* 34 (1976) 38.
4. J.D. Hunt: Solidification and Casting of Metals, p. 3, Metals Soc., London (1979).
5. G.P. Ivantzov: *Dokl. Akad Nauk SSSR* 58 (1947) 567.
6. G. Horvay and J.W. Cahn: *Acta Met.* 9 (1961) 695.
7. M.E. Glicksman and R.J. Schaefer: *J. Cryst. Growth* 1 (1967) 297.
8. M.E. Glicksman and R.J. Schaefer: *J. Cryst. Growth* 2 (1968) 239.
9. R. Trivedi: *Acta Met.* 18 (1970) 287.
10. G.E. Nash and M.E. Glicksman: *Acta Met.* 22 (1974) 1283.
11. J.S. Langer and Muller-Krumbhaar: *Acta Meth.* 26 (1978) 1681.
12. J.S. Langer and Muller-Krumbhaar: *Acta Met.* 26 (1978) 1689.
13. J.S. Langer and Muller-Krumbhaar: *Acta met.* 26 (1978) 1697.
14. M.H. Burden and J.D. Hunt: *J. Cryst. Growth* 22 (1974) 109.
15. T.F. Bowers, H.D. Brody and M.C. Flemings: *Trans. Met. Soc. AIME* 236 (1966) 624.
16. V. Laxmanan: *Acta Met.* 33 (1985) 1023, 1037.
17. W. Kurz and D.J. Fisher: *Acta Met.* 29 (1981) 11.
18. R. Trivedi: *J. Cryst. Growth* 49 (1980) 219.
19. D.G. McCartney and J.D. Hunt: *Met. Trans.* 15A (1984) 983.
20. L.H. Ungar and R.A. Brown: *Phys. Rev.* B29 (1984) 1367.
21. L.H. Ungar and R.A. Brown: *Phys. Rev.* B30 (1984) 3993.
22. L.H. Ungar and R.A. Brown: *Phys. Rev.* B31 (1985) 5931.
23. L.H. Ungar, M.J. Bennet and R.A. Brown: *Phys. Rev.* B31 (1985) 5923.
24. G.B. McFadden and S.R. Coriell: *Physica* 12D (1984) 253.
25. S.-C. Huang and M.E. Glicksman: *Acta Met.* 29 (1981) 701.
26. R. Trivedi: *J. Cryst. Growth* 73 (1985) 289.
27. K.A. Jackson and J.D. Hunt: *Trans. Met. Soc. AIME.* 236 (1966) 246.
28. D.G. McCartney: D. Phil. Thesis, Oxford Univ. (1981).
29. W.W. Mullins and R.F. Sekerka: *J. Appl. Phys.* 34 (1963) 323.
30. R.F. Sekerka: *J. Cryst. Growth* 3–4 (1968) 71.
31. M. Gunduz and J.D. Hunt: *Acta Met.* 33 (1985) 1651.
32. J.D. Hunt and K.A. Jackson: *Acta Met.* 13 (1965) 1212.

Applied Scientific Research 44: 27–42 (1987)

A model of a casting

S.C. FLOOD * & J.D. HUNT
Dept. of Metallurgy and Science of Materials, University of Oxford, Parks Road, Oxford OX1 3PH, UK; (present address: Alcan International Ltd., Banbury Laboratories, Southam Road, Banbury, Oxon OX16 7SP, UK)*

Abstract. A model of a casting is presented which describes the freezing of a mushy zone, growing with a dynamically calculated undercooling at the dendrite tips. Equiaxed grains are introduced ahead of a columnar front. The model resembles a combination of the Stefan and mushy zone problems.

Nomenclature

Symbol	Meaning
c_0	composition
C	heat capacity per unit volume
C_1, C_2, C_3	parameters
D	solutal diffusion coefficient
g	weight fraction
G	temperature gradient
h	heat transfer coefficient
H	heat content or enthalpy per unit volume
I	nucleation rate per unit volume
k	distribution coefficient
k_i	thermal conductivity ($i = s$ or l)
K_i	thermal conductivity of a single phase ($i = s$ or l)
L	latent heat per unit volume evolved at the front
m	liquidus slope
N	number of grain per unit volume
N^0	number of substrate particles per unit volume
r	extended radius of a grain
t	time
T	temperature at time t
v	velocity
x	distance
α	width of control volume immediately ahead of the columnar front
β	width of control volume immediately behind the columnar front
δ	boundary layer thickness
δt	time discretisation interval
δx	spatial discretisation interval
ΔH	latent heat per unit volume
ΔT	undercooling
χ	pseudo heat transfer coefficient
ω	extended volume fraction of grains
Ω	volume fraction of grains

Subscript	Meaning
e	east face of the control volume
E	centre of the control volume to the east
ex	of the mould
L	liquidus
l	liquid
0	Centre of the control volume
s	solid
w	west face of the control volume
W	centre of the control volume to the west

Superscript	Meaning
B	bulk
I	columnar front
m	modified
p	pure
*	equiaxed
′	at time $t + \delta t$
−	average

1. Introduction

Solidification involves heat and fluid flow and, for an alloy, the transport of solute as well. The modelling of castings has concentrated on heat flow and has tended to introduce solute and fluid flow by simple, empirical treatments, if at all.

Two classes of heat flow models can be identified:

1. Stefan-type models, in which the latent heat is liberated discontinuously at a sharp phase front and where, therefore, a Stefan 'jump' boundary condition (see equation (5)) must be applied [1,2]; and
2. Mushy zone models, in which the latent heat is evolved continuously over a temperature range [3,4] and a Stefan 'jump' condition is not involved because there is not a discontinuity in solid fraction across the casting.

The former are good representations of the solidification of a pure metal, while the latter are used to describe alloy solidification.

Macroscopic solidification models have tended to neglect the undercooling at the phase front. Recently, however, Flood and Hunt [5,6] have produced a model which combines the characteristics of the Stefan and mushy zone models and which is capable of calculating a variation in undercooling of a semi-solid front. A Stefan 'jump' boundary condition is applied in the vicinity of the dendrite tips and the semi-solid is treated as in a mushy zone model. The undercooling at the columnar front varies as a result of the heat flow and generation in the casting and equiaxed grains can exist in the bulk. Clyne has previously introduced a constant undercooling at a semi-solid front [7] and also has modelled a planar front with a varying undercooling [8] but the work described here is the first to include a varying undercooling at a semi-solid front and equiaxed grains ahead of it.

2. The model

The authors wanted to investigate, through the application of a mathematical model, whether the columnar-equiaxed transition can be explained in terms of the growth of equiaxed grains in the bulk and their thermal interaction with the columnar front. This required a model which included a dynamically calculated undercooling at the columnar dendrite tips because this undercooling is crucial in determining the velocity of the columnar dendrites and the growth rate of the equiaxed grains.

The key to the model is the relationship linking the velocity of a dendrite tip to its undercooling (this is applied to both the columnar and equiaxed dendrites):

$$v = \frac{C_1}{c_0}(\Delta T)^2 \tag{1}$$

where C_1 is a function of material constants [9]. The relationship is supported by both experimental measurements and an approximate analytical treatment of solute diffusion at the dendrite tip [9,10]. Modifications can be introduced with high temperature gradients, and other, more suitable growth velocity undercooling expressions can be used similarly to achieve closure under rapid solidification conditions [8].

The Scheil equation is used to describe the shape of the dendrites in the columnar semi-solid (this is a common technique, see references [11,7,12,5]) and the liquid volume fraction within an equiaxed grain. It is a function of temperature alone if a linear liquidus is assumed:

$$g_l = \left(\frac{T - T_L^p}{T_L - T_L^p} \right)^{1/(k-1)} \tag{2}$$

The assumptions behind the Scheil equation are discussed elsewhere [12]. The equation can be modified to account for solid state diffusion of solute (this is often important for interstitial solute, e.g. carbon in iron) and forward diffusion of solute down the interdendrite composition gradient [13,14].

The columnar dendrites are truncated Scheil shapes (see Fig. 1). The temperature at which the truncation occurs is calculated by iterating on the Stefan 'jump' condition (see later). The truncated Scheil and true dendrite shapes will deviate near the tips but will match within a short distance behind them. The discrepancy between the real and truncated Scheil shapes is not important because it occurs over only a small distance compared to the scale of the casting and because the correct quantity of latent heat is liberated after a very small distance behind the dendrite tips.

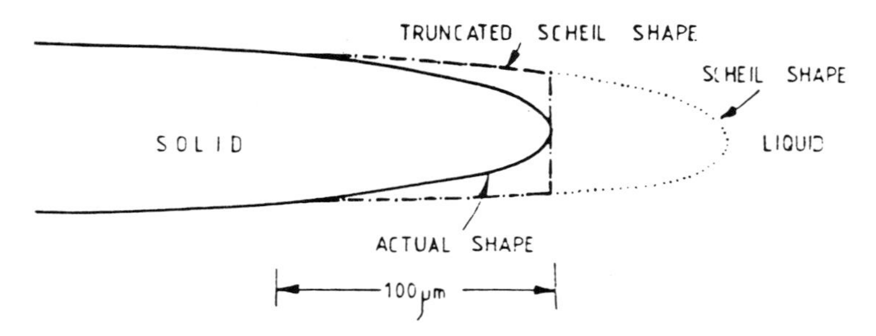

Fig. 1. The relationship between the complete Scheil, truncated Scheil and actual dendrite shapes.

The thermal fields of neighbouring grains are assumed to overlap because of the high thermal diffusivity of metals, and so the grains are isothermal and grow at the local bulk undercooling. A treatment has been developed to allow for a temperature difference between the grains and the bulk, but this is found to be negligible [6].

Impingement of the grains is accounted for by an Avrami-type treatment [15]: the concept of 'extended' volume, ω, as opposed to the actual volume, Ω, enables the kinetics of the growth of the equiaxed grains to be divorced from the geometrical complication of impingement. The increase in the actual volume fraction of grains, $d\Omega$, is related to the change in the 'extended' volume fraction, $d\omega$, by

$$d\Omega = (1 - \Omega)\, d\omega$$

Fluid flow is assumed not to affect heat flow in the semi-solid because of the high thermal diffusivity of metals and the small interdendritic spacings. In the bulk, convection is modelled by a temperature plateau ahead of a conducting boundary layer and quantities are described by bulk values. A small boundary layer corresponds to much convection in the bulk. In a recent version of the model, the size of the boundary layer is a function of the volume fraction equiaxed in the bulk.

In the first instance, nucleation of the grains was ignored. The influence of just the growth of the grains on the columnar-equiaxed transition was considered; grains were assumed always to be present in the bulk and they started to grow as soon as the local temperature fell below the liquidus. Later, calculations were performed with a temperature dependent nucleation rate [16,6]. Assuming heterogeneous nucleation in which substrate particles are consumed by the growing equiaxed grains, and that each substrate only nucleates one grain, the nucleation rate is given by:

$$I = N^0 \frac{(N^0 - N)}{N^0} (1 - \Omega) C_2 \exp\left(-C_3/(\Delta T^*)^2\right)$$

and typical values for C_2 and C_3 are $10^{29}\ \mathrm{s}^{-1}\ \mathrm{m}^{-3}$ and $46K^2$ respectively.

On the basis of an argument which considers the probability of a columnar dendrite being obstructed by an equiaxed grain in the bulk, the columnar-equiaxed transition is assumed to occur when $\Omega = 0.49$ immediately ahead of the front [6,17].

3. Mathematical formulation

The model can be applied to one, two or three dimensions. As an example, consider the one dimensional case (see Fig. 2). Solidification is modelled by modifying the Fourier heat conduction equation:

$$C^m\left(\frac{\partial T}{\partial t}\right) = \frac{\partial}{\partial x}\left(k_i\left(\frac{\partial T}{\partial x}\right)\right) + (1 - g_l)\frac{\partial \Omega}{\partial t}\Delta H \tag{3a}$$

where C^m is the modified heat capacity:

$$C^m = C + \Omega \Delta H \frac{\mathrm{d} g_l}{\mathrm{d} T} \tag{4}$$

with g_l defined by the Scheil equation (2). $\Omega = 1$ in the columnar semi-solid and is the local volume fraction of equiaxed crystals in the bulk. $\partial\Omega/\partial t = 0$ in the columnar zone. ΔH is the total latent heat per unit volume of liquid.

In a simple treatment of convection in the bulk, the diffusive term in (3a) is replaced by one involving a pseudo-heat transfer coefficient, χ, which is a function of the boundary layer width, δ:

$$C^m\left(\frac{\partial T}{\partial t}\right) = \chi\left(T_l^B - T^I\right) + (1 - g_l)\frac{\partial \Omega}{\partial t}\Delta H \tag{3b}$$

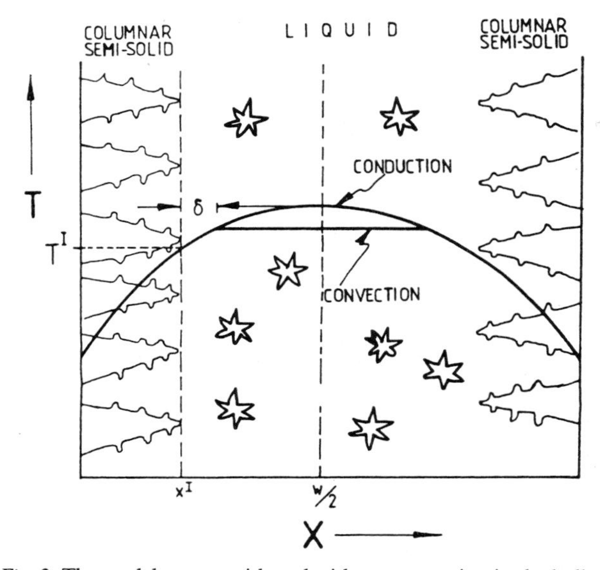

Fig. 2. The model system with and without convection in the bulk.

where T^I is the temperature of the columnar front and

$$\chi = \frac{k_l}{\delta}$$

The value to be used for δ is uncertain. The authors have used values ranging from 200 microns to the width of the casting; the true value must lie somewhere in this range. In more thorough treatments of convection, the heat transport is modelled within the boundary layer as well [6].

The boundary conditions are:

$$\frac{\partial T}{\partial x} = h(T - T_{ex}) \text{ at the edge}$$

where h is the heat transfer coefficient representing the resistance between the casting and the mould wall and T_{ex} is the mould temperature;

$$k_i \frac{\partial T}{\partial x} = 0 \text{ at the centre of the casting}$$

and the Stefan 'jump' condition:

$$\left.\begin{aligned} vL &= k_s^I G_s^I - k_l^I G_l^I && \text{with conduction in the bulk} \\ vL &= k_s^I G_s^I - \chi(T_l^B - T_l^I) && \text{with convection in the bulk} \end{aligned}\right\} \quad (5)$$

is applied at the columnar front, with v given by (1) and L by

$$L = (1 - g_l^I)\Delta H. \quad (6)$$

In the bulk:

$$\frac{\partial \Omega}{\partial t} = (1 - \Omega) N 4\pi \bar{r}^2 v^* = (1 - \Omega) N 4\pi \bar{r}^2 \frac{C_1}{c_0} (\Delta T^*)^2$$

where N is the number of grains per unit volume, $\bar{r}$ is the local average extended grain radius and ΔT^* is the local undercooling.

The thermal conductivity varies as:

$$k_i = K_l + (1 - g_l)(K_s - K_l).$$

See reference [6] for further details.

4. Computational procedure

A fully conservative (or strong) discretisation of the governing differential equation (3) is adopted; the conservative property of the scheme proved very useful when debugging the computer code.

In the one-dimensional case, a representative cross-section of the casting is split into a row of control volumes. The change of temperature over time at the centre of a box is calculated by an explicit finite difference scheme. Discretisation equations are used to step through time in small intervals δt; the box temperature at time $t + \delta t$ are obtained explicitly from those at time t.

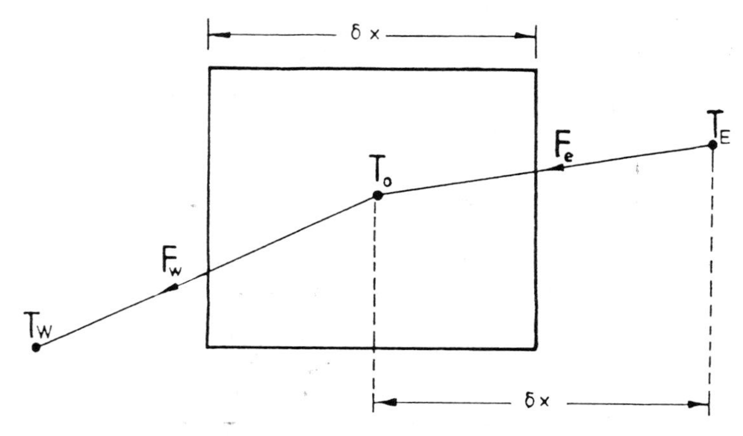

Fig. 3. A typical control volume.

The temperature dependency of the thermal conductivity is easily handled by employing the values for conductivity at the box faces. As examples of the form of the discretisation equations, consider respectively those for heat transport in the columnar semi-solid (see Fig. 3):

$$T_0' = T_0 + \frac{\delta t}{C_s^m(\delta x)^2}(k_w T_W - (k_w + k_e)T_0 + k_e T_E)$$

and the bulk liquid, with conduction:

$$T_0' = T_0 + \frac{\delta t}{C_s^m}\left[\frac{(k_w T_W - (k_w + k_e)T_0 + k_e T_E)}{(\delta x)^2} + (1 - g_l)\Delta H \delta \Omega_0\right]$$

where C_s^m is given by equation (4) and:

$$\delta \Omega_0 = (1 - \Omega_0) N 4 \pi r_0^2 v_0^* \delta t.$$

The columnar front is tracked directly. It begins to move inwards once the temperature at the edge falls below some predetermined value at which nucleation occurs. From then on, at the start of each time interval, the 'jump' condition at the columnar front is invoked iteratively to determine the temperature and velocity of the front. The distance moved by the front in an interval δt is $v \cdot \delta t$. The columnar front constitutes the common face of the last semi-solid and the first liquid boxes (see Fig. 4). In the early numerical schemes, the dimensions of these two boxes change as the front advances. To ensure numerical stability (for a given δt there is a minimum permitted control volume dimension [18]) and accuracy, it is necessary to amalgamate and divide respectively the boxes immediately ahead and behind the front. Smoother solutions are obtained in the later schemes by repeatedly shrinking and joining liquid boxes at the centre of the casting instead of immediately ahead of the front [6].

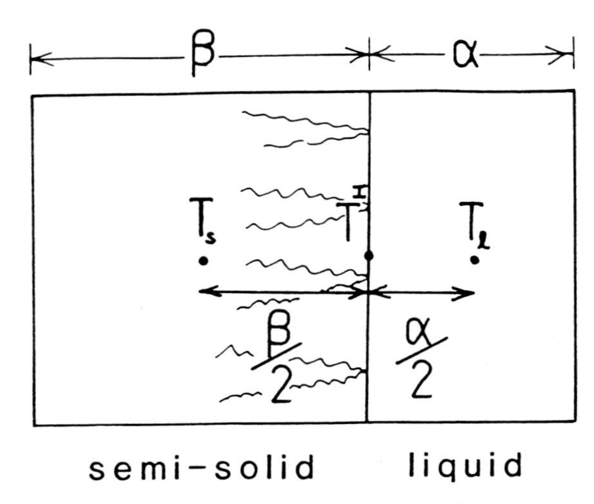

Fig. 4. The control volumes either side of the columnar front.

An explicit method is employed because the position of the front and its temperature are not known *a priori* at any stage in the calculation. An implicit scheme requires iterations at each time step to determine a consistent front position, temperature and velocity. It was thought that this would be very time-consuming but experience with the explicit iterative procedure used to satisfy the 'jump' condition suggests that convergence would be rapid. Consequently an implicit predictor-corrector implementation of the model is being developed, in which, additionally, a mapping is also used to immobilise the columnar front in computational space, hence avoiding the need for amalgamation and division of boxes.

Calculations have been performed in two dimensions with cylindrical control volumes and, in principle, can be extended to the third dimension.

5. Implementing the 'jump' condition

The truncation of the columnar dendrites at an undercooling causes a discontinuity in the latent heat liberation and requires the application of the Stefan 'jump' condition there. This boundary condition cannot be absorbed into the formulation by expressing the problem in the usual integral, control volume enthalpy approach [2,3] because the undercooling of the dendrite tips is not fixed *a priori* and equiaxed solid exists ahead of the front; it is impossible to define the enthalpy function in advance. The front has to be tracked directly and the parabolic velocity relation (1) provides the necessary closure for the temperature.

At the start of a time interval, the temperatures at the centres of the boxes either side of the columnar front, T_s and T_l, are known from the calculation of the previous time step. But the consistent value for T^I for the new time step has yet to be determined. The gradients at the front can be approximated by (see Fig. 4):

$$G_s = 2k_s^I \frac{(T^I - T_s)}{\beta}$$

$$G_l = 2k_l^I \frac{(T_l - T^I)}{\alpha}$$

If E_j is the error in T^I at the jth iteration then a value for T^I consistent with the 'jump' condition (5) can be obtained by iterating and finding progressively better values for T^I, according to the scheme:

$$E_j := \frac{(vL - k_s G_s^I + k_l G_l^I)}{pG_s^I - qG_l^I - \gamma_1 k_l^I + \gamma_s k_s^I - bL - \lambda v}$$

$$T_{j+1}^I := T_j^I + E_j$$

$$j := j + 1$$

until E_j is acceptably small. The symbols used in the expression for E_j are defined to be:

$$p = \frac{\mathrm{d}k_s^I}{\mathrm{d}T^I} \quad q = \frac{\mathrm{d}k_l^I}{\mathrm{d}T^I} \quad \gamma_1 = \frac{\mathrm{d}G_l^I}{\mathrm{d}T^I} \quad \gamma_s = \frac{\mathrm{d}G_s^I}{\mathrm{d}T^I} \quad b = \frac{\mathrm{d}v}{\mathrm{d}T^I} \quad \lambda = \frac{\mathrm{d}L}{\mathrm{d}T^I}.$$

Having found T^I, the velocity and latent heat at the front can be calculated using equations (1) and (6).

6. Results

Calculations were performed using physical parameters relating to Al–Cu alloys ($T_L^p = 933\text{K}$; $m = -2.6K\ \text{wt\%}^{-1}$; $k = 0.18$; $C = 3400\ \text{kJ K}^{-1}\ \text{m}^{-3}$; $\Delta H = 1.02 \times 10^9\ \text{J m}^{-3}$; $K_s = 180\ \text{W m}^{-1}\ \text{K}^{-1}$; $K_l = 98\ \text{W m}^{-1}\ \text{K}^{-1}$; $C_1 = 3 \times 10^{-4}\ \text{ms}^{-1}\ \text{K}^{-2}\ \text{wt\%}^{-1}$). Values of heat transfer coefficient were chosen to produce freezing times comparable to sand and metal mould castings, and some calculations with rapid solidification conditions have also been undertaken and are discussed briefly later. Unlike previous mushy zone models, the dynamically calculated undercooling at the front enables the model to produce cooling curves which exhibit recalescence. Most have been for one-dimensional castings (slit moulds) (see Fig. 5). The cooling curves show a higher bulk temperature during solidification when equiaxed grains are present owing to the liberation of latent heat ahead of the front. Also the columnar curves show a slight recalescence throughout freezing because of a deceleration of the front due to the increasing liberation of latent heat behind it as the semi-solid region increases. This last observation is peculiar to one-dimensional slit castings; in a cylindrical casting, the area of the columnar front decreases as it nears the centre of the casting and this causes a decreasing

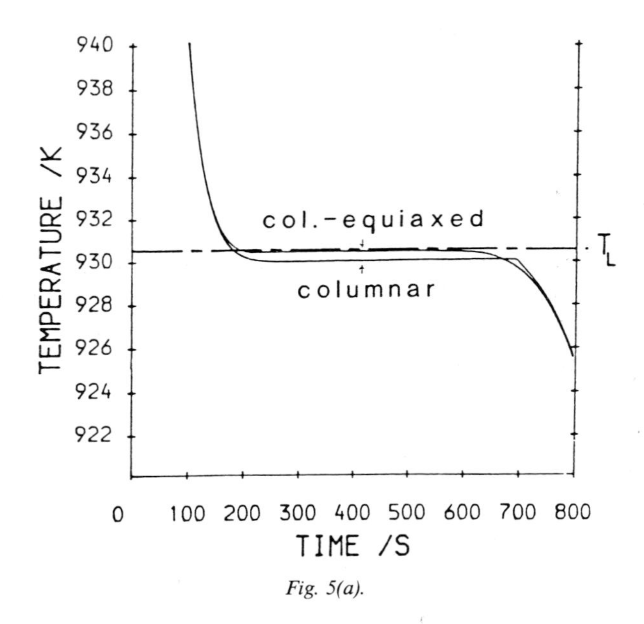

Fig. 5(a).

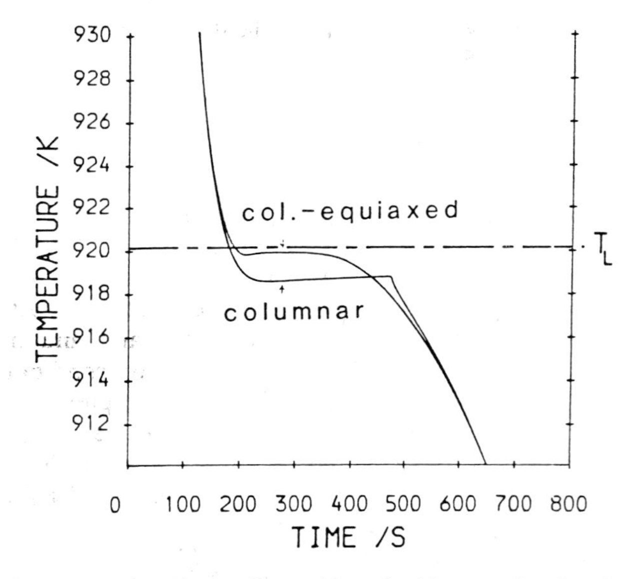

Fig. 5(b). Cooling curves for Al–Cu alloys with and without equiaxed grains ahead of the columnar front: (a) Al-1 wt% Cu (no convection); (b) Al-5 wt% Cu (no convection); and (c) Al-5 wt% Cu (convection in the bulk; $\delta = 0.2$ mm) (see p. 37). All cast at 973 K into a 100 mm wide slit mould. $h = 10^2$ Wm^{-2} K^{-1}.

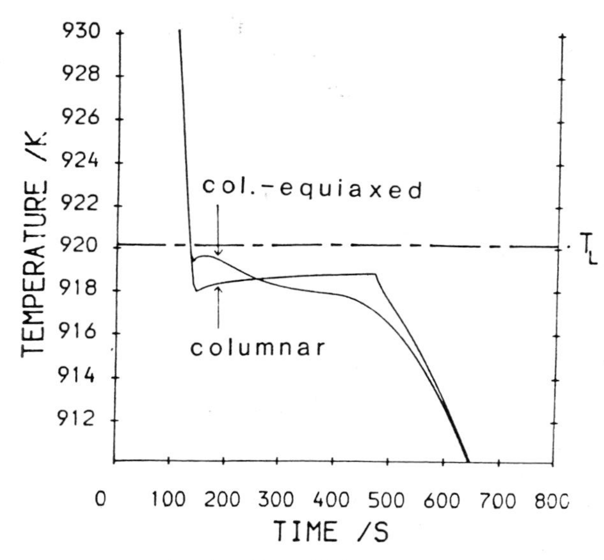

Fig. 5(c). See p. 36 for caption.

rate of latent heat production and, therefore, a drop in temperature (see Fig. 6).

Compare Figs. 5(a), (b) and (c). For a given rate of heat extraction, increasing the composition of an alloy causes an increase in the velocity and

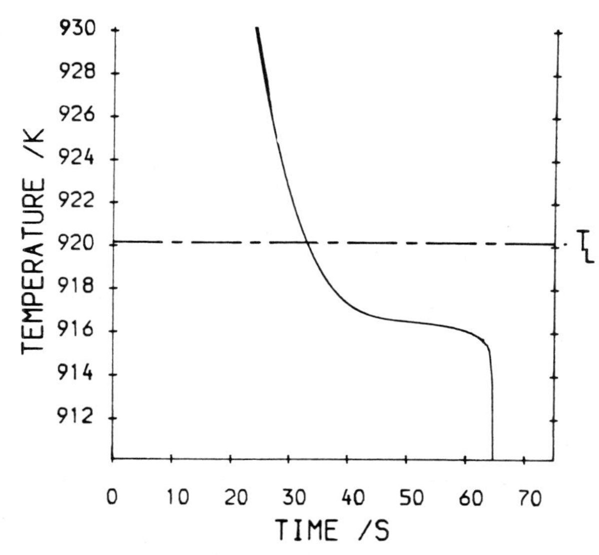

Fig. 6. Cooling curves for the columnar solidification of Al-5 wt% Cu alloy in a 100 mm diameter cylindrical mould. Poured at 973 K; $h = 10^3$ Wm^{-2} K^{-1}. No convection.

undercooling of the columnar front, and the temperature plateau in the cooling curve becomes shorter and further below the equilibrium liquidus temperature. This trend is a consequence of

(i) the compositional dependency of the velocity-undercooling relation (1) and
(ii) the variation in the Scheil solid fraction with composition.

Initially, the development of the equiaxed zone was investigated by placing nuclei ahead of the columnar front and allowing them to grow as soon as the local temperature fell below the liquidus. The scale and speed of equiaxed growth ahead of the columnar front depended only on the extent and degree of the undercooled liquid in the bulk:

(i) Increasing the composition promoted equiaxed growth by increasing the undercooling at the front (see Fig. 7).
(ii) Both decreasing the superheat and
(iii) decreasing the rate of heat extraction produced a lower temperature gradient in the bulk, thereby widening the undercooled layer and increasing the period in which the equiaxed grains could grow before the front reached them (see Figs. 7(a) and (b)).
(iv) Convection in the bulk accelerated the loss of the superheat and the onset of undercooling at the centre of the casting and therefore encouraged the growth of the equiaxed grains (see Fig. 7(c)).

In the calculations, when only the growth of the grains is considered, very small columnar ranges are produced. At moderate cooling rates, low tempera-

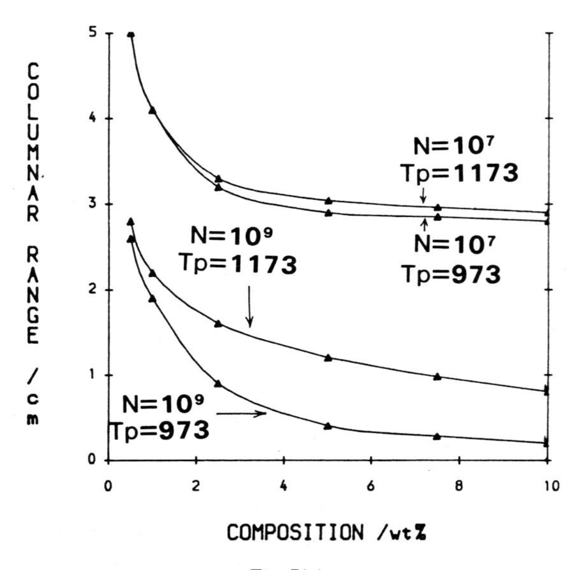

Fig. 7(a).

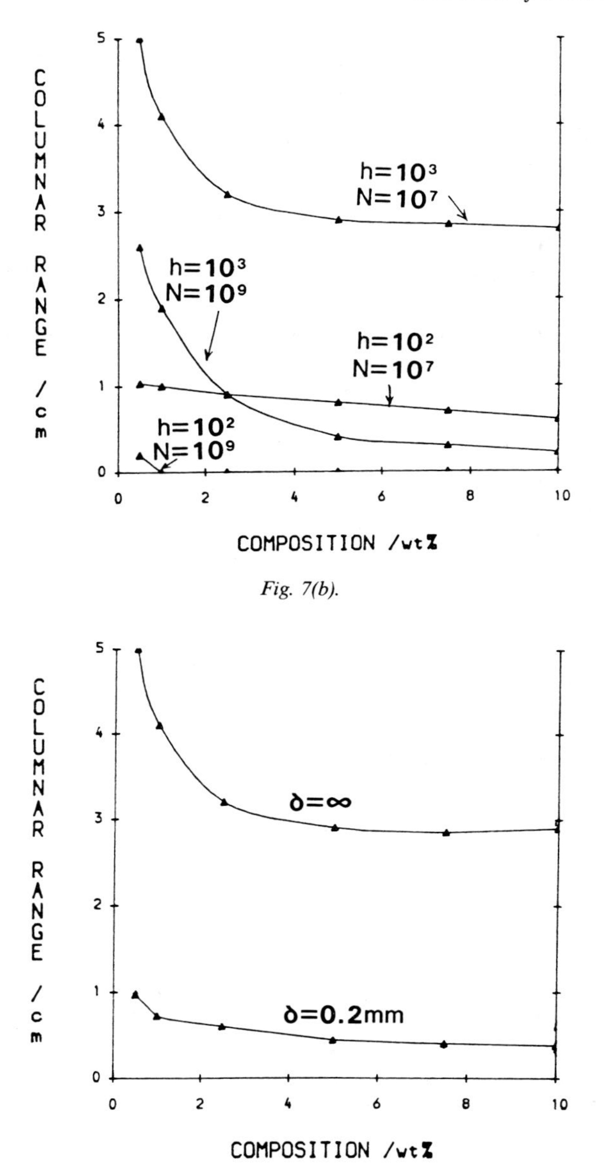

Fig. 7(b).

Fig. 7(c). Graphs of columnar range vs composition for Al–Cu alloys cast in a 100 mm wide slit mould. (a) The effect of grain density (N in m^{-3}) and pouring temperature (T_p in K). $h = 10^3$ Wm^{-2} K^{-1}; 10^9 grains m^{-3}. No convection. (b) The effect of heat transfer coefficient (h in Wm^{-2} K^{-1}) and grain density (N in m^{-3}). Poured at 973 K. No convection. (c) The effect of convection (δ = boundary layer width). Poured at 973 K; $h = 10^3$ Wm^{-2} K^{-1}; 10 grains m^{-3}.

ture gradients are quickly attained throughout the bulk and equiaxed growth soon occurs at the centre; even with high cooling rates which produce steeper gradients, growth immediately ahead of the front is fast. The difference between the undercooling and, therefore, the tip velocities, of the columnar and equiaxed dendrites is not very great and so obstruction of the front occurs readily.

In practice, the equiaxed growth appears not to be so dominant. Hence, in reality, factors other than the growth of the equiaxed grains must also determine macrostructure, such as the buoyancy of the grains and their convective motion in the liquid. These effects would tend to reduce the presence of equiaxed grains ahead of the columnar front. The buoyancy of the grains is being investigated currently.

As expected, the model shows that the superheat is quickly removed owing to the low Stefan number and high thermal conductivity of metals – the rate-determining process during solidification is the removal of the latent as opposed to the sensible heat ($L/C = 300$ K). In the calculations (see Fig. 7(a)), the effect of superheat on the growth of the columnar and equiaxed grains is significant only at high rates of heat extraction and in large castings because only under these circumstances can an appreciable temperature difference be maintained between the edge and centre of a casting. At low cooling rates and in small castings, the temperature gradient is shallow and the growth of the grains is unaffected by the initial pouring temperature. In practice, however, the effect of superheat can be dramatic, especially in small castings [19,20]. From the model, therefore, it appears not to be possible to explain satisfactorily the influence of superheat in terms of the growth of the grains: it is suggested that, as is usually proposed, the variation with superheat of the equiaxed zone and grain size is due to the delay in the bulk becoming undercooled affecting the survival of chill nuclei (the Big Bang mechanism [20]).

When a temperature dependent nucleation rate is included, the onset of equiaxed growth is delayed until the critical nucleation undercooling is attained locally causing larger equiaxed ranges and smaller equiaxed zones. If the columnar front does not exceed the critical nucleation undercooling then neither does the bulk, and no equiaxed grains are nucleated. Hence, under certain circumstances, the columnar range can be dramatically decreased by increasing the cooling rate: at a low cooling rate the columnar front might never attain the critical nucleation undercooling but on increasing the cooling rate, the front undercooling will increase and may exceed the critical value and thus permit nucleation in the bulk. It is suggested that the model could be used to investigate the efficiency of a grain refiner, thus extending the work of Maxwell and Hellawell [21].

Rapid solidification was modelled: two high values for the heat transfer coefficient ($h = 10^4$ and 10^5 W m^{-2} K^{-1}) and a thin width of casting (100 microns) were used. Nucleation undercoolings of 50 and 100 K were assumed (see Fig. 8). Rapid recalescence and a corresponding decrease in the columnar front velocity were noted. Shortly after nucleation of solid at the edge, the liquid ahead of the front is at a lower temperature and thus latent heat is conducted both into the solid and also the liquid; after the liquid has

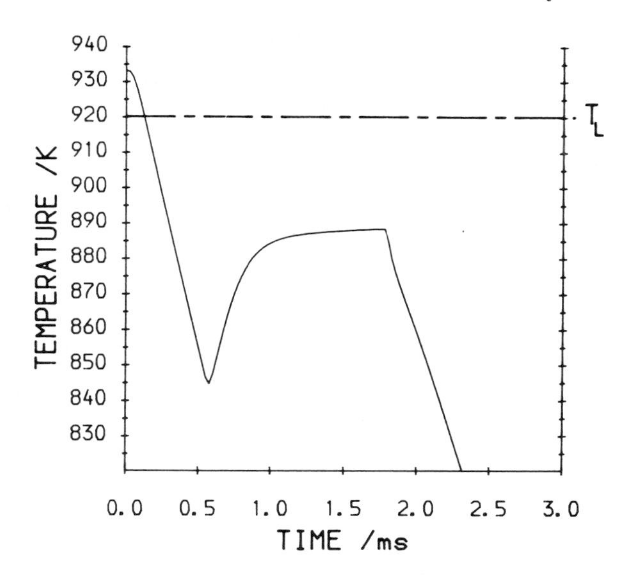

Fig. 8. Cooling curve for the rapid columnar solidification of Al-5 wt% Cu alloy. 100 um wide ribbon initially at 973 K. $h = 10^5$ Wm^{-2} K^{-1}.

recalesced, though, heat is extracted through the solid only and the front velocity falls.

Conclusions

A model for the solidification of a casting has been presented in which the columnar mass is represented as a mushy zone truncated at an undercooling which is free to vary. Equiaxed grains can exist and grow ahead of the front. Because of the freely varying undercooling and the presence of grains in the bulk, it is not possible to prescribe an enthalpy function and treat the solidification as in the usual integral, control volume enthalpy-based method. The front has to be tracked directly. Because of the truncation of the Scheil shape at the columnar tips, there is a discontinuity in the evolution of the latent heat at the columnar front: this introduces the Stefan 'jump' condition into the problem. This condition is not present in normal mushy zone models because of the continuous form assumed for the variation of bulk solid fraction with temperature. The current model, therefore, is original in so far as it is a combination of the Stefan and mushy zone models.

The model can produce realistic cooling curves and has been used to investigate the growth of equiaxed grains and the development of the equiaxed zone ahead of a columnar front. The expected trends with superheat, composition, convection and cooling rate have been reproduced. However, the superheat effect is not as pronounced in the model as that seen in practice. This is attributed to the failure of the model to consider the Big Bang mechanism for the production of chill nuclei. The model predicts small columnar ranges and this is probably due to its neglect of the buoyancy and convective motion of the equiaxed growth ahead of the front.

Although the model requires further development to include the effects of movement of equiaxed grains, it forms a useful basis for the investigation of the evolution of the macrostructure of a casting and the interplay of the various casting parameters.

Acknowledgements

The Authors should like to thank Professor Sir Peter Hirsch FRS and the Oxford University Computing Service for the provision of office and computing facilities respectively, and one of them (SCF) is also grateful to the Royal Society of London, the United Kingdom Science and Engineering Research Council, and Alcan International Ltd for financial support.

References

1. J. Stefan: *Ann. Phys. u. Chem.* (ed. by Wiedemann) N.F., 42 (1891) 269–281.
2. N. Shamsundar: In: *Moving Boundary Problems*, ed. by D.G. Wilson, A.D. Solomon and P.T. Buggs (1981) 165–185. Pubd by Academic Press Inc., New York, U.S.A.
3. P.N. Hansen: In: *Solidification and Casting of Metals* (1979) 350–356. Pubd by Metals Society, London, U.K.
4. R.E. Marrone, J.O. Wilkes and R.D. Pehlke: *Cast Metals Research J.* 9 (1970) 184.
5. S.C. Flood and J.D. Hunt: In: *Proc. Conf. on Modeling of Casting and Welding Processes*, Henniker, New Hampshire, U.S.A. (1983) 207–218.
6. S.C. Flood: D. Phil Thesis, University of Oxford (1985).
7. T.W. Clyne: In: *Proc. 2nd Intern. Conf. on Numerical Methods in Thermal Problems*, Venice, Italy (1981) 240–256.
8. T.W. Clyne: *Metallurgical Transactions* 15B (1984) 369–381.
9. M.H. Burden and J.D. Hunt: *J. Crystal Growth* 22 (1974) Part I: 99–108, Part II: 109–116.
10. M. Tassa and J.D. Hunt: *J. Crystal Growth* 34 (1976) 38–48.
11. I. Jin and J.G. Sutherland: In: *Solidification and Casting of Metals* (1979) 256–259. Pubd by Metals Society, London, U.K.
12. R. Elliot: In: *Eutectic Solidification Processing* (1983) 247–251. Pubd by Butterworths & Co., London, U.K.
13. T.W. Clyne and W. Kurz: *Metallurgical Transactions* 12A (1981) 965–971.
14. J.D. Hunt: In: *Solidification and Casting of Metals* (1979) 3–9. Pubd by Metals Society, London, U.K.
15. J.W. Christian: In: *The Theory of Transformations in Metals and Alloys*. Part I (1975) 17–18. Pubd by Pergamon Press, Oxford, U.K.
16. J.W. Christian: ibid., 450.
17. J.D. Hunt: *Mat. Sci. Eng.* 65 (1984) 75–83.
18. B. Carnahan, H.A. Luther and J.O. Wilkes: In: *Applied Numerical Methods* (1969) 449–450. Pubd by John Wiley, New York, U.S.A.
19. B. Chalmers: In: *Principles of Solidification* (1964) 257–259, 265–268. Pubd by John Wiley, New York, U.S.A.
20. R. Morando, H. Biloni, G.S. Cole and G.F. Bolling: *Metallurgical Transactions* 1 (1970) 1407–1412.
21. I. Maxwell and A. Hellawell: *Acta Metallurgica* 23 (1975) 901–909.

Applied Scientific Research 44: 43–49 (1987)

A model for the numerical computation of microsegregation in alloys

A.J.W. OGILVY [1] & D.H. KIRKWOOD [2]

[1] *Osprey Metals Ltd., Neath, South Wales, UK;* [2] *Department of Metallury, Sheffield University, Sheffield, UK*

Abstract. A development of the Brody-Flemings model for the prediction of dendritic microsegregation in alloys is proposed. The original model considered one-dimensional back diffusion into thickening platelike dendrite arms of fixed spacing. The present modification allows for the dendrite arm coarsening which is observed to occur during solidification and considers microsegregation in both binary and multi-component alloy systems. It is shown that numerical calculations of microsegregation using finite difference techniques based on this model give good agreement with experiment.

Nomenclature

$B = (1-f)/(1-(1-k)f)$, correction factor for fast diffusing species
c concentration in solid, wt.%
c^0 average concentration, wt.%
c^l concentration in liquid, wt.%
D diffusion coefficient in solid
f fraction solid ($= X/L$)
k equilibrium partition coefficient
L half dendrite arm spacing
t time
W cooling rate
x direction coordinate normal to dendrite plate
X distance solidified
β liquidus gradient
subscript i refers to the ith solute element

Introduction

The first attempts to predict microsegregation quantitatively [1,2] assumed that within a characteristic volume element, related to a dendrite spacing or the grain size, the liquid region during solidification is completely mixed, whereas negligible diffusion occurs in the solid and local thermodynamic equilibrium is maintained at the moving interface described by a constant equilibrium partition coefficient. This model gives rise to the so-called Scheil equation: $c = kc^0(1-f)^{k-1}$.

With the introduction of microprobe analysis it soon became apparent that the assumption of negligible solid-state diffusion was untenable and Brody and Flemings [3] proposed a model involving one-dimensional back diffusion into growing dendrites of plate-like morphology. Apart from this modification, the previous assumptions of interface equilibrium and complete solute mixing in the liquid are maintained.

An analytical solution of the differential equation was derived from the model by assuming the dendrite growth rate and approximating for the composition gradient at the solid interface. This has been criticised however [4] as being an invalid approximation when diffusion is important. A general disadvantage of analytical treatments of microsegregation is their inability to deal with the complexity of real systems in which partition coefficients, diffusion coefficients and dendrite spacings all change with time or temperature during solidification. These problems are readily overcome using numerical techniques to solve the basic differential equations.

The partial differential equation derived from the Brody-Flemings model of one-dimensional back diffusion into plate-like dendrites in a binary alloy system is:

$$c^l(1-k)\,\mathrm{d}X/\mathrm{d}t = D(\partial c/\partial x)_X + \mathrm{d}c^l/\mathrm{d}t(L-X)$$

where L is half the appropriate dendrite arm spacing (see Fig. 1). Using an explicit finite difference method to calculate diffusion and the above equation to control the interface movement, Brody and Flemings were able to compute the solute distribution throughout the solidification process, and compare their predictions with measurements of microsegregation in Al-Cu alloys solidified under a range of conditions [5]. In order to obtain agreement with experiment it was necessary to employ dendrite arm spacings approximately one third of the measured value. This large discrepancy has been attributed later by Flemings [6] to the simple geometries assumed, to the use of inaccurate diffusion data or to coarsening effects. It is believed that diffusion data in the Al-Cu system is now well enough established not to be considered a source of significant error [7]. It is clear however that dendrite morphologies can be extremely complex and change during growth, particularly in highly alloyed systems [8] and under conditions of rapid growth and low temperature gradient. It would not be expected that the model proposed in this paper could be satisfactorily applied under these conditions.

The observation that the spacing of secondary (and higher order) arms increases during solidification was made by Kattamis et al. [9], who proposed that this coarsening occurred by smaller arms melting away at the expense of their larger neighbours by a diffusion process driven by surface tension. It has also been shown that coalescence of adjacent arms to form a single larger arm can occur driven by the same surface tension forces [10,11]. These coarsening processes have a important effect on the calculation of dendritic microsegregation as will be shown.

A fuller discussion of the assumptions involved in different microsegregation models has been published recently [12].

Microsegregation in binary alloys involving arm coarsening

The present model makes the same assumptions as Brody and Flemings, except that the average dendrite arm spacing is allowed to increase with time during solidification according to some given relationship. Two adjacent dendrite plates are considered as in Fig. 1, in which the centre to centre spacing is $2L$. Within this we select an element bounded by a plate centre (at $x = 0$) and the centre of the liquid region (at $x = L$) for the calculation on solute redistribution. A schematic diagram of solute distribution within the element is given in Fig. 2, representing a point in time during the solidification process. The solute rejected by the interface movement must be absorbed partly by back diffusion into the solid, and partly absorbed into the liquid raising its uniform composition. This distribution is expressed in the first two terms of the right-hand side of the solute balance equation:

$$c^l(1-k)\,\mathrm{d}X/\mathrm{d}t = D(\partial c/\partial x)_X + \mathrm{d}c^l/\mathrm{d}t(L-X) + (c^l - c^0)\,\mathrm{d}L/\mathrm{d}t. \quad (1)$$

The end term represents the increase in the size of the element due to arm coarsening, which brings in liquid of average composition that requires to be raised to the composition of the existing liquid. It is necessary that liquid of average composition be added to conserve the overall composition in the element: the physical meaning of this is that somewhere in the system a dendrite arm is melting and part of this solid together with its associated liquid is transferred at average composition to the element. This dilution effect is responsible for a reduction in microsegregation.

If we can assume that the cooling rate and the liquidus slope of the alloy system are constants, then we may substitute $\mathrm{d}c^l/\mathrm{d}t = W/\beta$ into eqn. 1 to obtain:

$$\mathrm{d}X/\mathrm{d}t = \frac{\beta D(\partial c/\partial x)_X + W(L-X) + \beta(c^l - c^0)\,\mathrm{d}L/\mathrm{d}t}{c^l(1-k)}. \quad (2)$$

Back diffusion of solute into the solid dendrite plate may be calculated using a finite difference formulation of the diffusion equation: $\partial c/\partial t = D\,\partial^2 c/\partial x^2$, and employing the boundary condition $\partial c/\partial x = 0$ at $x = 0$. The movement of the interface δX in the time step δt is obtained from the finite difference formulation of eqn. 2, in which $(\partial c/\partial x)_X$ is approximates from compositions in the solid near the interface at unequally spaced nodes by a method due to Crank [13]. The increase in δL during the time step is obtained from the given functional relationship referred to above. The increase in liquid composition δc^l is now obtained by back substitution in eqn. 1, from which the interface composition of the solid is in turn calculated. These calculations can be repeated, adjusting the diffusion and partition coefficients as appropriate for

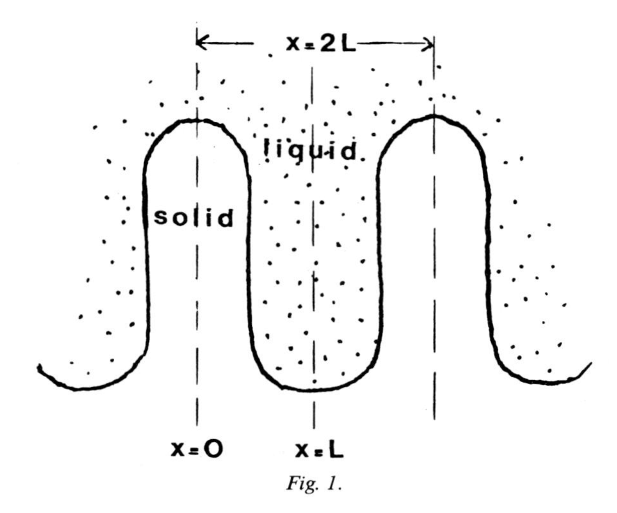

Fig. 1.

the temperature at each step until complete solidification is achieved, that is when either the liquid disappears or it attains eutectic composition. An earlier paper [14] provides further details on the finite difference computation.

In many situations the use of a constant cooling rate during solidification may not be appropriate and a reformulation of eqn. 2 in terms of constant heat extraction given by Howe [15] may be more realistic.

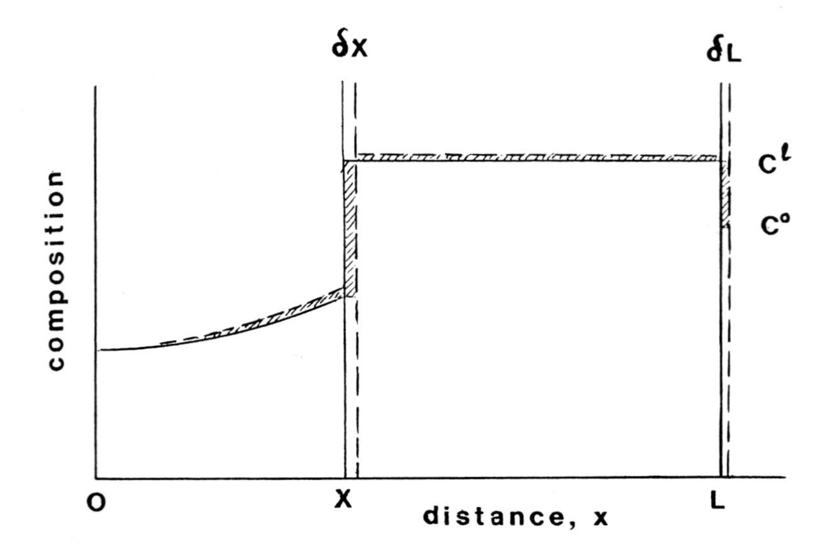

Fig. 2.

Microsegregation in ternary and higher component systems

In a ternary system a solute balance at the moving interface may be written for both diffusing solute species, 1 and 2:

$$c_1^l(1-k_1)\,\mathrm{d}X/\mathrm{d}t = D_1(\partial c_1/\partial x)_X + \mathrm{d}c_1^l/\mathrm{d}t(L-X) + (c_1^l - c_1^0)\,\mathrm{d}L/\mathrm{d}t$$

$$c_2^l(1-k_2)\,\mathrm{d}X/\mathrm{d}t = D_2(\partial c_2/\partial x)_X + \mathrm{d}c_2^l/\mathrm{d}t(L-X) + (c_2^l - c_2^0)\,\mathrm{d}L/\mathrm{d}t.$$

We may also write for the cooling rate: $W = \beta_1\,\mathrm{d}c_1^l/\mathrm{d}t + \beta_2\,\mathrm{d}c_2^1/\mathrm{d}t$. Multiplying the two upper equations by the appropriate β and adding to eliminate $\mathrm{d}c^l/\mathrm{d}t$ from each, we have:

$$\mathrm{d}X/\mathrm{d}t = \big[\beta_1 D_1(\partial c_1/\partial x)_X + \beta_2 D_2(\partial c_2/\partial x)_X + R(L-X) + \{\beta_1(c_1^l - c_1^0) + \beta_2(c_2^l - c_2^0)\}\,\mathrm{d}L/\mathrm{d}t\big]\big[\beta_1 c_1^l(1-k_1) + \beta_2 c_2^l(1-k_2)\big]^{-1}.$$

It is clear from the form of this equation that it may be generalised for higher component alloys:

$$\mathrm{d}X/\mathrm{d}t = \frac{\Sigma\beta_i D_i(\partial c_i/\partial x)_X + R(L-X) + \Sigma\beta_i(c_i^l - c_i^0)\,\mathrm{d}L/\mathrm{d}t}{\Sigma\beta_i c_i^l(1-k_i)}.$$

The finite difference procedure is carried out exactly as in the binary case; having determined δX for a given time step, δc^l may be obtained for each solute element by back substitution in the first equations.

Microsegregation in an alloy containing a fast diffusing solute element

When explicit finite difference equations are used to solve the diffusion equation, a condition for the mathematical stability of the solution is that we must choose a time step δt such that $D\delta t/\delta x^2 \leqslant 1/2$ [16]. This requires that where we have a fast diffusing species, the time step will be determined by it and will necessarily be small. Therefore a large number of computer calculations are needed to complete the solidification and the time taken for the overall microsegregation calculation can be unacceptable.

In many practical situations one (or more) elements may diffuse so fast as to establish partial equilibrium between the solid and liquid phases, for example interstitial elements such as carbon in steels. In such cases the compositional gradients of the solute in each phase is negligible, and we may write for the solute balance at the moving interface:

$$c_i^l(1-k_i)\,\mathrm{d}X/\mathrm{d}t = k_i X\,\mathrm{d}\frac{\mathrm{d}c_i^l}{\mathrm{d}t}/\mathrm{d}t + \mathrm{d}c_i^l/\mathrm{d}t(L-X) + (c_i^l - c_i^0)\,\mathrm{d}L/\mathrm{d}t.$$

Eliminating $\mathrm{d}c_i^l/\mathrm{d}t$ in all the solute balance equations as before, we have:

$$\mathrm{d}X/\mathrm{d}t = \frac{\Sigma\beta_i D_i(\partial c_i/\partial x)_X + R(L-X) + \Sigma\beta_i(c_i^l - c_i^0)B_i\,\mathrm{d}L/\mathrm{d}t}{\Sigma\beta_i c_i(1-k_i)B_i}.$$

Table 1. Predicted effect of back diffusion on microsegregation in Al-5wt.%Cu alloy solidified at a cooling rate of 1 K/s.

Assumptions	Cmin./Co
Fixed dendrite arm spacing	0.25
Dendrite coarsening	0.34
Dendrite coarsening Constant D (liquidus value)	0.46

where $B_i = 1$ for slow diffusing solute elements (e.g. in substitutional solution), and $B_i = (1-f)/(1-(1-k_i)f)$ for fast diffusing solute elements (e.g. in interstitial solution).

This equation may also be recast in an alternative form where the assumption of a constant rate of heat extraction is more appropriate (see Howe [15]).

Discussion

Table 1 shows the results of numerical calculations using the present model to predict microsegregation in Al-5wt%Cu alloys. They are expressed as the ratio of the minimum concentration at the centre of the dendrite arm to the average concentration, which was measured by Bower et al. [5] to be ~ 0.30 and by Bennett [17] to be ~ 0.35 under a variety of solidification conditions. Using a fixed arm spacing (the final spacing measured after solidification) clearly results in an overestimation of the degree of microsegregation as previously shown by Brody and Flemings [3]. Using the coarsening relationship: $L = 8t^{0.31}$ (where$L$ is in μm. and t in seconds) obtained from experimental measurements of secondary arm spacings during solidification in this alloy [11], the prediction of the model is in good agreement with the measured values of microsegregation. The third computation demonstrates the sensivity of the calculation to the choice of diffusion coefficient. A constant value has been used, calculated at the liquidus temperature of the alloy, and this clearly predicts too low an amount of microsegregation.

The use of this model for predicting microsegregation in more complex alloys has been made by Howe and is reported in this volume [15].

References

1. G.H. Gulliver: *Metallic Alloys*, Charles Griffin, London, (1922).
2. E. Scheil: *Z. Metallk.* 34 (1942) 70.
3. H.D. Brody and M.C. Flemings: *Trans AIME* 236 (1966) 615.
4. T.W. Clyne and W. Kurz: *Metall. Trans.* 12A (1981) 965.
5. T.F. Bower, H.D. Brody and M.C. Flemings: *Trans. AIME* 236 (1966) 624.
6. M.C. Flemings: *Solidification Processing*, McGraw-Hill, New York (1974).
7. C.J. Smithells, *Metals Reference Book*, 5th edn., Butterworths, London (1976)

8. B.A. Rickinson and D.H. Kirkwood: *Proc. Conf. on Solidification and Casting of Metals*, Sheffield, July 1977, Metals Society, London (1979), p. 44.
9. T.Z. Kattamis, J.C. Coughlin and M.C. Flemings: *Trans. Metall. Soc. AIME* 239 (1967) 1504.
10. K.H. Chien and T.Z. Kattamis: Z. Metallkunde 61 (1970) 475
11. K.P. Young and D.H. Kirkwood: *Metall. Trans.* 6A (1975) 197.
12. D.H. Kirkwood: *Mat. Sci. and Eng.* 65 (1984) 101.
13. J. Crank: *Quart. J. Mech. App. Math.* 10 (1957) 220.
14. D.H. Kirkwood and D.J. Evans: Conf. on 'The Solidification of Metals', Iron and Steel Inst., London (1978) ISI P110.
15. A.A. Howe: *Appl. Sci. Res.* 44 (1987) 51–59.
16. G.D. Smith: *Numerical Solution of Partial Differential Equations*, 2nd edn., Clarendon Press, Oxford (1978).
17. D.A. Bennett: Ph.D. thesis, Univ. of Sheffield (1978).

Applied Scientific Research 44: 51–59 (1987)

Development of a computer model of dendritic microsegregation for use with multicomponent steels

A.A. HOWE
British Steel Corporation, Swinden Laboratories, Rotherham, UK

Abstract. Progress is reported in the development of a computer program for modelling the process of microsegregation in multicomponent steels. The program is based on the single phase binary and ternary models described in the previous paper by Ogilvy and Kirkwood but has been extended to consider general, multicomponent alloys. The cooling rate required in the program formulation is currently determined according to a constant, set rate of heat extraction, but any, independently derivable rule could be employed. The peritectic reaction from ferritic to austenitic solidification encountered in a great many steels has been included in the formulation with the relevant, three-phase compositions decreed in pseudo-binary fashion by appropriate carbon equivalents. The onset of the austenite-cementite eutectic is similarly described. Manganese sulphide precipitation is described by removal of the elements from the residual liquid in appropriate ratio so as not to exceed a prescribed solubility product. Such devices can be fashioned similarly for other phase and precipitate developments as required.

Nomenclature

A	Back diffusion term, m wt% s^{-1}
B	Correction factor for fast – diffusing species
C	Concentration in the solid, wt%
Cl	Concentration in the liquid, wt%
C_0	Bulk concentration, wt%
Cp	Concentration in the liquid during the peritectic, wt%
D	Diffusivity in the solid, $m^2 s^{-1}$
f	Fraction solid ($= X/L$)
H	Volumetric latent heat, Jm^{-3}
K	Partition coefficient
L	Half dendrite arm spacing, m
M	Atomic weight, (a.m.u.)
$\dot{Q}$	Heat extraction rate, $Jm^{-3}s^{-1}$
R	Atomic weight ratio
t	time, s
T	Temperature, k
W	Cooling rate, ks^{-1}
x	Distance coordinate normal to dendrite plate
X	Distance solidified, m
Z	Correction increment, wt%
β	Liquidus gradient, k wt$\%^{-1}$
θ	Heat capacity, $Jm^{-3}k^{-1}$

Subscripts:

c	carbon
i	"i"th element
Mn	manganese
S	sulphur

Superscripts:

bcc	body centred cubic delta ferrite
fcc	face centred cubic austenite

1. Introduction

The various forms of segregation resulting from solidification are of great importance and dictate the degree to which the material has to be worked subsequent to casting before the required physical property specification can be met. The microsegregation between the dendrite arms is, of itself, important in this respect but is usually an accepted feature and it is the presence of macrosegregation, i.e. larger scale compositional inhomogeneity, which is more frequently described as a problem. It must be remembered, however, that the source of virtually all the macrosegregation in dendritic solidification is this inherent microsegregation, a quantitative understanding of which is therefore central.

A finite difference formulation for microsegregation with temperature dependent diffusivity had been developed at BSC but was deemed inferior to the model due to Kirkwood and Ogilvy [1] because the former required a prescribed, interdendritic growth law and did not include the process of secondary arm coarsening. The latter model was made available to BSC but required modification and extension for proper use with steels.

2. Modification of basic solute balance formulation

The ternary description available at the onset of this work [2] was insufficient even for so-called 'plain carbon' steels and it was therefore necessary to develop a general, multicomponent formulation. Analysis of the source equations revealed that this extension could be readily achieved, assuming that the cooling rate expression can be extended linearly for any number of components, i.e.:

$$W = \sum \frac{\delta Cl_i}{\delta t} \cdot \beta_i .$$

A further aspect was the inherent assumption that this cooling rate was constant through solidification, which was inappropriate for practical casting processes. Again, however, analysis of the source equations indicated that this was not a problem. The cooling rate must be supplied for each program iteration but does not have to be constant. At present, this is derived from an assumption of a constant rate of volumetric heat extraction:

$$\dot{Q} = H \cdot \frac{\mathrm{d}f}{\mathrm{d}t} + \theta \cdot \frac{\mathrm{d}T}{\mathrm{d}t}$$

($f = X/L$ where both X and L are functions of t)

$$W = \frac{1}{\theta} \cdot \left\{ \dot{Q} - \frac{H}{L} \cdot \left(\frac{\mathrm{d}X}{\mathrm{d}t} - f \cdot \frac{\mathrm{d}L}{\mathrm{d}t} \right) \right\}.$$

The basic, simultaneous solution of the solute balance equations for the growth rate is as follows:

$$\frac{dX}{dt} = \frac{L(1-f)\cdot\frac{\dot{Q}}{\theta} + \frac{dL}{dt}\cdot\left(f(1-f)\cdot\frac{H}{\theta} + \sum\beta_i\cdot(Cl_i - Co_i)\right) + \sum\beta_i\cdot A_i}{(1-f)\cdot\frac{H}{\theta} + \sum\beta_i\cdot Cl_i(1-K_i)}$$

where $A_i = D_i \cdot \frac{\delta C_i}{\delta x}$ for substitutional elements or $L\cdot f\cdot K_i\cdot\frac{dCl_i}{dt}$ for interstitial elements.

The value of dCl_i/dt for the interstitial elements in the term, A_i, has to be that derived from the previous iteration. The following, improved formulation can be made which fully eliminates this term from the expression. This corresponds to equation 3 in the previous paper suitably modified for the constant heat extraction assumption:

$$\frac{dX}{dt} = \left[L(1-f)\cdot\frac{\dot{Q}}{\theta} + \frac{dL}{dt}\cdot\left\{f(1-f)\cdot\frac{H}{\theta} + \sum\beta_i\cdot(Cl_i - Co_i)\cdot B_i\right\}\right.$$
$$\left. + \sum^{\text{sub}}\beta_i\cdot A_i\right]\left[(1-f)\cdot\frac{H}{\theta} + \sum\beta_i\cdot Cl_i(1-K_i)\cdot B_i\right]^{-1}$$

where B_i is the 'correction factor' for interstitial elements:

$B_i = (1-f)/(1-(1-K_i)f)$

and the superscript, sub, refers to summation for the substitutional elements, only. It is hoped that this revised formulation will solve some of the instability problems at high fractions solid.

3. Compound and eutectic precipitation

The region of the Fe–C–Cr ternary considered in Ogilvy's research work [2] included terminal solidification upon attainment of the austenite-cementite eutectic, with the relevant carbon content in the residual liquid dependent on the chromium content. Similarly, other elements would be expected to influence the carbon content at the eutectic. Provided carbon remains the 'dominant' element in the residual liquid, this can be assessed in pseudo-binary fashion by use of carbon equivalents. Relevant coefficients are employed for cast irons in determining whether a particular composition is hypo- or hyper-eutectic, and these can be applied to the residual, interdendritic liquid composition. The values employed are taken from reference [3] except that the chromium coefficient is replaced by the quadratic equation employed by Ogilvy. A similar approach could be followed for stainless steels if a treatment using both chromium and nickel equivalents, instead of those for carbon, is employed.

The eutectic will terminate solidification at a constant temperature only if it involves each component element but this simple termination would be a reasonable approximation because, even with the highest carbon steels of interest, the eutectic would only be encountered in a very low volume fraction of residual liquid.

The most important compound in terms of this work to be precipitated from the liquid is manganese sulphide, MnS. A great many steels precipitate this compound, albeit in small quantities, towards the end of solidification, but it is not considered to be an adequate approximation to assume that solidification terminates with its development. Consequently, a routine has been introduced to remove manganese and sulphur from the residual liquid in appropriate ratio as it precipitates but which allows this liquid to continue its solidification with decreasing temperature. A maximum solubility product of 0.506 (wt%) [2] in the liquid is employed, as determined by Schwerdtfeger [4]. The resultant formulation is as follows:

$$Cl_{\mathrm{Mn}} \rightarrow Cl_{\mathrm{Mn}} - R \cdot Z, \quad Cl_{\mathrm{S}} \rightarrow Cl_{\mathrm{S}} - Z$$

where $R = M_{\mathrm{Mn}}/M_{\mathrm{S}}$ and

$$Z = \left((Cl_{\mathrm{Mn}} + R \cdot Cl_{\mathrm{S}}) - \sqrt{(Cl_{\mathrm{Mn}} + Cl_{\mathrm{S}})^2 - 4R \cdot (Cl_{\mathrm{Mn}} \cdot Cl_{\mathrm{S}} - 0.506)} \right) / 2R.$$

4. The peritectic reaction

A great many steels undergo the delta-ferrite/austenite peritectic reaction during solidification (Fig. 1) whereas the supplied model [1] only considered single phase solidification. Two additional routines were therefore required: one to determine the relevant phase compositions bounding the reaction, and another to handle the physical progress of the reaction.

In a similar fashion to the case of the eutectic reaction (section 3), the liquid composition at which the peritectic is encountered is treated in a pseudo-binary manner with the use of carbon equivalents. Unfortunately, no such coefficients specific to this reaction appeared to be available from other sources and the values employed (Table 1) were generated for this study from very limited data (20 casts of carbon and low alloy steels). The respective compositions in the ferrite and austenite were determined simply from that in the liquid multiplied by the relevant partition coefficients. The following equation can therefore be generated relating the fraction solid of bcc, delta-ferrite to the overall fraction solid, bulk carbon content and peritectic liquid carbon content:

$$f^{\mathrm{bcc}} = \frac{\left(1 - Co_{\mathrm{c}}/Cp_{\mathrm{c}} - f\left(1 - K_{\mathrm{c}}^{\mathrm{fcc}}\right)\right)}{\left(K_{\mathrm{c}}^{\mathrm{fcc}} - K_{\mathrm{c}}^{\mathrm{bcc}}\right)}.$$

The peritectic reaction was assumed to follow equilibrium which, unlike the case in many alloy systems, is a reasonable assumption in steels where its

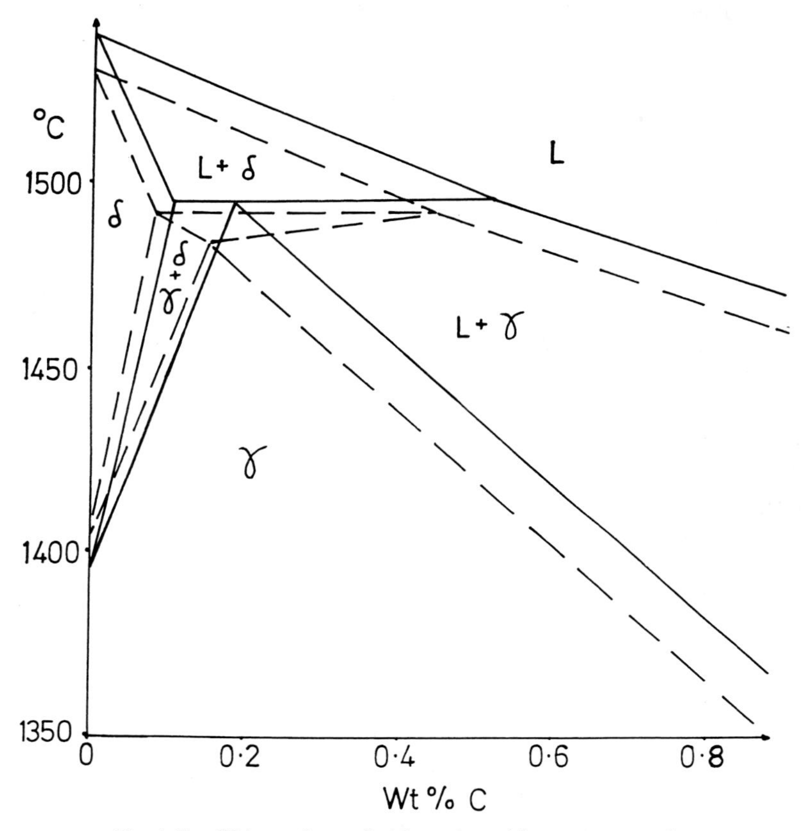

Fig. 1. Fe–C binary & pseudo-binary in multicomponent steel.

progress is dominated by rapidly diffusing carbon atoms. The back diffusion term, A_i, from the solute balance equation was restructured for carbon and nitrogen for progress through the peritectic, with uniform concentration in each phase. The expression thus generated is most readily understood from the schematic description, Fig. 2:

$$A_i = L \cdot \left(\frac{\mathrm{d}f^{\mathrm{bcc}}}{\mathrm{d}t} \cdot Cl_i \left(K_i^{\mathrm{fcc}} - K_i^{\mathrm{bcc}} \right) + \frac{\mathrm{d}Cl_i}{\mathrm{d}t} \cdot \left(K_i^{\mathrm{bcc}} \cdot f^{\mathrm{bcc}} + K_i^{\mathrm{fcc}} \left(f - f^{\mathrm{bcc}} \right) \right) \right)$$

where subscript 'i' is restricted to interstitial elements.

Modification consistent with that described at the end of section 2 may be required if the peritectic occurs or persists towards the end of solidification, but its incorporation for nitrogen is problematic. (The problem does not occur for carbon in the present, peritectic formulation because its liquid content is controlled independently of the equations employed for the other elements; neither does it affect the formulation for the other elements because its liquidus gradient, β, is zero during this process.)

Table 1. Carbon equivalent coefficients, E_i.

	C	Si	Mn	P	S	Cr	Mo	Ni	Cu	V
Peritectic	(1)	−0.123	+0.04	–	+0.06	−0.018	−0.05	+0.08		
Eutectic	(1)	+0.31	−0.03	+0.33	–	*	–	+0.05	+0.075	−0.10

$E_{Cr} = -0.015Cr + 0.00216Cr^2$ (ref. [2]).
$E = C + \Sigma E_i C_i$.

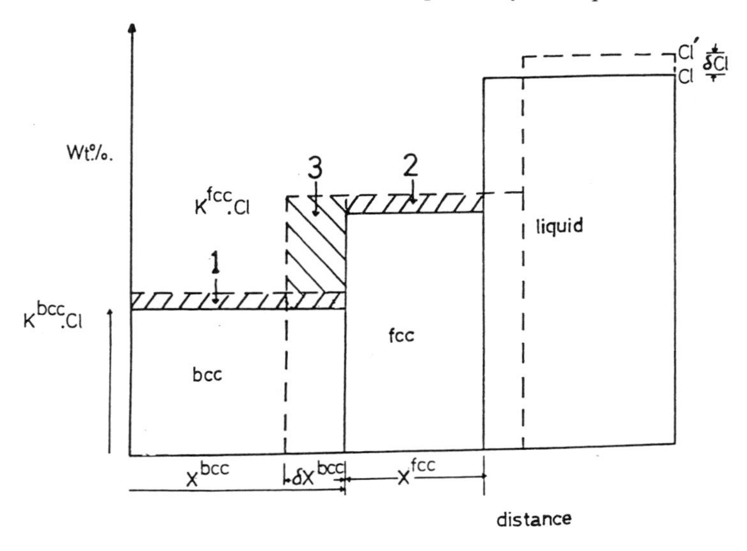

Fig. 2. Schematic representation of solute balance for interstitial elements during the peritectic reaction. Effective back-diffusion term in time δt comprises the three labelled areas, $1+2+3$: $\delta Cl \cdot (K^{bcc} \cdot X^{bcc}) + \delta Cl\ (K^{fcc} \cdot X^{fcc}) + \delta X^{bcc} \cdot Cl \cdot (K^{fcc} - K^{bcc})$. ———: position/concentration before solidification increment; — — —: position/concentration after solidification increment.

The diffusive behaviour of the 'slow' elements in substitutional solid solution is, at present, assumed to be unaffected by the solid state phase change, but superior formulations are being considered.

5. Initial operation and results

Some stability problems have been experienced with the program at high fractions solid. Depending on the composition, heat extraction rate and, notably, the mathematical description of the dendrite coarsening process, the program would not reliably achieve 100% solid under stable conditions. The schematic dendrite in Fig. 3 suggests that the greater inaccuracy in description of the dendrite should be at low fractions solid. The calculated results are, however, very insensitive to the morphological description at low fractions solid but surprisingly sensitive to representation of the coarsening process at completion of solidification. Further work is in hand to restructure or refine the iterative calculations under these conditions to make the program more robust in operation. It is hoped, for example, that the revised formulation described in section 2 will help in this respect. In the meanwhile, certain observations can be made but the program has yet to be proven and shown to be sufficiently reliable for confident application.

In steels of low, overall content of highly segregating species, the calculated solidus (an indicator of the "net" level of microsegregation) appears to be in

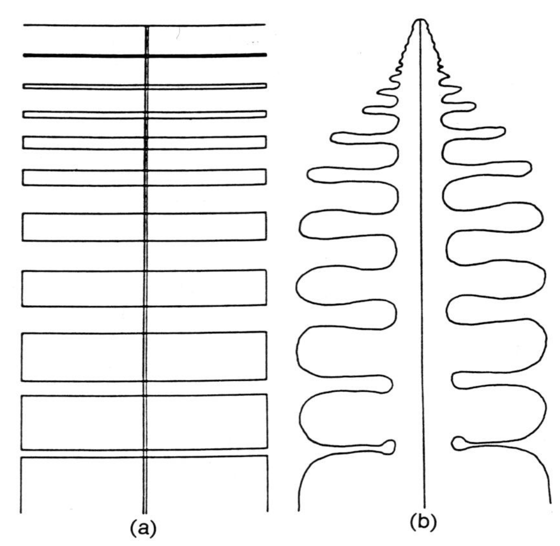

Fig. 3. Schematic dendrite structure implicit in 'MISEG' program. (a) *A 'MISEG' Dendrite*: Thickening and coarsening of secondary arms from a finite initial separation. No contribution from primary arms to solute redistribution. (b) *A More Realistic Representation*: Thickening and coarsening of secondary arms from zero length and separation at dendrite tip. Major contribution from primary arms to solute redistribution at early stages of solidification.

good agreement with that measured by thermal analysis experiments. Agreement is poor, however if significant amounts of strongly partitioning species such as carbon are present, in that the computed results fall substantially below such measurements, Table 2. (The experimental examples in Table 2 are taken from published work [5]) Indeed, the calculated results appear to reflect a real effect of persistent, highly segregated films of too small a volume fraction to be detected by thermal analysis. For example, several steels in reference [5] contained interdendritic austenite/carbide/phosphide eutectic which would only be encountered at well over 100 °C below the quoted, experimental solidus. Moreover, these films are now believed to be largely responsible for the total lack of measureable ductility down to temperatures previously described as substantially sub-solidus [6,7].

Table 2. Comparison of experiment [5] and computed solidification temperatures.

Temperature, °C	Steel 202 (low C)		Steel 310 (low C, high Cr)		Steel 205 (medium C)	
	Exp.	Comp.	Exp.	Comp.	Exp.	Comp.
Liquidus	1515	1514	1500	1498	1498	1498
Peritectic (max)	1475	1471	N/A	N/A	1480	1482
Solidus	1440	1447	1440	1438	1425	1392

The steel designations are those used in ref. [5], and are not international specifications.

The expected effects of manganese and sulphur on the solidus are observed, whereby a presence of manganese sufficient to remove the bulk of the sulphur from the residual liquid through precipitation of MnS markedly raises the calculated solidus temperature.

6. Conclusions

Significant progress has been made in the adaptation and extension of the Ogilvy-Kirkwood microsegregation model for multicomponent steels, allowing for eutectic termination and compound precipitation, and the peritectic phase transformation. The program is still under development in order to render it more 'robust' in operation, and requires more rigorously derived source data on multicomponent equilibria and the dendrite side-arm coarsening process. Initial results, however, are promising and it is expected that BSC will soon have a powerful, flexible tool for a quantitative understanding of the segregation process.

7. Acknowledgement

The author would like to thank Dr R. Baker, Director of Research, BSC, for permission to publish this paper.

References

1. A.J.W. Ogilvy and D.H. Kirkwood: A model for the numerical computation of microsegregation in alloys. *Appl. Sci. Res.* 44 (1987) 43–49.
2. A.J.W. Ogilvy: PhD Thesis, Sheffield University (1983).
3. H.T. Angus: *Cast Iron*, Butterworths (1976).
4. K. Schwerdtfeger: *Arch. Eisenhutten*. 41 (1970) 923–937.
5. *A Guide to the Solidification of Steels*, Jernkontoret, Stockholm (1977).
6. M.C.M. Cornelissen: paper presented at 'Progress in Modelling of Solidification Processes', Sheffield University (1986) (Inst. of Metals).
7. E. Schmidtmann and F. Rakoski: *Stahl. u. Eisen* 103 (18) (1983) 881–882.

Applied Scientific Research 44: 61–92 (1987)

Finite element simulation of solidification problems

R.W. LEWIS & P.M. ROBERTS

Institute for Numerical Methods in Engineering, University College of Swansea, University of Wales, Singleton Park, Swansea SA2 8PP, United Kingdom

Abstract. The modelling of liquid-solid phase change phenomena is extremely important in many areas of science and engineering. In particular, the solidification of molten metals during various casting methods in the foundry, provides a source of important practical problems which may be resolved economically with the aid of computational models of the heat transfer processes involved. Experimental design analysis is often prohibitively expensive, and the geometries and complex boundary conditions encountered preclude any analytical solutions to the problems posed. Thus the motivation for numerical simulation and computer aided design (CAD) systems is clear, and several mathematical/computational modelling techniques have been brought to bear in this area during recent years.

This paper reports on the application of the finite element method to solidification problems, principally concerning industrial casting processes. Although convective heat transfer has been modelled, the work herein considers only heat conduction, for clarity. The heat transfer model has also been coupled with thermal stress analysis packages to predict mechanical behaviour including cracking and eventual failure, but this is reported elsewhere.

Following the introduction, the mathematical and computational modelling tools are described in detail, for completeness. A discussion on the handling of the phase change interface and latent heat effects is then presented. Some aspects of the solution procedures are examined next, together with special techniques for dealing with the mold-metal interface. Finally, some numerical examples are presented which substantiate the capabilities of the finite element model, in both two and three dimensions.

Nomenclature

c = heat capacity
$\boldsymbol{C}$ = capacitance matrix
f = time function
$\boldsymbol{F}$ = loading term
h = heat convection coefficient
H = specific enthalpy
$|J|$ = Jacobian determinant
$|\hat{J}|$ = patch approximation to $|J|$
k = thermal conductivity
$\boldsymbol{K}$ = conductance matrix
L = latent heat
$\hat{\boldsymbol{n}}$ = unit outward normal
N_i = nodal shape function
q = known heat flux
R_i = nodal heat capacity
S = phase change interface
t = time
T = temperature
$\hat{T}$ = known boundary temperature
$\boldsymbol{T}$ = vector of nodal temperatures
T_a = ambient temperature
T_c = solidification temperature
T_L = liquidus temperature
T_0 = initial temperature
T_s = solidus temperature
$\boldsymbol{x}$ = space coordinates
α = interface heat transfer coefficient
γ = iteration parameter
Γ = boundary of domain
δT = solidification range
Δt = timestep magnitude
∇ = vector gradient operator
ε = convergence tolerance
θ = timestepping parameter
Θ = known vector in alternating-direction formulation
λ = Laplace modifying parameter
(ξ, η) = local space coordinates
ρ = density
τ = time limit
$\phi(\xi)$ = shape function factor
$\psi(\eta)$ = shape function factor
Ω = domain of interest

1. Introduction

The growing use of computational modelling techniques in recent years reflects the potential economic benefits they offer to many industrial processes, including casting methods in the foundry. Numerical simulations of the thermal, and mechanical, behaviour of both castings and moulds, will permit designers to create robust, effective products, whilst simultaneously minimising process costs and waste. To achieve a practical simulation package, a heat transfer analysis program is required at the core of an overall flexible CAD system, incorporating pre- and post-processing graphics and other geometric modelling tools. With such a system the designer can reliably predict hot spots which could result in shrinkage porosity; alterations in geometric configuration or to process parameters will eliminate such risks and ensure directional solidification in advance of any casting being poured. The desired mechanical properties of the finished casting can be controlled by predicting, via numerical experimentation, the effects of using differing mould materials and the placement of various chills and/or cooling lines. Waste associated with risers can be significantly reduced by optimising their magnitude and location.

This paper focusses on the mathematical model which is the basis of the numerical heat transfer analysis program. The development of a non-linear heat conduction simulator which uses the finite element method is presented. Finite elements are employed because of their facility in modelling complex domain configurations, and handling of non-linear boundary conditions. For completeness some details of the discretisation process are given together with a brief description of the mathematical model, i.e. differential equation and boundary conditions, involved.

Various timestepping methods are considered for use in conjunction with the Galerkin semi-discrete weighted residual form of the differential equation, and the treatment of the non-linear terms which arise is also discussed. The representation of latent heat effects is an important feature in the modelling of solidification problems. The tracking of the phase change boundary is an important numerical problem, which is resolved in our work by use of the enthalpy method which incorporates latent heat in a smooth manner.

Techniques for improving the efficiency of the simulation program are then looked at, including an alternating-direction implicit, and a 'mixed' implicit/explicit solution algorithm coupled with a direct profile solver. A quadratic conductivity approximation is also presented which speeds up matrix generation.

A crucial component in simulating casting solidification is the handling of the heat transfer at the interface between the mould and the metal, a critical process parameter acutely affecting the solidification rate. The heat transfer coefficient can vary in response to type and thickness of die coat, to the pressure, temperature and cooling rate of the metal, and due to the formation of an air gap. A coincident node technique is illustrated for dealing with the air gap problem. A short description of an infinite element approach for representing heat losses to an external medium is also given.

The numerical results from a series of simulations of practical problems are then shown. The problems solved for include the techniques presented earlier, and involve both two and three-dimensional meshes.

2. Heat transfer models

2.1. *Mathematical models*

The mathematical model consists of the differential equation governing the conduction of heat through a medium, coupled with appropriate boundary conditions. The Fourier law of conservation of energy may be expressed as

$$\nabla \cdot (k\nabla T) = \rho c\dot{T}; \quad (\boldsymbol{x}, t) \in \Omega \times [0, \tau]. \tag{1}$$

The form of boundary condition is usually of two types, i.e. Dirichlet (known temperature) or Neuman (known heat flux), or can be a mixture of both. If we denote the distinct boundary regions Γ_1, Γ_2 of the domain of Ω (see Fig. 1) where Dirichlet and Neuman conditions are known, then we have

$$T(\boldsymbol{x}, t) = \hat{T}(\boldsymbol{x}, t); \quad (\boldsymbol{x}, t) \in \Gamma_1 \times [0, \tau] \tag{2}$$

and

$$k \nabla T \cdot \hat{\boldsymbol{n}} + q + h(T - T_a) = 0; \quad (\boldsymbol{x}, t) \in \Gamma_2 \times [0, \tau] \tag{3}$$

where $\Gamma_1 \cap \Gamma_2 = \phi$.

Since we are dealing with a transient problem we must also specify the initial state, or temperature distribution, giving the initial condition

$$T(\boldsymbol{x}, 0) = T_0(\boldsymbol{x}); \quad \boldsymbol{x} \in \Omega. \tag{4}$$

In situations involving a change of phase, the classical Stefan problem is used

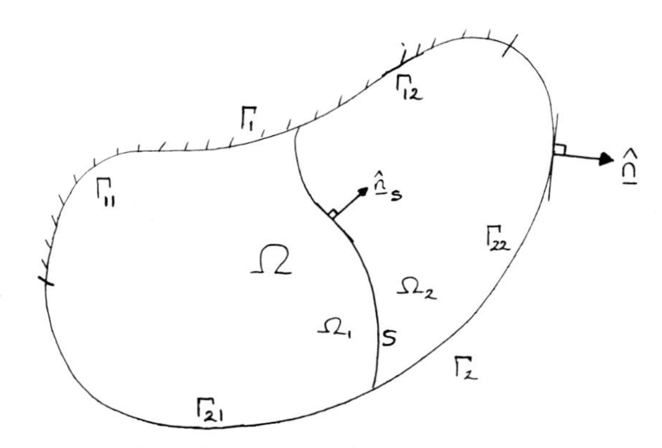

Fig. 1. Domain of interest and phase change interface.

to describe the conduction in a domain comprising two separate phases. Thus on the subdomains Ω_1, Ω_2 we have ($\Omega_1 \cup \Omega_2 = \Omega$, see Fig. 1)

$$\nabla \cdot (k_i \nabla T_i) = \rho_i c_i \dot{T}_i; \quad (\boldsymbol{x}, t) \in \Omega_i \times [0, \tau] \tag{5}$$

$$T_i(\boldsymbol{x}, t) = \hat{T}_i(\boldsymbol{x}) \cdot f_i(t); \quad (\boldsymbol{x}, t) \in \Gamma_{1i} \times [0, \tau] \tag{6}$$

$$k_i \nabla T_i \cdot \hat{\boldsymbol{n}} + q_i + h_i(T_i - T_{ai}) = 0; \quad (\boldsymbol{x}, t) \in \Gamma_{2i} \times [0, \tau] \tag{7}$$

$$T_i(\boldsymbol{x}, 0) = T_{0i}(\boldsymbol{x}); \quad \boldsymbol{x} \in \Omega_i. \tag{8}$$

Furthermore, since the interface S is moving, we must have the interface condition

$$k_1 \nabla T_1 \cdot \hat{\boldsymbol{n}}_s - k_2 \nabla T_2 \cdot \hat{\boldsymbol{n}}_s = \rho L \dot{S}. \tag{9}$$

In the above form, the Stefan problem requires a front tracking algorithm to determine the interface S at any given time – this requires special timestepping or deforming mesh techniques. However, a 'weak' form of the above problem can be posed by defining the enthalpy as

$$H = \int_0^T \rho c \, \mathrm{d}T \tag{10}$$

or

$$\rho c = \frac{\mathrm{d}H}{\mathrm{d}T}. \tag{11}$$

Then Eq. (1) (or Eq. (5)) becomes

$$\nabla \cdot (k \nabla T) = \frac{\mathrm{d}H}{\mathrm{d}T} \dot{T} \tag{12}$$

or

$$\nabla \cdot (k \nabla T) = \dot{H}. \tag{13}$$

Thus in Eq. (11) we are only solving for temperature; the front tracking problem has been 'removed', and the latent heat L will be included in the definition of enthalpy H (see below).

2.2. Numerical models

2.2.1. Spatial discretisation. The finite element method has been chosen to discretise the domain Ω in space, rather than by using finite differences, on the basis of its inherent flexibility in handling complicated shapes and boundary conditions. Thus the domain Ω (2- or 3-dimensional) is divided up into distinct regions or elements Ω_e such that

$$\bigcup_{e=1}^{M} \Omega_e = \Omega.$$

Nodal points are then distributed at element vertices, along the element edges, and possibly in the element interiors, as in Fig. 2. The usual element

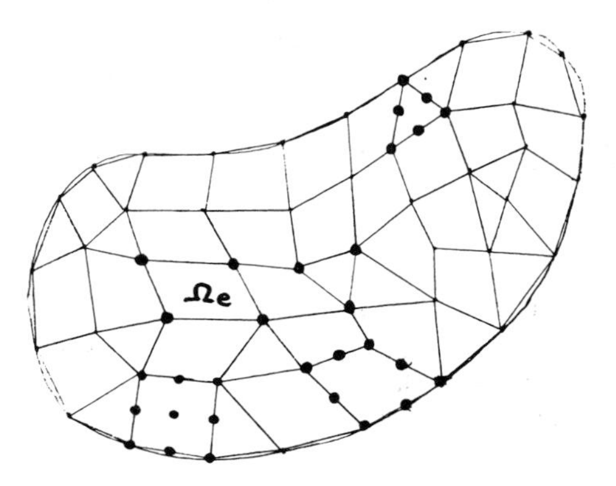

Fig. 2. Schematic discretisation of general domain into finite elements.

shapes consist of triangles and quadrilaterals, with tetrahedra or parallelepipeds in 3-dimensions. The application of the finite element method to the heat conduction equation is well documented, and indeed appears as a standard example in nearly all finite element textbooks [1]. Therefore, only a brief description is warranted here. We approximate the temperature distribution by a weighted series of locally supported polynomial functions – the weights/amplitudes correspond to the temperatures at the nodal points or 'nodes'. Thus we have the approximation

$$\tilde{T}(\boldsymbol{x}, t) = \sum_{i=1}^{N} N_i(\boldsymbol{x}) T_i(t) \tag{14}$$

where N_i are the basis, or shape functions and T_i the nodal temperatures. Substituting this approximation into the Galerkin weighted residual form of the differential equation, Eq. (12), we obtain the semi-discrete system

$$\boldsymbol{C}\dot{\boldsymbol{T}} + \boldsymbol{K}\boldsymbol{T} = \boldsymbol{F} \tag{15}$$

where $\boldsymbol{T}$ is the vector $\{T_i\}$ of unknown nodal temperatures. The capacitance matrix $\boldsymbol{C}$, the conductance matrix $\boldsymbol{K}$, and the thermal loading term $\boldsymbol{F}$ are defined as follows:

$$C_{ij} = \int_\Omega \frac{\mathrm{d}H}{\mathrm{d}T} N_i N_j \,\mathrm{d}\Omega \tag{16}$$

$$K_{ij} = \int_\Omega \nabla N_j \cdot (k \nabla N_i) \,\mathrm{d}\Omega + \int_{\Gamma_2} h N_i N_j \,\mathrm{d}\Gamma_2 \tag{17}$$

$$F_i = -\int_{\Gamma_2} (q - hT_a) n_i \,\mathrm{d}\Gamma_2. \tag{18}$$

The appearance of the boundary integral terms in Eqs. (17), (18) takes into account the flux boundary condition Eq. (3). The symmetric form of $\boldsymbol{K}$ reflects the implicit use of a Green's identity ('integration by parts') to obtain the weak form of the residual equation. Given the integrals of Eqs. (16)–(18), we still have to evaluate them numerically. Analytical integration can only be performed on simple, regular elements, and thus recourse to numerical integration is necessary. To facilitate this, the irregular element configurations are mapped onto regular domains, typically $[-1, 1]^n$, via isoparametric transformations [1]. The isoparametric mappings use the same functions N_i which form the basis of the approximation space. Thus boundary curves of the same order as the polynomial functions N_i can be represented exactly. Having mapped each element onto the same convenient domain we then have to choose a quadrature rule which is sufficiently accurate, but which does not require excessive, even redundant, computation. A popular choice meeting these requirements is that of the Gauss-Legendre quadrature rule which places the sampling points at the zeroes of Legendre polynomials. In one dimension a single point integrates exactly a straight line, two points a cubic polynomial, three points a quintic polynomial and so on. 'Standard' Gauss-Legendre integration rules have been employed throughout in the heat transfer programs developed, although for 3-dimensional elements cheaper schemes with fewer points but similar accuracy could have been utilised.

2.2.2. Timestepping algorithms. So far we have only considered the spatial discretisation of our governing partial differential equation, hence giving the 'semi-discrete', ordinary, differential system equation (15)

$$\boldsymbol{C}\dot{\boldsymbol{T}} + \boldsymbol{K}\boldsymbol{T} = \boldsymbol{F}$$

in which the time derivative remains. We now have to address ourselves to the question of temporal discretisation. Many possible timestepping algorithms are possible for this first order equation. However, we have decided on the use of a straightforward two-level scheme which can vary between explicit (no matrix inversion, timestep limited by stability) and implicit (matrix inversion, unconditional stability) solution strategies. The scheme is frequently referred to as the 'θ-method' after the usual notation used for the free parameter. The algorithm can be expressed in the form

$$\boldsymbol{C}\frac{(\boldsymbol{T}^{n+1} - \boldsymbol{T}^n)}{\Delta t} + \theta \boldsymbol{K}\boldsymbol{T}^{n+1} + (1-\theta)\boldsymbol{K}\boldsymbol{T}^n = \boldsymbol{F}; \quad \theta \in [0, 1]$$

or, more concisely

$$[\boldsymbol{C} + \theta\, \Delta t \boldsymbol{K}]\boldsymbol{T}^{n+1} = [\boldsymbol{C} + (\theta - 1)\ \Delta t \boldsymbol{K}]\boldsymbol{T}^n + \Delta t \boldsymbol{F} \tag{19}$$

where the superscript denotes the time level, i.e. $\boldsymbol{T}^n = \boldsymbol{T}(\boldsymbol{n}\ \Delta t)$. Note that $\theta = 0$ corresponds to an explicit scheme, requiring no matrix inversion providing $\boldsymbol{C}$ is diagonalised or 'lumped', and $\theta = 1$ corresponds to a fully implicit scheme. Both these schemes have first order, $0(\Delta t)$, accuracy. The choice $\theta = \frac{1}{2}$ corresponds to the well known Crank-Nicolson method, which is second order, $0(\Delta t^2)$, accurate but can be suspect to spurious oscillations, or 'noise', for larger timesteps – it is marginally stable. It is a two-level scheme since the unknown values at t^{n+1} are determined completely from the previous, known, values at t^n. Other values for θ can of course be used, $\theta = \frac{2}{3}$ has been recommended, having been derived from finite element timestepping – where the time domain is notionally discretised by 'shape' functions in the time 'direction'. Values of $\theta \geqslant \frac{1}{2}$ lead to an unconditionally stable scheme – i.e. errors in the solution do not grow, but for $\theta < \frac{1}{2}$ the time step magnitude, Δt, must be kept within the Courant limit – proportional to the element size and diffusion term k, and inversely proportional to the capacitance term ρc or $(\mathrm{d}H/\mathrm{d}T)$.

The 'θ-method' has been used extensively in the examples to be presented later, but other algorithms have been investigated during the development of the heat conduction program [2,3]. These includes the three level scheme of Lees [4], which can be expressed as

$$\boldsymbol{C}\frac{(\boldsymbol{T}^{n+2} - \boldsymbol{T}^{n})}{2\,\Delta t} + \boldsymbol{K}\frac{(\boldsymbol{T}^{n+2} + \boldsymbol{T}^{n+1} + \boldsymbol{T}^{n})}{3} = \boldsymbol{F}$$

or

$$\left[\boldsymbol{C} + \frac{2\,\Delta t}{3}\boldsymbol{K}\right]\boldsymbol{T}^{n+2} = \left[-\frac{2\,\Delta t}{3}\boldsymbol{K}\right]\boldsymbol{T}^{n+1} + \left[\boldsymbol{C} - \frac{2\,\Delta t}{3}\boldsymbol{K}\right]\boldsymbol{T}^{n} + \Delta t\boldsymbol{F}. \tag{20}$$

The three-level recurrence relation (20) solves for $\boldsymbol{T}$ at t^{n+2}, given *two* previous values $\boldsymbol{T}^{n+1}$, $\boldsymbol{T}^{n}$. It is unconditionally stable, but requires some form of start-up e.g. by use of the above two level scheme, or by using $\boldsymbol{T}^1 = \boldsymbol{T}^0$ and a very small initial timestep (the use of different timesteps Δt between $t^n \rightarrow t^{n+1}$ and $t^{n+1} \rightarrow t^{n+2}$ requires some modification to Eq. (20)). Although stable, our experience has shown that solutions exhibited oscillatory, 'noisy' behaviour – this was, however, considerably attenuated by redefining $\boldsymbol{T}^n$ in equation (20) to be the average [3]

$$\bar{\boldsymbol{T}}^{n} \rightarrow \frac{(\boldsymbol{T}^{n+1} + \boldsymbol{T}^{n} + \boldsymbol{T}^{n-1})}{3}.$$

Another timestepping algorithm which has been utilised is that of a Laplace modified scheme [5], used in conjunction with the finite element alternating-direction method [2,6,7]. More details regarding the solution strategy of the alternating-direction technique will be given below. However, it is pertinent to include the Laplace modified algorithm in this section on timestepping procedures. Applied to Eq. (12), a first order Laplace-modified algorithm, within a Galerkin weighted residual form, gives

$$\begin{aligned}
&\int_\Omega (T^{n+1} - T^{n}) \cdot N\,|\hat{J}|\,\mathrm{d}\Omega + \lambda\,\Delta t\int_\Omega \nabla'(T^{n+1} - T^{n}) \cdot \nabla' N\,|\hat{J}|\,\mathrm{d}\Omega \\
&\quad + \lambda^2\,\Delta t^2\int_\Omega \frac{\partial^2}{\partial x\,\partial y}(T^{n+1} - T^{n})\frac{\partial^2 N}{\partial x\,\partial y}\,|\hat{J}|\,\mathrm{d}\Omega \\
&\quad = -\Delta t\int_\Omega \frac{k}{\mathrm{d}H/\mathrm{d}T}\nabla T^{n} \cdot \nabla N\,|J|\,\mathrm{d}\Omega
\end{aligned} \tag{21}$$

where the domain is transformed onto a union of rectangular (cuboids in 3-dimensions) regions. On the left hand side it can be seen that we are solving for *differences* in temperature over the interval $[t^n, t^{n+1}]$. The term λ – the Laplace modifier – is an approximation to the diffusive term k. For constant coefficients, and ignoring the higher order, $O(\Delta t^2)$, term, setting $\lambda = 0, \frac{1}{2}k, k$ reduces Eq. (20) to explicit, Crank-Nicolson, and implicit schemes respectively. The perturbation term, $\alpha\lambda^2$, is essential to the formulation of the alternating-direction finite element scheme, and its form closely resembles that which can be derived from the analogous finite difference method. $|J|$ denotes the Jacobian of the mesh transformation onto the regular form, and $|\hat{J}|$ denotes its 'patch' approximation [6], discussed below.

2.2.3. Nonlinear problems. Although tacitly assumed, our notation so far has ignored the possible dependencies of the coefficients k, ρ, c, H on the independent variables space, time and temperatures. Thus Eq. (12) should be written as

$$\nabla \cdot (k(\boldsymbol{x}, t, T)\, \nabla T) = \frac{\mathrm{d}H}{\mathrm{d}T}(\boldsymbol{x}, t, T) \cdot \dot{T}. \tag{22}$$

Variations in space and time do not present any special difficulties – apart from discontinuities. More serious, however, is the dependence of the coefficients k, $(\mathrm{d}H/\mathrm{d}T)$ on the actual solution, i.e. the temperature distribution. We are faced with the problem of how to determine the coefficients, i.e. what temperature values should be used. The matrix problem is given by (ignoring spatial/temporal coefficient dependencies)

$$\boldsymbol{C}(\boldsymbol{T})\dot{\boldsymbol{T}} + \boldsymbol{K}(\boldsymbol{T})\boldsymbol{T} = \boldsymbol{F}(\boldsymbol{T}). \tag{23}$$

Put into the timestepping algorithm given by Eq. (19), we have

$$[\boldsymbol{C}(\boldsymbol{T}) + \theta\, \Delta t \boldsymbol{K}(\boldsymbol{T})] \cdot \boldsymbol{T}^{n+1} = [\boldsymbol{C}(\boldsymbol{T}) + (\theta - 1)\, \Delta t \boldsymbol{K}(\boldsymbol{T})] \cdot \boldsymbol{T}^n + \Delta t \boldsymbol{F}(\boldsymbol{T}) \tag{24}$$

or, simply,

$$\boldsymbol{L}(\boldsymbol{T}) \equiv \boldsymbol{P}(\boldsymbol{T})\boldsymbol{T} - \boldsymbol{G}(\boldsymbol{T}) = 0. \tag{25}$$

For such nonlinear problems some form of iterative solution procedure is required (linear problems can be considered as only requiring *one* iteration!). Taking a truncated Taylor series expansion of Eq. (25) we obtain the well known Newton-Raphson iteration scheme

$$\begin{cases} \boldsymbol{T}_0^n = \boldsymbol{T}^n \\ \boldsymbol{T}_{k+1}^{n+1} = \boldsymbol{T}_k^n + \Delta \boldsymbol{T}_k \quad k = 0, 1, 2 \ldots \\ \Delta \boldsymbol{T}_k = -\left(\dfrac{\mathrm{d}}{\mathrm{d}\boldsymbol{T}} \boldsymbol{L}(\boldsymbol{T}_k^n)\right)^{-1} \boldsymbol{L}(\boldsymbol{T}_k^n) \end{cases} \tag{26}$$

where k denotes the number of iterations. The 'Jacobian' matrix (or 'tangential' stiffness matrix) requires the calculation of derivatives. Only in special circumstances can these derivatives be determined analytically – for 'real' problems the only recourse is for approximate numerical derivatives, which represents a substantial computing cost. Furthermore, the Jacobian matrix has to be reconstructed at every iteration. Thus, although the Newton-Raphson scheme gives quadratic convergence rates and is guaranteed to converge (provided the initial 'guess' $\boldsymbol{T}_0^n$ is 'close enough' to the solution $\boldsymbol{T}^{n+1}$), it is very expensive to operate. Modified Newton-Raphson schemes, which iterate directly using the first Jacobian constructed, only require one matrix construction per time step, but the convergence characteristics are frequently little better than direct iteration which costs less.

In our experience the successive substitution procedure has proved adequate, but Newton-Raphson type schemes may be required for highly non-linear problems to achieve acceptable convergence characteristics.

The scheme we have employed is based on direct, successive substitution iteration. For reasonable timesteps, $\boldsymbol{T}^n$ is often close enough to $\boldsymbol{T}^{n+1}$ to ensure convergence by the direct approach. The algorithm adopted is of a 'predictor-corrector' form, viz (from Eq. (24)) [8].

Predictor:

$$[\boldsymbol{C}(\boldsymbol{T}^n) + \Delta t\theta \boldsymbol{K}(\boldsymbol{T}^n)]\,\boldsymbol{T}_*^{n+1} = [\boldsymbol{C}(\boldsymbol{T}^n) + \Delta t(\theta - 1)\boldsymbol{K}(\boldsymbol{T}^n)]\,\boldsymbol{T}^n + \Delta t\boldsymbol{F}(\boldsymbol{T}^n)$$

or

$$[\boldsymbol{C}^n + \Delta t\theta \boldsymbol{K}^n]\boldsymbol{T}_*^{n+1} = [\boldsymbol{C}^n + \Delta t(\theta - 1)\boldsymbol{K}^n]\,\boldsymbol{T}^n + \Delta t\boldsymbol{F}^n \tag{27}$$

Corrector:

$$\left[\boldsymbol{C}(\boldsymbol{T}_p^{\bar{n}}) + \Delta t\theta \boldsymbol{K}(\boldsymbol{T}_p^{\bar{n}})\right]\boldsymbol{T}_{p+1}^{n+1} = \left[\boldsymbol{C}(\boldsymbol{T}_p^{\bar{n}}) + \Delta t(\theta - 1)\boldsymbol{K}(\boldsymbol{T}_p^{\bar{n}})\right]\boldsymbol{T}^n + \Delta t\boldsymbol{F}(\boldsymbol{T}_p^{\bar{n}})$$

or

$$\left[\boldsymbol{C}_p^{\bar{n}} + \Delta t\theta \boldsymbol{K}_p^{\bar{n}}\right]\boldsymbol{T}_{p+1}^{n+1} = \left[\boldsymbol{C}_p^{\bar{n}} + \Delta t(\theta - 1)\boldsymbol{K}_p^{\bar{n}}\right]\boldsymbol{T}^n + \Delta t\boldsymbol{F}_p^{\bar{n}} \tag{28}$$

where $p = 0, 1, 2, \ldots$ (i.e. p is the iteration counter)

$$\boldsymbol{T}_p^{\bar{n}} = \gamma \boldsymbol{T}_p^{n+1} + (1-\gamma)\boldsymbol{T}^n;\ \ \gamma \in (0, 1] \tag{29}$$

$$\boldsymbol{T}_0^{n+1} = \boldsymbol{T}_*^{n+1}.$$

The weighting parameter γ, which allows for variation in the choice of temperature values for coefficient calculation, is chosen to take the same value as θ and in most of our practical tests both are set at $\frac{1}{2}$ giving improved accuracy and convergence.

With this predictor-corrector scheme, we control the timestep by the number of iterations (corrections) required to reach convergence. Convergence is deemed to have been reached when $\| T_{p+1}^{n+1} - T_p^{n+1} \| < \varepsilon$, for some p and a user specified tolerance ε. If the number of iterations required becomes excessive then the timestep Δt is reduced; conversely, if very few iterations are needed then Δt may be increased. This straightforward, direct iteration, method has proven to be successful in all our tests, and is very economical compared with Newton type iteration techniques.

3. Latent heat representation and phase change

The above heat conduction model must take into account the effect of latent heat if it is to be used to simulate solidification. The latent heat L may be used to determine the progression of the phase change interface S in the formulation of the Stefan problem Eq. (5)–(9). As we have noted this formulation requires some form of sophisticated front tracking algorithm. However there is a simpler approach we can follow [9]. The latent heat effect is approximated by a sharp increase in heat capacity within a narrow temperature range about the solidification point, T_c, as illustrated schematically in Fig. 3. The temperature range δT from the solidus temperature, T_s, to the liquids value, T_L, can be extremely small so that for many substances the heat capacity curve $\rho c(T)$ exhibits a discontinuous Dirac function at T_c. Since we are using numerical integration, which is only adequate for relatively smooth curves, the direct calculation of ρc as a coefficient will lead to very poor results.

However, the integral of the heat capacity, the physical quantity known as the enthalpy H, as given by Eq. (10)

$$H = \int_0^T \rho c \, \mathrm{d}T$$

is a smooth function of temperature in the phase change zone. Thus, it is preferable to use the enthalpy curve $H(T)$, rather than the specific heat

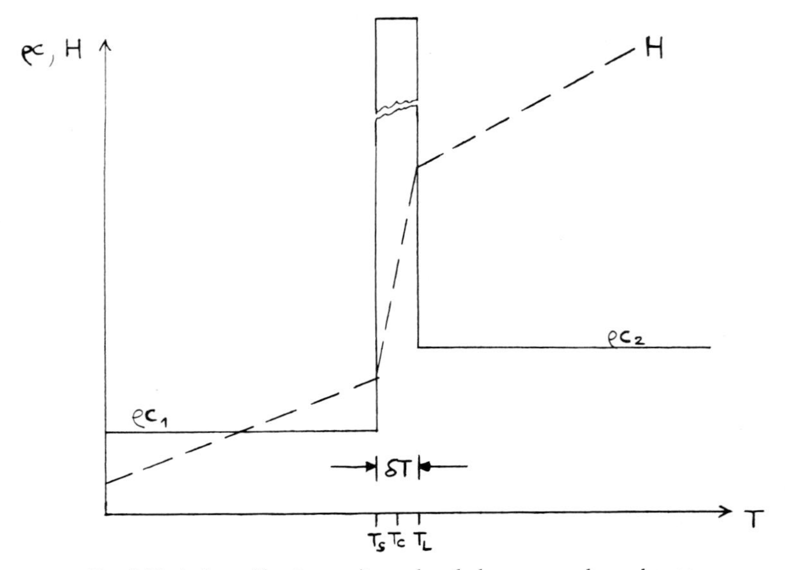

Fig. 3. Variation of heat capacity and enthalpy near a phase change.

capacity $\rho c(T)$ curve. Hence we interpolate for the enthalpy in the same way as for the temperature, i.e.

$$H = \sum_{i=1}^{N} N_i(\boldsymbol{x}) H_i(t). \tag{30}$$

The enthalpy values are then utilised in deriving the temperature derivative $(\mathrm{d}H/\mathrm{d}T)$ for the integrand of entries in the capacitance matrix , see Eq. (16). Several averaging techniques have been used to determine this derivative term, such as

$$\frac{\mathrm{d}H}{\mathrm{d}T} \cong \frac{1}{3} \sum_{i=1}^{3} \left\{ \frac{\partial H}{\partial x_i} \bigg/ \frac{\partial T}{\partial x_i} \right\} \tag{31}$$

or

$$\frac{\mathrm{d}H}{\mathrm{d}T} \cong \frac{\left\{ \sum_{i=1}^{3} (\partial H/\partial x_i)^2 \right\}^{1/2}}{\left\{ \sum_{i=1}^{3} (\partial T/\partial x_i)^2 \right\}^{1/2}}. \tag{32}$$

However, a simple backward difference approximation [10]

$$\left(\frac{\mathrm{d}H}{\mathrm{d}T} \right)^n \cong \frac{H^n - H^{n-1}}{T^n - T^{n-1}} \tag{33}$$

has proven to be successful, giving excellent results for reasonable timestep sizes. This last scheme is computationally quicker than the averaging approach, but does require the storage of previous enthalpy values at nodes. The interpolation of H onto its nodal values saves a significant amount of computing since H does not have to be evaluated from the $H(T)$ curve at every integration sampling point – the practice of interpolating for coefficients using the same basis functions as for the unknown variables is sometimes referred to as a 'group formulation'.

The use of the enthalpy method maintains a correct heat balance provided the timestep is not so large as to involve great temperature changes at the nodes, over one time increment, when δT is very small. For many metals and alloys δT is relatively large so that the above technique works well, as will be illustrated by the examples given below.

4. Novel solution techniques

In this section we detail some of the methods employed to increase the computing efficiency of the overall solution process associated with the finite element method of discretisation, as applied to solidification problems and casting techniques in particular.

4.1. Alternating-direction finite element method

The use of alternating-direction implicit (ADI) solution algorithms is well known to finite difference modellers due to their efficiency both in solving transient problems, and as a rapid iteration scheme for steady-state, elliptic, problems. An alternating-direction (AD) finite element scheme has been utilised by the present authors to compare its efficiency with standard solution methods [2]. The AD method involves the solution of matrix systems congruent to its finite difference namesake, but the derivation of these systems is not analogous. The finite difference approach is via operator splitting, i.e. solving or 'sweeping' along orthogonal directions, resulting in reduced bandwidth systems – typically tridiagonal matrices for standard central difference approximations.

However, the AD finite element algorithm relies on the tensor product form of the shape functions N associated with the Lagrangian 'family' of elements, i.e. the shape functions can be factored into directional parts, $N_i(\xi, \eta) = \phi_i(\xi)\psi_i(\eta)$. Using this fact, and introducing some reasonable numerical artifices, we can factor the overall system matrix into directional components, e.g. $K = K_x K_y K_z$, where each of K_x, K_y, K_z has very narrow bandwidth (tridiagonal for linear elements) for suitable nodal orderings.

We have already met the associated timestepping scheme, a Laplace modified algorithm in Eq. (21). The scheme given there could use standard finite element methodology to give a perfectly acceptable discrete system. However, to understand how it represents an AD formulation it is necessary to detail the underlying 'philosophy' of the discretization process. Such detail can be found in references [2,5,6] – only a brief sketch of the technique will be given here.

Firstly, the AD technique is based on regular, rectangular in two dimensions, meshes. However, unions of such meshes can be handled, and coupled with the isoparametric transformation a great variety of mesh configurations can be undertaken. To facilitate description, a simple rectangle is taken, as in Fig. 4. The general region Ω_g, Fig. 4(a), in 'global' coordinates (x, y) is mapped onto the rectangular mesh Ω, Fig. 4(b), in coordinates (X, Y). Thereafter each element is mapped onto $[-1, 1]^2$ in (ξ, η) coordinates for numerical integration (Gauss-Legendre). The reason for the appearance of the 'patch' approximation $|\hat{J}|$ to the actual Jacobian $|J|$, for the transformation of Ω_g onto Ω, is that $|J|$ is not of the tensor product from required to factorise the integrals of (on the left hand side of) Eq. (21) into directional components. Following [6], we adopt the following patch approximation

$$|\hat{J}|_{ij} = \sqrt{|J|_i} \cdot \sqrt{|J|_j}$$

where $|J|_i$ is the mean value of $|J|$ at node i, with the average being determined over the patch of elements intersecting at node i. The system Eq. (21) can be written in matrix notation as

$$\bar{\boldsymbol{K}}(\boldsymbol{T}^{n+1} - \boldsymbol{T}^n) = \Delta t \Theta^n \tag{35}$$

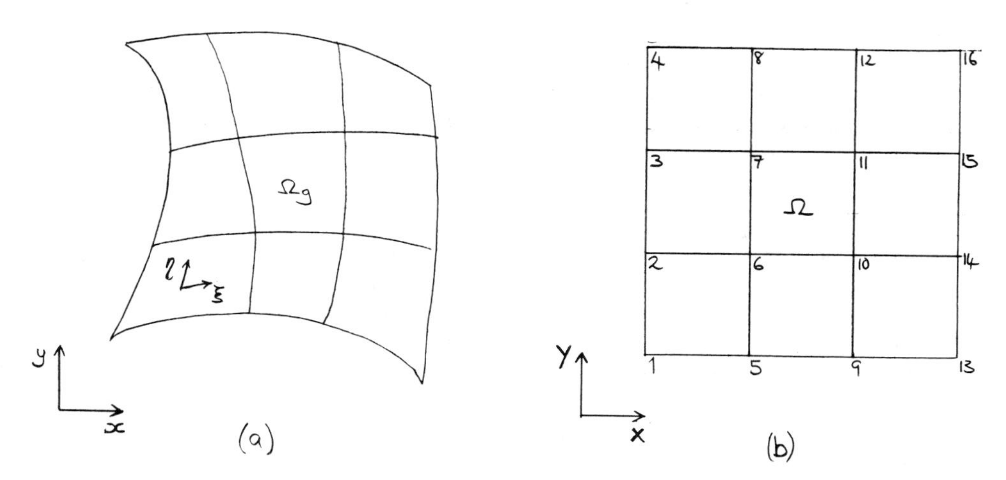

Fig. 4. Finite element mesh for alternating-direction formulation (a) general region Ω_g, (b) isoparametric base mesh Ω.

where

$$\Theta^n = \left\{ - \int_\Omega \frac{k^n}{(\mathrm{d}H/\mathrm{d}T)^n} \nabla T^n \cdot \nabla N \,|\, J \,|\, \mathrm{d}\Omega \right.$$

$$\left. + \int_\Omega (\,|\, J \,|-|\, \hat{J} \,|\,) \left(\frac{\boldsymbol{T}^n - \boldsymbol{T}^{n-1}}{\Delta t} \right) N \,\mathrm{d}\Omega \right\}. \tag{36}$$

The term involving $(\,|\,J\,|-|\,\hat{J}\,|\,)$ is an error-correction term which can be omitted for rectangular meshes, or for meshes in which the elements are not badly distorted. Now, because of the form of $|\,J\,|$ (used throughout in the left hand side of equations (21), (35)) the system matrix can be factored as

$$\bar{\boldsymbol{K}} = \boldsymbol{D}^{1/2} \boldsymbol{K} \boldsymbol{D}^{1/2} \tag{37}$$

where $\boldsymbol{D} = \mathrm{diag}\{\,|\,J\,|_i\}$. Furthermore the tensor product form of the shape functions, and the inclusion of the $0(\Delta t^2)$ perturbation term enables the following factorisation, with $N_i(X, Y) = \phi_i(X)\psi_i(Y)$, in two dimensions,

$$K_{ij} = \int_{\Omega_{ij}} \left(N_i N_j + \lambda \,\Delta t \,\nabla N_i \cdot \nabla N_j + \lambda^2 \,\Delta t^2 \frac{\partial^2 N_i}{\partial X\, \partial Y} \frac{\partial^2 N_j}{\partial X\, \partial Y} \right) \mathrm{d}X \,\mathrm{d}Y$$

$$= \int_{\Omega_{ij}^x} (\phi_i \phi_j + \lambda \,\Delta t \phi_i' \phi_j') \,\mathrm{d}X \cdot \int_{\Omega_{ij}^y} (\psi_i \psi_j + \lambda \,\Delta t \psi_i' \psi_j') \,\mathrm{d}Y \tag{38}$$

where $\phi_i' = \dfrac{\partial \phi_i}{\partial X}$ and $\psi_i' = \dfrac{\partial \psi_i}{\partial Y}$. You will notice that the perturbation term is necessary to 'complete the square' in the integrand.

Thus,

$$\boldsymbol{K} = \boldsymbol{K}_x \cdot \boldsymbol{K}_y \tag{39}$$

where K_x, K_y have very narrow band structure – tridiagonal for four-noded bilinear elements, or pentadiagonal for nine-noded biquadratic elements and so on. The system given by Eq. (35) may be solved as follows. Firstly Θ^n is determined at each time step – this involves standard finite element integrals. Secondly, we solve for $\boldsymbol{v}^n$ from $\boldsymbol{K}_x\boldsymbol{v}^n = D^{-1/2}\,\Delta t\Theta^n$, and then for $\boldsymbol{w}^n$ from $K_y\boldsymbol{w}^n = \boldsymbol{v}^n$. Finally we have $\boldsymbol{T}^{n+1} = \boldsymbol{T}^n + D^{-1/2}\boldsymbol{w}^n$. The solution stages can be performed rapidly using well known algorithms, and furthermore, K_x, K_y are block structured making for efficient vectorisation. K_x, K_y are also time independent and so need only be constructed once.

The above *AD* algorithm has been successfully used to solve solidification problems on relatively simple geometries; but further program development will enable the potential savings to be realised for quite complex configurations.

4.2. Mixed implicit/explicit solution

Explicit time stepping requires no matrix inversion, and is thus very quick to operate, but is limited by timesteps constrained by stability criteria. Implicit schemes, on the other hand, are unconditionally stable and can take quite large timesteps (but is still first order, $0(\Delta t)$, accurate); however, they require matrix inversion/factorisation which can be time consuming for problems, as in three dimensions, which involve large bandwidths. The mixed implicit-explicit technique, pioneered by Hughes and Liu [11,12], seeks to combine the advantages of each scheme whilst minimising their deficiencies. Hence, explicit solution can be carried out cheaply in areas of the domain where the solution is changing slowly, whereas, to maintain stability, implicit treatment of thermally 'active' regions is required [13].

Taking the first order system given in Eq. (15)

$$\boldsymbol{C}\dot{\boldsymbol{T}} + \boldsymbol{K}\boldsymbol{T} = \boldsymbol{F}$$

we can classify the elements of the mesh as being explicit ($\theta = 0$, $\boldsymbol{C}$ 'lumped' in Eq. (19)) or implicit ($0.5 \leqslant \theta \leqslant 1.0$ in (19)), giving

$$\left.\begin{aligned} \boldsymbol{K} &= \boldsymbol{K}_I + \boldsymbol{K}_E \\ \boldsymbol{C} &= \boldsymbol{C}_I + \boldsymbol{C}_E \\ \boldsymbol{F} &= \boldsymbol{F}_I + \boldsymbol{F}_E \end{aligned}\right\}. \tag{40}$$

Setting $\boldsymbol{V} = \dot{\boldsymbol{T}}$, and $\tilde{\boldsymbol{T}}^{n+1}$ as an intermediate nodal temperature vector, we can express Eq. (16) in the form, for a single element,

$$\boldsymbol{C}\boldsymbol{V}^{n+1} + \boldsymbol{K}_I\boldsymbol{T}^{n+1} + \boldsymbol{K}_E\tilde{\boldsymbol{T}}^{n+1} = \boldsymbol{F}^{n+1} \tag{41}$$

where

$$\tilde{\boldsymbol{T}}^{n+1} = \boldsymbol{T}^n + \Delta t\boldsymbol{V}^n(1-\theta) \tag{42}$$

and

$$\boldsymbol{T}^{n+1} = \tilde{\boldsymbol{T}}^{n+1} + \Delta t \boldsymbol{V}^{n+1}\theta. \tag{43}$$

Substitution of Eq. (43) into (41) results, after some manipulation, in the form

$$[\boldsymbol{C} + \Delta t\theta \boldsymbol{K}_I]\boldsymbol{T}^{n+1} = \left[\Delta t\theta\left(\boldsymbol{F}^{n+1} - \boldsymbol{K}_E\tilde{\boldsymbol{T}}^{n+1}\right) + \boldsymbol{C}\tilde{\boldsymbol{T}}^{n+1}\right]. \tag{44}$$

For an explicit element, i.e. $\theta = 0$, Eq. (42) becomes

$$\boldsymbol{T}^{n+1} = \boldsymbol{T}^n + \Delta t \boldsymbol{V}^n$$

but from Eq. (16)

$$\boldsymbol{V}^n = \boldsymbol{C}^{-1}[\boldsymbol{F}^n - \boldsymbol{K}\boldsymbol{T}^n]$$

and hence

$$\boldsymbol{C}\tilde{\boldsymbol{T}}^{n+1} = \boldsymbol{C}\boldsymbol{T}^n + \Delta t[\boldsymbol{F}^n - \boldsymbol{K}\boldsymbol{T}^n].$$

This equation is exactly the same given by the predictor of Eq. (27), again with $\theta = 0$, for $\boldsymbol{T}_*^{n+1}$. Setting $\theta = 0$ in Eq. (44) gives

$$\boldsymbol{C}\boldsymbol{T}^{n+1} = \boldsymbol{C}\tilde{\boldsymbol{T}}^{n+1} = \boldsymbol{C}\boldsymbol{T}^n + t[\boldsymbol{F}^n - \boldsymbol{K}\boldsymbol{T}^n]$$

where now $\boldsymbol{C}$, $\boldsymbol{K}$, $\boldsymbol{F}$ are determined using some intermediate 'new' temperature $\boldsymbol{T}^{\bar{n}} = \gamma\tilde{\boldsymbol{T}}^{n+1} + (1-\gamma)\boldsymbol{T}^n$. Once more, this is identical to the form of the corrector Eq. (28). Similar arguments show the algorithm of Eqs. (41)–(44) is identical to the predictor-corrector form for the implicit case $0.5 \leqslant \theta \leqslant 1.0$.

Thus, an implicit-explicit algorithm can be implemented by specifying θ *for each element* [13], rather than globally, within the existing predictor-corrector solution algorithm. However, to take full advantage of the presence of explicit elements, which only contribute to the diagonal entries of the system ($\boldsymbol{C}$ is diagonalised for explicit elements), an active column profile solver is used. Essentially this performs Gaussian elimination only for those matrix entries which have off-diagonal terms in the equation. Thus, with many explicit elements, significant reductions in processing time can be achieved. One such case could be the use of explicit elements to represent the area defined by the sand mould, since it is not the heat transfer in the mould, but that in the cast which is of primary concern.

4.3. *Quadratic conductivity approximation*

The governing Eq. (1) is nonlinear, and, as we have noted, this necessitates some form of iterative solution procedure. This requires the re-formulation of the system matrices $\boldsymbol{C}(\boldsymbol{T})$ and $\boldsymbol{K}(\boldsymbol{T})$ for each iteration due to their temperature dependence. This is particularly expensive for finite elements since this would involve, adopting a simplistic approach, costly re-integration of all the terms of Eqs. (16)–(18).

However, we can partly alleviate this computational burden by making assumptions about the form of the temperature dependent coefficients –

namely the conductivity $k(T)$ and the heat capacity $(\mathrm{d}H/\mathrm{d}T)$. We have already discussed the interpolation of the enthalpy H using the finite element shape functions and nodal values H_i, as in Eq. (30). If we denote the heat capacity $(\mathrm{d}H/\mathrm{d}T)$ by R, and assume a finite element discretisation for R, i.e.

$$R(\boldsymbol{x}, t) \cong \sum_{i=1}^{N} N_i(\boldsymbol{x}) R_i. \tag{45}$$

Then the capacitance matrix, after lumping by summation of rows, becomes

$$C_{ii} = \sum_{j=1}^{N} R_j \int_{\Omega} N_i N_j \, \mathrm{d}\Omega$$

or

$$\boldsymbol{C} = \boldsymbol{I}(\boldsymbol{C}_0 \boldsymbol{R}) \tag{46}$$

where

$$\boldsymbol{C}_{0ij} = \int_{\Omega} N_i N_j \, \mathrm{d}\Omega \tag{47}$$

and $\boldsymbol{R}$ is the vector of heat capacities (derived from enthalpy values $\boldsymbol{H}$). The matrix $\boldsymbol{C}_0$ can be formed once at the start of the problem by numerical integration; subsequent calculations of the diagonalised capacitance matrix can be obtained much more rapidly by matrix multiplication as in Eq. (46).

We now apply similar 'group' formulation ideas to the conductance matrix $\boldsymbol{K}(\boldsymbol{T})$. Samonds et al. [14] recently put forward a 'quadratic conductivity approximation' whereby the finite element integrals involving k can be evaluated initially, with subsequent evaluations of $\boldsymbol{K}$ by matrix multiplication involving the nodal temperatures.

Hence, we approximate by a quadratic function, i.e.

$$k(T) = a + bT + cT^2 \tag{48}$$

where a, b, c are material constants. If we ignore the boundary condition contribution, for clarity, then the integral term which gives us the product $\boldsymbol{K}(\boldsymbol{T}) \cdot \boldsymbol{T}$ in Eq. (15) over a single element becomes

$$I_{k^e} = \int_{\Omega^e} \nabla n \cdot k(T) \, \nabla T \, \mathrm{d}\Omega^e \tag{49}$$

which on substitution of the quadratic form Eq. (48) yields, after some use of vector operator identities,

$$I_{k^e} = a \int_{\Omega^e} \nabla N \cdot \nabla T \, \mathrm{d}\Omega^e + \frac{b}{2} \int_{\Omega^e} \nabla N \cdot \nabla (T^2) \, \mathrm{d}\Omega^e + \frac{c}{3} \int_{\Omega^e} \nabla N \cdot \nabla (T^3) \, \mathrm{d}\Omega^e. \tag{50}$$

The next simplifying assumption is to interpolate the square and cube of the temperature onto the same finite element basis as T, i.e.

$$T^2 = \sum_{i=1}^{N} N_i T_i^2, \quad T^3 = \sum_{i=1}^{N} N_i T_i^3 \tag{51}$$

which gives us

$$I_k = a\boldsymbol{K}_0\boldsymbol{T} + \tfrac{1}{2}b\boldsymbol{K}_0\boldsymbol{T}^2 + \tfrac{1}{3}c\boldsymbol{K}_0\boldsymbol{T}^3 = \boldsymbol{K}_0\left(a\boldsymbol{T} + \frac{b}{2}\boldsymbol{T}^2 + \frac{c}{3}\boldsymbol{T}^3\right) \tag{52}$$

where

$$\boldsymbol{K}_0 = \int_\Omega \nabla N_i \cdot \nabla N_j \, \mathrm{d}\Omega. \tag{53}$$

The conductance matrix $\boldsymbol{K}_0$ need only be evaluated, by numerical integration, once at the initialisation stage of computation.

If we apply this procedure to explicit elements in the mixed implicit-explicit formulation then the $\boldsymbol{K} \cdot \boldsymbol{T}$ contribution to the right hand side vector can be rapidly established by the matrix multiplications of Eq. (52) for all iterations.

A similar procedure could be utilised for the implicit elements, using a $\boldsymbol{K}_0$ on the left hand side for $\boldsymbol{K}$, subtracting $\Delta t\theta(\frac{1}{2}b\boldsymbol{K}_0\boldsymbol{T}_p^{\bar{n}^2} + \frac{1}{3}c\boldsymbol{K}_0\boldsymbol{T}^{\bar{n}^3})$ from the right hand side of Eqs. (27), (28). However, since the consistent capacitance matrix has to be evaluated, and it does not cost much more to calculate the conductivity matrix at the same time, it is doubtful whether the reduction in matrix assembly costs will justify the loss of accuracy for implicit elements.

The quadratic conductivity approximation given above has demonstrated its effectiveness in reducing matrix generation times, particularly for explicit timestepping. One important consideration is the need for piecewise quadratic approximations should the conductivity of the melt vary sharply on solidification, except for elements which contain more than one phase. However, the approximation given here has proved adequate in our test problems, resulting in much faster runs, especially for three dimensional meshes.

The quadratic conductivity technique shown here is thus justified for materials where $k(t)$ can be reasonably approximated by a quadratic curve. Notable exceptions include cases where $k(T)$ exhibits near discontinuous behaviour.

4.4. Mould-metal interface models – coincident node technique

A critical problem in the simulation of casting methods is the treatment of the heat transfer between metal and mould at the relevant interface. Samonds et al. [15] have investigated various schemes for handling the mould-metal interface, including the use of thin elements at the interface, and the use of a coincident node technique, to be described in this section. The coincident node approach is, numerically, virtually equivalent to the use of thin elements, but

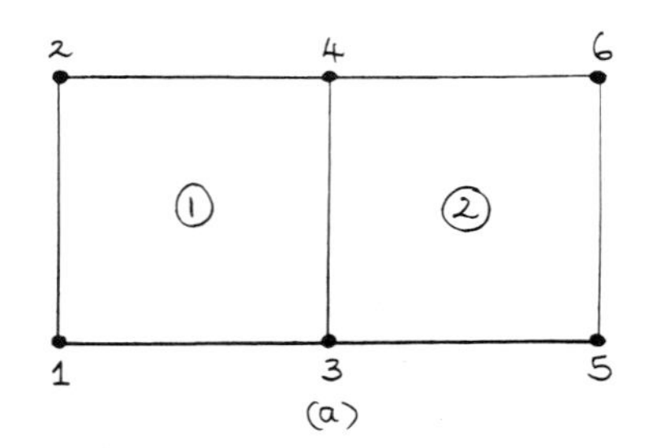

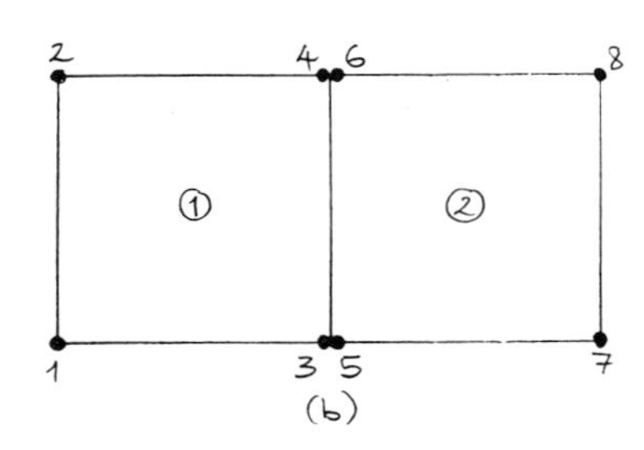

Fig. 5. Linear elements with common interface (a) standard connections, (b) coincident interface nodes.

offers greater flexibility in mesh design, and also significant savings in computer time.

Consider two bilinear elements which are adjacent as shown in Fig. 5(a). The nodal connectivities for these elements can be listed as

Element	Nodes			
1	1	3	4	2
2	3	5	6	4

Upon assembly of the system matrices, with this element topology, a condition of perfect conduction will exist between elements 1 and 2 because of the common interface defined by nodes 3 and 4.

If the elements are considered as disjoint, e.g. by the introduction of a spatially coincident node at each interface node, as shown in Figure 5(b), the nodal connections for the elements become

Element	Nodes			
1	1	3	4	2
2	5	7	8	6

In this case standard matrix assembly will assume a condition of perfect insulation between these elements since they have no nodes in common.

Neither perfect conduction nor insulation is adequate to describe the heat transfer between metal and mould. Common modelling practice is to assume a convective type heat transfer, as given by the expression

$$q_I = h(T_{\text{METAL}} - T_{\text{MOULD}}) \tag{54}$$

where q_I is the heat flux across the interface, and h is a controlling coefficient. Experimental values for h and its variation with time, or temperature, as solidification progresses are available. We now seek to incorporate such a heat transfer procedure within the coincident node formulation. This is quite easily effected by making the following addition to each element conductivity matrix (i.e. for each element abutting the interface) [8],

$$\boldsymbol{K}^e = \boldsymbol{K}^e + \int_{\Gamma_I} \alpha N_i \left(N_j - \tfrac{1}{2} N_m\right) \mathrm{d}\Gamma_I \quad i,\ j = 1, 2, \ldots N_I \tag{55}$$

where N_I is the number of nodes on the interface boundary Γ_I; m is the number of the node coincident with node j, and α is the interface heat transfer coefficient (c.f. h in Eq. (54)). The integrated contribution of Eq. (55) must be applied to elements on both sides of the interface, and the $N_i N_m$ cross term has a factor of $\frac{1}{2}$since it will appear twice during the integration.

Thus, given accurate measurement of the transfer parameter α we can expect good results from this coincident node approach, which is flexible during mesh design and has demonstrated its efficiency through our numerical tests.

4.5. Infinite element heat loss calculations

When dealing with external boundaries which are distant from the main area of interest it is impractical to enmesh the external region with finite elements. The associated computational overhead would be great even for a truncated external mesh. A viable alternative is to discretise the entire external region as if it were of infinite dimension, using 'infinite elements' [16,17] which map the infinite regions onto the usual prototype domain for numerical integration. Although perhaps this may be of limited use in casting simulation, it has important practical applications in many areas, including heat transfer, and the infinite element technique has been used successfully by the present authors in the context of heat losses from geothermal or petroleum reservoirs to the surrounding rock strata [16].

Zienkiewicz et al. [17] put forward an infinite element method based on a mapping of a semi-infinite region onto $[-1, 1]^n$ – the usual finite element domain for Gauss-Legendre integration. The one-dimensional mapping is schematically illustrated in Fig. 6(a), with a two-dimensional infinite 'strip' element shown in Fig. 6(b), and its use in conjunction with a *finite* element mesh given in Fig. 6(c).

Referring to Fig. 6(a), the one dimensional mapping, which is at the heart of the mapped infinite element method can be expressed as

$$x = \left| -\frac{\xi}{(1-\xi)} \right| x_0 + \left| 1 + \frac{\xi}{(1-\xi)} \right| x_2 \tag{56}$$

or

$$x = N_0(\xi)x_0 + N_2(\xi)x_2.$$

This map transforms the region $[x_1, x_3)$ (x_3 at infinity) onto the interval $-1 \leqslant \xi \leqslant 1$ with $x_1 = (x_0 + x_2)/2$. Note that $N_0(\xi) + N_2(\xi) \equiv 1$, and so the map is not affected by a shift of origin.

This transformation can be easily coupled with existing shape function mappings (e.g. isoparametric) to derive 'strip' elements of the form illustrated by Fig. 6(b). It should be noted that all that is required to utilise these infinite elements into existing finite element programs is some minor modifications to

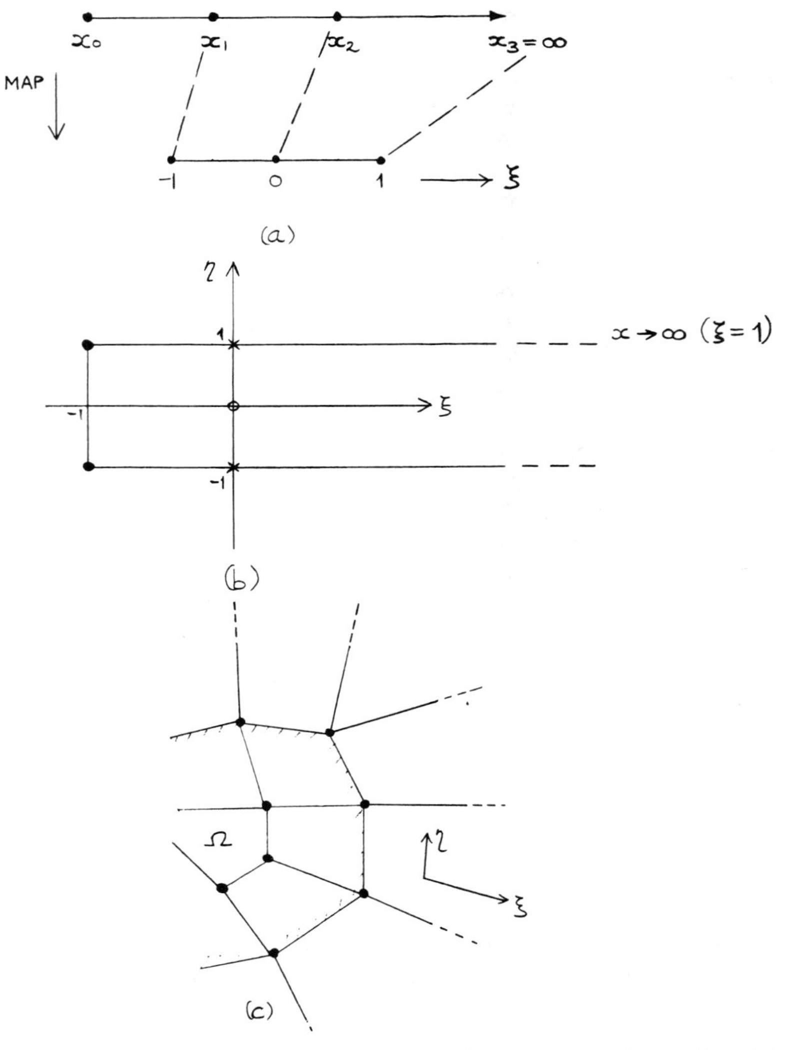

Fig. 6. Infinite elements: (a) one dimensional mapping, (b) strip element, (c) coupling with finite element mesh.

the routine which calculates the Jacobian matrix determinant for numerical integration.

5. Numerical examples

In this section we present some of the test problems which have been simulated using the finite element heat conduction program with its various enhancements discussed above. All the problems involve solidification, with most representing real casting projects.

The first example, however, deals with a corner freezing problem and is intended to demonstrate the effectiveness of the alternating-direction solution algorithm. The remaining examples will use the predictor-corrector timestepping procedure.

The mesh which discretises the square region is illustrated in Fig. 7, and comprises 9-noded biquadratic Lagrangian elements – giving pentadiagonal matrices in the AD formulation. The region is initially set at the solidus temperature $-0.15\,°\mathrm{C}$, with material properties $\rho c = 1.0$, $k = 1.08$, and latent heat $L = 70.26$. The phase change temperature interval δT (see Fig. 3) is taken as $2°$, and a constant timestep $\Delta t = 0.05$ was employed. At $t = 0$ the faces $x = 0$, $y = 0$ are instantaneously set at $-45\,°\mathrm{C}$ to initiate freezing. The outer faces $x = 4$, $y = 4$ are assumed to be perfectly insulated. Figure 8 compares the temperature history profiles at $x = y = 1$ for the AD algorithm against both the predictor-corrector (Crank-Nicolson), and Lees timestepping schemes. The curves show very good agreement and temperature contours from all these methods are almost indistinguishable, showing the AD formulation to be as accurate as standard solution methods. However, the CPU time required per timestep by the techniques was

Alternating-direction algorithm = 4.02 s
Lees' algorithm = 10.22 s
Crank-Nicolson algorithm = 10.10 s.

Thus the AD method is 2.5 times faster than the usual timestepping methods,

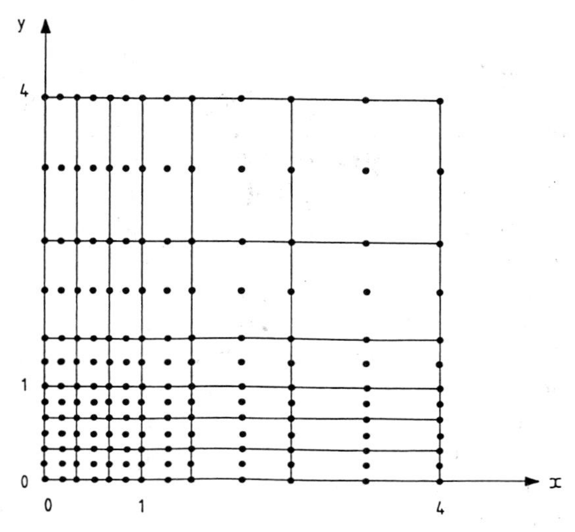

Fig. 7. Corner freezing problem – finite element mesh

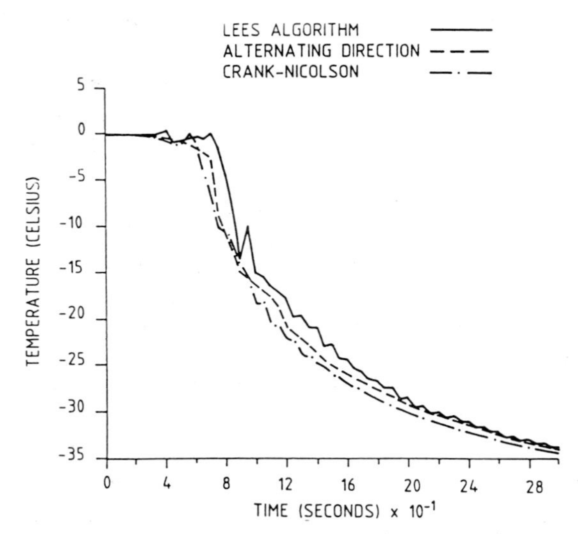

Fig. 8. Temperature histories at point $x = y = 1$.

this figure should rise to about 10 for vector processor machines. The above times are for an ICL 2966 computer, a scalar machine with virtual memory capability.

The AD algorithm has also been successfully utilised for problems involving non-rectangular domains; a wedge shaped domain with similar physical properties and problem definition to the above example has been successfully simulated [2].

The AD code, however, requires further development to model general mesh configurations and to cater for three dimensional meshes. An existing three-dimensional heat conduction program, incorporating the predictor-corrector timestepping scheme, has been amended to investigate the various procedures for improved casting simulation discussed in the previous section.

To demonstrate the effectiveness of the implicit-explicit method an experimental casting of a tapered slab of aluminium bronze in a resin bonded silica sand mould was simulated [13]. Table 1 gives the thermal properties of the metal, where the latent heat figure is a weighted average of the constituent pure metals. Table 2 gives the properties of the sand mould – two sets of conductivity curves were investigated. The cross-sectional finite element mesh, shown in Fig. 9, consists of 56, 8-noded quadratic, isoparametric elements, with the sand elements shaded. The initial temperature of the molten metal was 1156 °C, with the initial mould-metal interface set at the liquidus temperature, 1080 °C. No thermal resistance, i.e. perfect conduction, was assumed at the interface, and the latent heat was taken to evolve linearly over the solidification range. The external boundary, of the sand elements, was fixed at room temperature, 20 °C, with the initial temperature being the same

Table 1. Material properties – aluminium bronze.

Composition (%): Copper 80.5	
Aluminium 9.5	
Iron 4.5	
Nickel 5.5	
Solidification Range : 1,050* – 1,080 °C ($\delta T = 30$ °)	
Latent Heat, L : 2.37×10^5 J/kg	
Specific Heat, c : 452.2 J/kg °C	
Density, ρ : 7,600 kg/m^3	
Conductivity, $k(T)$:	
Temperature (°C)	k(J/sm° C)
200	54.42
400	46.05
600	43.95
800	41.86
1000	41.86
1200	41.86

throughout the sand mould. For the riser, the top boundary was fixed at the initial metal temperature, 1156°, with the sides assumed to be perfectly insulated.

The initial timestep was taken as 1.0 s, but this was altered by the predictor-corrector controlling algorithm, with an iteration tolerance of 0.5 °C. The first run assumed $\theta = \gamma = \frac{1}{2}$ (Crank-Nicolson, implicit) for all elements, and Fig. 10 depicts the movement of the solidus temperature contour (1050 °C) at 90s intervals, for the high sand conductivity case. The shape and progression of the solidification front agrees well with physical expectations, even for the coarse mesh used. A similar numerical experiment on a slab which is not tapered, as in Fig. 11, shows that a large proportion of the slab solidifies almost simultaneously, unlike the even progress of the solidification for the tapered case. This suggests feeding problems in this region yielding shrinkage porosity, an effect which occurs in practice.

Table 2. Material properties – sand.

Density, ρ : 1,444.7 kg/m^3	
Conductivity, $k(T)$: (High) $0.6606 - 2.084 \times 10^{-4}T + 7.741 \times 10^{-7}T^2$	
(J/sm° C) (Low) $0.2828 - 1.681 \times 10^{-4}T + 1.352 \times 10^{-7}T^2$	
Specific heat, $c(T)$:	
Temperature (°C)	c(J/kg °C)
200	975.7
400	1092.9
600	1151.5
800	1159.9
1000	1176.7

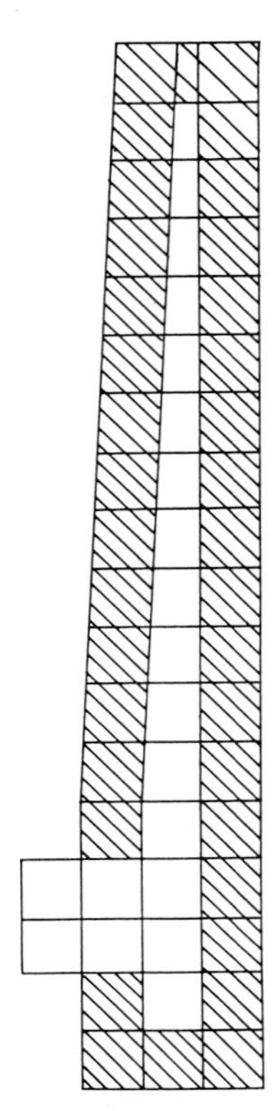

Fig. 9. Tapered slab casting – mesh with sand elements shaded.

The problem was run again, this time making all the sand elements explicit ($\theta = 0$), i.e. 64% of the mesh is solved by explicit timestepping. When using the active column profile solver this resulted in a reduction of 40% in the running time, with little qualitative difference from the results already shown. Thus the efficiency of the implicit-explicit scheme, coupled with the profile solver, can provide substantial savings in casting simulation.

The tapered slab problem was solved for once again, but this time employing the quadratic conductivity approximation to achieve even further processing time reduction. Figure 12 shows the temperature history profiles, at selected points along the slab, with consistent capacitance matrices (lumped for explicit elements) and a piecewise linear determination of k from the set of values given in Tables 1, 2 at each integration point. Figure 13 shows the same profiles, but using the quadratic conductivity approximation and the diagonalised capacitance technique [14]. There is very close agreement between the two sets of curves – a maximum difference of only 2%. However the assembly time of the quadratic approximation case was 46% that of the standard method, and, since the assembly time makes up 85% of the total solution cost,

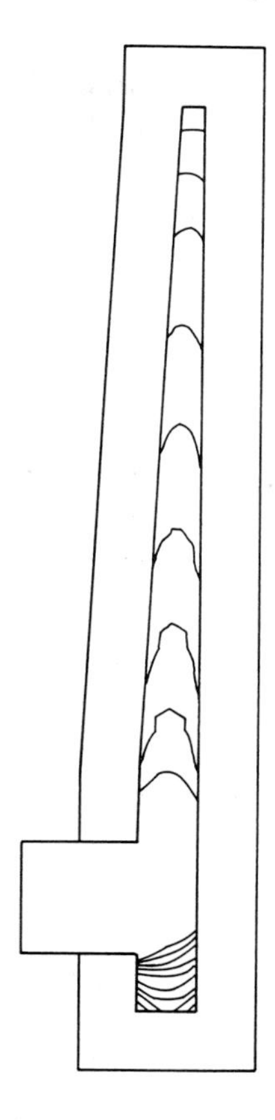

Fig. 10. Tapered slab casting – solidification front history.

thus giving an overall saving of 39% in total processing time. This is for the predictor-corrector algorithm running in mixed implicit-explicit mode, with explicit elements in the sand mould.

To demonstrate the three dimensional simulation capability, investigations were carried out [15] into a production gravity die casting of an aluminium-silicon-copper alloy (BS.1490 : LM4) with material properties given in Table 3.

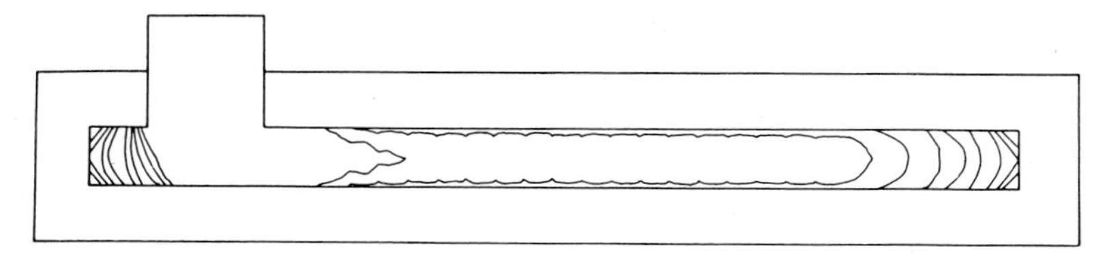

Fig. 11. Untapered slab – solidification front history.

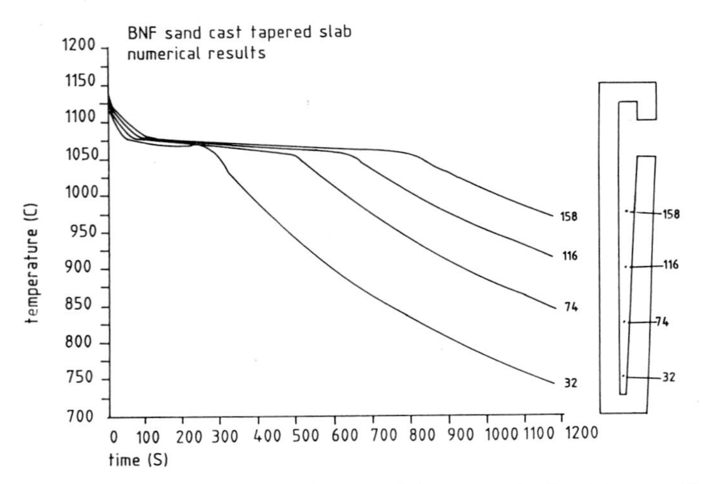

Fig. 12. Temperature history curves for tapered slab – standard matrix assembly.

Half of the die assembly, with its finite element mesh, is shown in Fig. 14, with the cast shaded – the internal geometry/mesh of the cast is shown in Fig. 15. The die composed of Grade 17 grey iron, the material properties for which can be found in Table 4.

Thermocouples were placed at various positions both in the die and in the cavity. The resulting temperature histories at these points are given in Fig. 16, for a melt which was poured at approximately 760 ° C.

The finite element mesh consisted of 374 linear brick (8-noded) elements and 645 nodes. The coincident node method was employed at the interface,

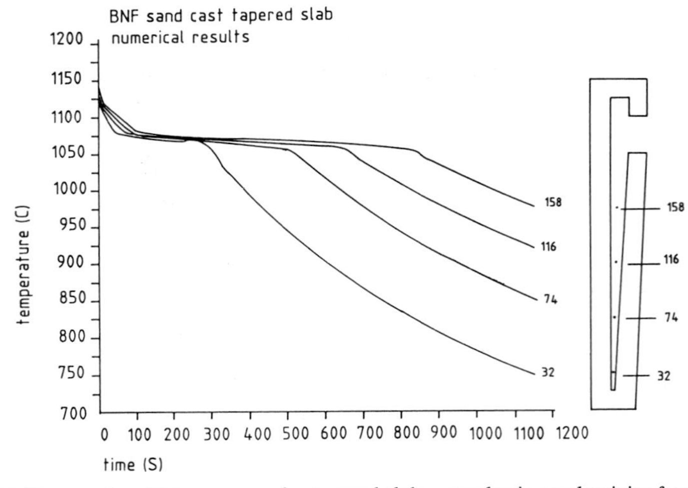

Fig. 13. Temperature history curves for tapered slab – quadratic conductivity function.

Table 3. Material properties – alloy BS.1490 : LM4.

Composition (%):		
Aluminium	89.41	
Silicon	5.70	
Copper	3.00	
Iron	0.70	
Manganese	0.37	
Zinc	0.32	
Titanium	0.20	
Nickel	0.16	
Magnesium	0.06	
Lead	0.05	
Tin	0.03	

Solidification Range : 525° – 625° C ($\delta T = 100°$)
Latent Heat, L : 92.3 cal/g
Density, ρ : 2.67 g/cm^3
Conductivity, $k(T)$, Specific Heat, $c(T)$:

Temperature (° C)	k(cal/sec-cm-° C)	c(cal/g – ° C)
225	0.393	0.234
325	0.400	0.247
425	0.390	0.268
525	0.390	0.301
625	0.442	0.301
725	0.442	0.301

with a heat transfer coefficient $\alpha = 5.0 \times 10^{-2}$ cal/sec-cm^2-° C, reduced after 19 seconds by 75% to model the formation of an air gap. Around the riser we took $\alpha = 3.0 \times 10^{-2}$ cal/sec-cm^2-° C, reduced after 25 seconds by 60%. These values were based upon previous experimental and computational work [15].

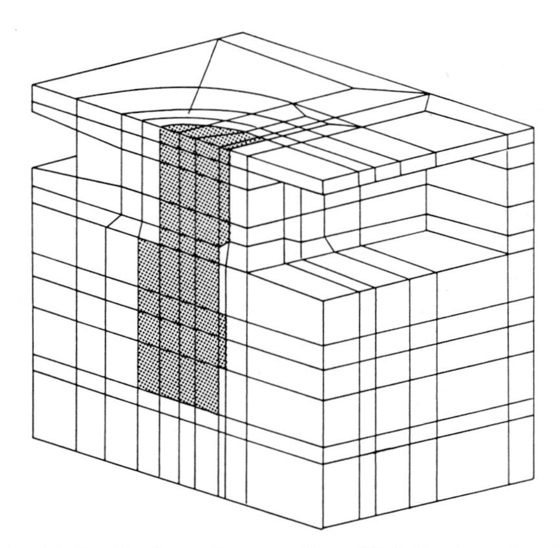

Fig. 14. Gravity die casting assembly – 3D finite element mesh.

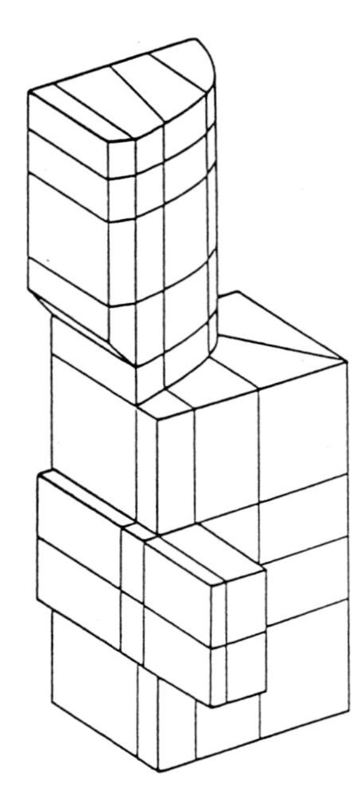

Fig. 15. Finite element mesh of cast plus riser.

The bottom of the die and the plane of symmetry were assumed to be perfectly insulated, whereas free convection was assumed at the remaining external faces. The temperature histories calculated at the thermocouple positions are shown in Fig. 17. These results were obtained using all implicit

Table 4. Material properties – grade 17 grey iron.

Composition (%):		
	Iron	93.55
	Carbon	3.18
	Silicon	2.03
	Manganese	0.56
	Phosphorus	0.59
	Sulphur	0.09

Density, ρ: 7.23 g/cm^3

Quadratic conductivity approximation:

$k(T) \simeq [1{,}622 \times 10^{-1} - 2.477 \times 10^{-4} T + 2.002 \times 10^{-7} T^2]$ cal/sec-cm-°C

Conductivity, $k(T)$, Specific Heat, $c(T)$:

Temperature (°C)	K(cal/sec-cm-°C)	c(cal/g-°C)
225	0.118	0.13
325	0.101	0.14
425	0.092	0.15
525	0.089	0.16
625	0.084	0.18
675	0.082	0.19

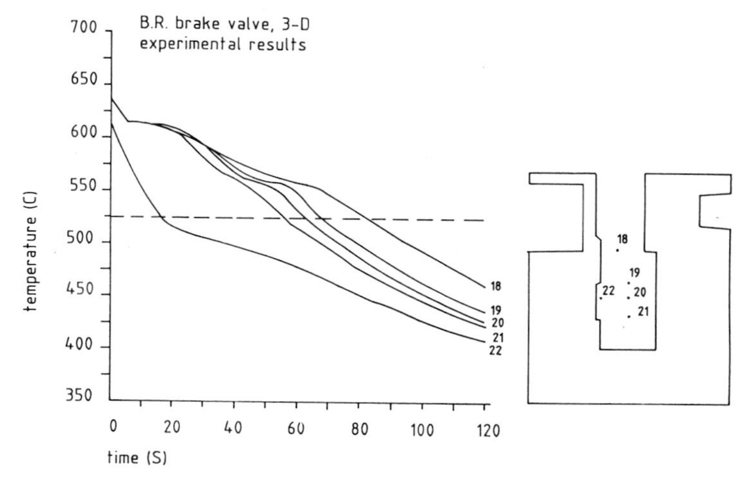

Fig. 16. Experimental temperature histories for die casting.

elements and standard conductivity matrix assembly. Comparing Figs. 15, 17 we note a good general agreement between the two sets of curves, except a numerical prediction of re-heating in the projecting wing of the casting after the air gap has been formed. Using an implicit-explicit formulation with all the die elements explicit lead to instability. Instead, a single layer 'coating' of implicit die elements around the cast was employed – making 64% of the elements explicit. The results of this mixed form, with the quadratic conductivity approximation, are given in Fig. 18 – note that the difference from the

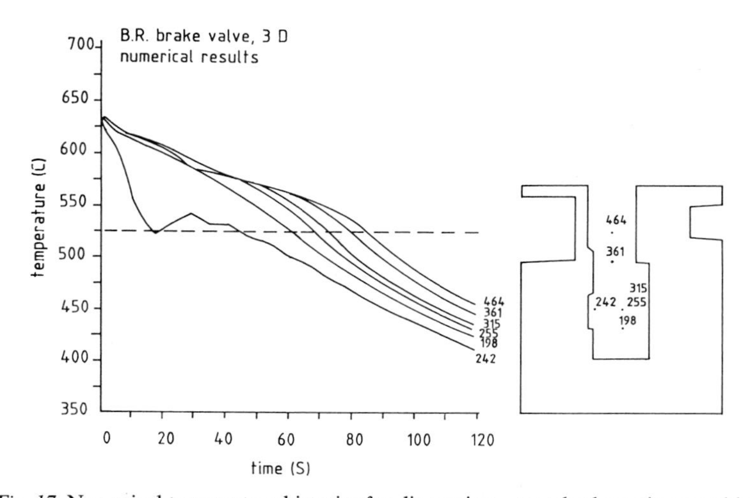

Fig. 17. Numerical temperature histories for die casting – standard matrix assembly.

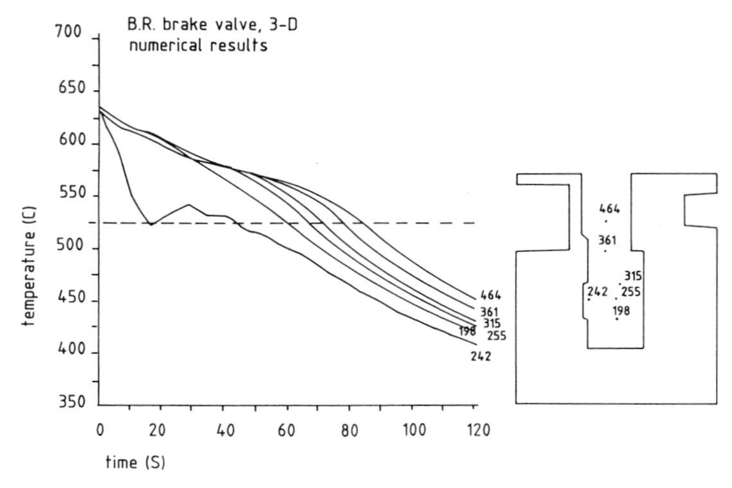

Fig. 18. Numerical temperature histories for die casting – explicit time stepping and quadratic conductivity function.

curves of Fig. 17 is negligible. However the assembly time was reduced by 40%, and the mixed timestepping algorithm yielded a reduction of 63% compared with the fully implicit run.

The use of the mixed implicit-explicit formulation, coupled with a quadratic conductivity approximation, thus effectively *halved* the CPU time required to model the gravity die cast simulation. Figure 19 depicts the movement of the 575 °C contour, which represents a solidification fraction of 80% – taken as the point at which the liquid metal feeding has ceased. It is apparent that the simulation predicts an underfed region in the casting.

Finally, the close agreement between experimental and numerical results demonstrates the effectiveness of the coincident node technique for modelling the mould-metal heat transfer at the interface. Further work needs to be carried out to model the air gap formation, in particular by the variation of α as above and in other test problems investigated [8].

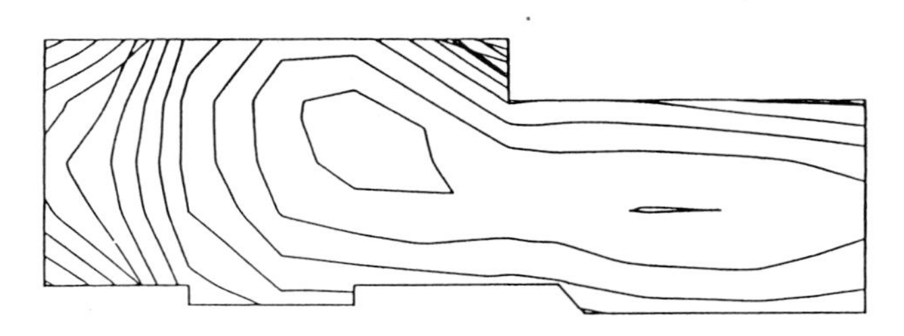

Fig. 19. ISO-solidification contour history – 80% at 3-second intervals.

6. Summary

The development of a finite element program for modelling heat conduction with phase change has been presented. The program has been used successfully to model a wide range of solidification problems including the casting processes illustrated by the examples given. Efforts to improve the efficiency of the finite element algorithm have succeeded in achieving substantial reductions in computer costs. Specifically the mixed implicit-explicit timestepping technique has been incorporated to speed up solution times, and the use of a quadratic conductivity approximation, which has significantly reduced the matrix assembly time associated with finite elements. The reduction of the computing overhead thus achieved makes for much cheaper numerical simulation of casting methods, particularly for three-dimensional discretisation, which is nearly always necessary due to the complex configurations involved.

A flexible technique for modelling the critical heat transfer processes associated with the mould-metal interface has been developed – namely the coincident node method. It has already established its viability, and its adaptability, via the variation of the heat transfer coefficient, make it a useful artifice in casting simulation. Further work needs to be carried out, however, to properly take into account the nature of air gap formation and how it affects the heat transfer coefficient.

It is hoped that further research and development of our simulation capability in this area, will make computer modelling of casting processes a practical foundry tool with which to optimise process and design features without recourse to much more expensive experimental analysis. Thus numerical simulation is envisaged as playing a prominent role in making industrial forming processes more economical and hence more effectual in a highly competitive market.

Our future objectives for our finite element model must include the simulation of convective heat transfer, where appropriate, and the coupling of the above heat transfer model with a comprehensive stress analysis package including crack propagation. A shorter term aim, however, is a comparison of cooling zones in the cast incorporating a model for analysing crystalline growth during solidification.

Once these objectives have been achieved a powerful simulation program will be available which can predict most, if not all, of the difficulties presently met in casting processes.

Acknowledgements

The authors wish to acknowledge the substantial contribution to the work presented in this paper made by Dr M. Samonds, Dr K. Morgan and Mr R. Symberlist.

References

1. O.C. Zienkiewicz and K. Morgan: *Finite Elements and Approximation*, 1st edn., p. 328, John Wiley & Sons Ltd., New York (1983).
2. R.W. Lewis, K. Morgan and P.M. Roberts: Application of an alternating-direction finite element method to heat transfer problems involving a change of phase, *Numerical Heat Transfer*, Vol. 7 (1984).
3. R.W. Lewis, K. Morgan and P.M. Roberts: Determination of thermal stresses in solidification problems. In: J.F.T. Pittman, O.C. Zienkiewicz, R.D. Wood and J.M. Alexander (eds) *Numerical Analysis of Forming Processes*, Chap. 15, John Wiley & Sons Ltd., (1984).
4. M. Lees: A linear three-level difference scheme for quasi-linear parabolic equations, *Math. Comp*. 20 (1966) 516–622.
5. J. Douglas Jr. and T. Dupont: Alternating-direction Galerkin methods on rectangles. In: B. Hubbard (ed.), *Proc. Numerical Solution of Partial Differential Equations*, Vol. 2, pp. 133–214, Academic Press, New York (1977).
6. L.J. Hayes: Implementation of finite element alternating-direction methods on non-rectangular regions, *Int. J. Numerical Methods in Engineering* 16 (1980) 35–49.
7. P.M. Roberts: The development of alternating-direction finite element methods for enhanced oil recovery simulation, Ph.D. Thesis, University College Swansea, University of Wales, Swansea (1984).
8. M. Samonds: Finite element simulation of solidification in sand mould and gravity die castings, Ph.D. Thesis, University College Swansea, University of Wales, Swansea (1985).
9. G. Comini, S. Del Guidice, R.W. Lewis and O.C. Zienkiewicz: Finite element solution of non-linear heat conduction problems with special reference to phase change, *Int. J. Numerical Methods in Engineering* 8 (1974) 613–624.
10. K. Morgan, R.W. Lewis and O.C. Zienkiewicz: An improved algorithm for heat conduction problems with phase change, *Int. J. Numerical Methods in Engineering* 12, (1978) 1191–1195.
11. T.J.R. Hughes and W.K. Liu: Implicit-explicit finite elements in transient analysis: implementation and numerical examples, *ASME. J. Appl. Mech.* 45 (1978) 375–378.
12. W.K. Liu: Development of mixed time partition procedures for thermal analysis of structures, *Int. J. Numerical Methods in Engineering*, 19, (1983) 125–140.
13. M. Samonds, K. Morgan and R.W. Lewis: Finite element modelling of solidification in sand castings employing an implicit-explicit algorithm. *Applied Mathematical Modelling*, 9 (1985) 170–174.
14. M. Samonds, R.W. Lewis, K. Morgan and R. Symberlist: Efficient three-dimensional finite element metal casting simulation employing a quadratic conductivity approximation, *Proc. Fourth Int. Conf. on Numerical Methods in Thermal Problems*, Part 1, pp. 64–77, Pineridge Press, Swansea (1985).
15. M. Samonds, R.W. Lewis, K. Morgan and R. Symberlist: Finite element modelling of the mould-metal interface in casting simulation with coincident nodes on thin elements. In: R.W. Lewis, K. Morgan, J.A. Johnson, W.R. Smith (eds) *Computational Techniques in Heat Transfer*, Vol. 1, Pineridge Press, Swansea (1985).
16. R.W. Lewis, K. Morgan and P.M. Roberts: Infinite element modelling of heat losses during thermal recovery processes, paper SPE 13515, *8th SPE Symp. Reservoir Simulation*, Dallas, Texas (1985).
17. O.C. Zienkiewicz, C. Emson and P. Bettess: A novel boundary infinite element, *Int. J. Numerical Methods in Engineering*, Vol. 19, pp. 393–404 (1983).

Applied Scientific Research 44: 93–109 (1987)

On the sensitivity of numerical simulations of solidification to the physical properties of the melt and the mould

T.J. SMITH, [1,*] A.F.A. HOADLEY [2] & D.M. SCOTT [2]
[1] *Department of Applied Mathematics and Theoretical Physics, University of Cambridge, Silver Street, Cambridge CB3 9EW, UK;*
[2] *Department of Chemical Engineering, University of Cambridge, Pembroke Street, Cambridge CB2 3RA, UK.*
* Present affiliation: *GST Professional Services Limited, 8 Green Street, Willingham, Cambridge CB4 5JA, UK.*

1. Introduction

Heat transfer problems involving solidification have been studied extensively for over a century. These problems are encountered in a wide range of industrial processes such as food processing, energy storage, and crystal production. Also, solidification is an important aspect of many natural phenomena, for example, ground freezing, and the formation of igneous rocks.

In many of these situations the process is continuous or the time scale is very long and hence the solidification rate can be considered to be constant. In other cases the geometry of the problem is not important. The exception is the casting process where solidification governs the shape, the properties, the yield and the production rate of a given product. It is perhaps not surprising that the foundry industry has been the major driving force behind the relatively recent advances in the computer modelling of solidification [1–4].

With the exception of some of the new rapid solidification techniques, casting processes rely on a mould or die to convey the heat from the casting while giving the product its desired shape. The materials used for the mould or die vary considerably. The casting of complicated shapes is usually achieved with a sand or clay mould which is destroyed in order to remove the product after solidification. Permanent moulds made from oxidation resistant materials are frequently used when casting non-ferous metals such as aluminium. The properties of the mould have a considerable influence on the solidification rate and need to be modelled just as closely as the actual casting itself. In the case of permanent moulds reliable thermal data are available for most materials, although some uncertainty is introduced when a powder is used to coat the inside surfaces to aid separation. Sand moulds give quite a different situation. The mould usually consists of at least two layers of different materials. The backing sand which provides the mould with its strength is usually bound together, but not saturated, with a water, clay mixture, while cores may use organic binders. A pattern plate is rammed into the sand to produce the desired shape, leading to density variations within the mould. Hartley and

Babcock [5] conducted experiments with sand-based moulds which indicate that variables such as particle size, ramming density, moisture and binder content all have an effect on thermal properties such as the enthalpy and conductivity. As foundries find these variables difficult to monitor and control, there must always be considerable uncertainty surrounding sand mould data.

For many castings there exist resistances to heat transfer other than through the casting and the mould. A full or partial air gap may often form between the casting surface and the mould due to solid-state contraction of the casting or expansion of the mould. The exact effects of this are very difficult to define accurately. The first problem is that the formation of this gap is time dependent [6]. In addition the heat transfer across such a gap is very complex. Point-point contacts, radiation between surfaces, and convection and conduction withing the air may all be significant modes of heat transfer. It is common to use a time dependent heat transfer coefficient to model this phenomenon [4,7]. However, its value is very dependent on the mould and casting geometry.

Heat transfer to the atmosphere is particularly important in metallic mould castings [8], but it is also a significant factor in large sand mould castings where there is sufficient time for the heat to penetrate to the mould surface prior to complete solidification. The heat at the mould surface is usually dissipated by convection and radiation, and should be taken into account when modelling casting solidification.

Finally, natural convection flow is known to take place within the melt during the early stages of solidification. Buoyancy-induced circulation occurs as a result of a significant temperature difference between the pouring temperature and solidification temperature. This temperature difference is known as the melt superheat. During the period when there is significant superheat convection is the dominant mode of heat transfer between the liquid and the solid interface. This convecting period is important for a number of reasons. In alloy solidification, convection is the means by which solute rejected at the dendrite tip is remixed with the bulk liquid. This may lead to microsegregation [7,9]. It is also important in influencing the grain size and orientation [10] and the important transition point between columnar and equiaxed dendrite growth [11]. Finally in continuous casting, convection within a superheated melt is believed to be one of the most likely causes of 'shell breakout' [12], which occurs if insufficient metal has solidified during the time it is in contact with the mould.

As can be seen, solidification in castings is affected by a number of complex interacting processes. Consequently, if a failure occurs, it is often difficult to determine the cause, particularly as the sensitivity of the solidification process to the various controlling parameters is often difficult to determine. In this respect numerical simulation does not always offer an adequate solution as it is often difficult to specify accurately the physical properties of the system. In order to have confidence in the results of a numerical simulation it is necessary to know the accuracy with which the physical properties must be specified and the sensitivity of the numerical results to errors in these parameters. Here it is interesting to note that in the simulation of sand castings the physical properties of the metal include temperature dependent effects while the mould material is often characterised by single global parameters. This is despite the fact that density and water content variations within the mould lead to strongly temperature dependent thermal properties [13].

The objective of this paper is an examination of the sensitivity of conduction controlled solidification in castings to the physical properties and geometry of the melt and mould. An assessment of the accuracy to which these parameters need to be measured for use in numerical simulations of the casting process can then be provided.

A review of analytical solutions to simplified one-dimensional realisations of the castings process is given. This is to identify those parameters of greatest significance.

An experimental study in a two-dimensional geometry using wax is described. Such a study provides data with which to compare analytical and numerical solutions, and also emphasises any departures of the data from the analytical solutions due to multidimensional effects.

The numerical simulation of solidification in the simple geometry studied experimentally is carried out, to demonstrate that such simulations can be accurate, and to provide a basis for extending the analytical results to multidimensional geometries.

It was found that the initial period of solidification was particularly sensitive to the properties of the mould; this is of particular relevance to the production of thin walled castings. It was also found that some widely used analytical formulae for solidification development can be inaccurate during the early stages of solidification, and a modification is proposed.

Discrepancies between the experiment and the numerical simulation led to the discovery of a solid-state phase transition just below the freezing point of n-eicosyne. This substance was thought to have well-known properties, and the discovery highlights the need for accurate data for solidification simulation.

2. Analytical background

The development of solidification in casting is a complex, non-linear, three-climensional process, and as such a general analytical treatment is not possible. However, analytical solutions can be obtained for certain simplified one- and two-dimensional problems. While these may have little direct practical application, they are useful in that they allow the functional dependence of solidification development on the problem invariants to be evaluated. It also permits a more comprehensive interpretation of both experimental data and numerically derived solutions. Here, a number of one-dimensional analytical solutions are discussed which correspond to different parameterisations of the mould. Some preliminary conclusions on the influence of the mould on solidification development can then be drawn. These are investigated in greater depth both experimentally and numerically in subsequent sections.

The fundamental problem considered in this section is the development of solidification in a melt occupying the semi-infinite half space $x > 0$. It is assumed that at $t = 0$ the temperature in the melt is uniform at the solidification temperature, T_s, and that there is no convection within the melt. Conduction controlled solidification is initiated by instantaneously reducing the mould temperature to $T_0 < T_f$. For $t > 0$ the temperature distributions in the liquid melt, T_l, the solid melt, T_s and the mould, T_m, are then governed by the equations

$$\rho_m C_m \frac{\partial T_m}{\partial t} = \frac{\partial}{\partial x}\left(k_m \frac{\partial T_m}{\partial x}\right), \quad -\infty < x < 0 \tag{1a}$$

$$\rho_s C_s \frac{\partial T_s}{\partial t} = \frac{\partial}{\xi}\left(k_s \frac{\partial T_s}{\partial x}\right), \quad 0 < x < a \tag{1b}$$

$$\rho_l C_l \frac{\partial T_l}{\partial t} = \frac{\partial}{\partial x}\left(k_l \frac{\partial T_l}{\partial x}\right), \quad a < x < \infty \tag{1c}$$

in which ρ denotes density, C denotes specific heat and k denotes thermal conductivity with subscripts m, s and l refering to the mould, solid and liquid melt respectively.

At the solid-liquid interface, $x = a$, the temperature is continuous, $T_l(a, t) = T_s(a, t)$. In addition, the release of latent heat, L, by the solidifying melt requires that

$$\rho_l L \frac{da}{dt} = k_l \frac{\partial T_l}{\partial x} - k_s \frac{\partial T_s}{\partial x}. \tag{2}$$

If there is no thermal resistance across the solid-mould interface, the temperature is also continuous, $T_s(0; t) = T_m(0; t)$. If there is some resistance due to poor thermal contact, the flux across the interface is given for $x = 0$, $t > 0$ by

$$k_s \frac{\partial T_s}{\partial x} = h(T_s - T_m) \tag{3}$$

in which h is a suitable heat transfer coefficient.

Solutions of these equations are easily obtained for the two extreme cases of the mould acting as an ideal heat sink or as a barrier to heat transfer. The former corresponds to conditions existing in many die-castings while the latter is common in many sand-castings.

If the physical properties of the mould and melt are such that $\rho_m C_m \gg \rho_s C_s \approx \rho_l C_l$ and $k_m \gg k_s \approx k_l$ and there is no thermal resistance between the melt and the mould, the mould acts as an ideal heat sink such that $T_m(x; t) = T_0$, $x < 0$, $t > 0$ and the significant region of heat transfer is confined to the solidified melt. This represents the fundamental Stefan problem, the solution for solidification development being given by [14]

$$a = 2\lambda(\alpha_s t)^{1/2} \tag{4}$$

in which $\alpha_s = k_s/(\rho_s C_s)$ is the thermal diffusivity of the solid, λ is a constant given by the solution of the transendental equation

$$\frac{e^{-\lambda^2}}{\lambda \operatorname{erf} \lambda} = \frac{\pi^{1/2}}{S} \tag{5}$$

$S = L/C_s(T_f - T_0)$ is the Stefan number and erf is the error function. Asymptotic analysis of equation (5) in the limit $S \to 0$, leads to $\lambda \sim (S/2)^{1/2}$, applicable to many metals for which $S \ll 1$.

Solidification development according to equation (4) has been observed in a number of experiments at times greater than some critical time, t_c. At times less than t_c a significant departure from this behaviour has been noted. It is generally considered that this departure is due to the inability of a physical system to cope with the infinite heat flux across the boundary $x = 0$ at $t = 0$ demanded by the analytical solution. In practice, it is argued that for times less than t_c the heat flux into the mould is a constant with a magnitude determined by the physical properties of the system.

In many castings there is often imperfect thermal contact between the solid melt and the mould. This is often due to the formation of an airgap caused by thermal contraction of the solid melt as it cools. Under these conditions, the thermal resistance is significant but is not the sole controlling influence on solidification development. An approximate analytical solution for this situation was derived by Adams [15] who showed that

$$\frac{\mathrm{d}a}{\mathrm{d}t} = \frac{2S}{Z}\left(\frac{h\alpha_s}{k}\right)\left\{1 + \left[\left(1 + \frac{4S}{3}\right)\left(\frac{Z^3 - 1}{Z^3}\right)\right]^{1/2}\right\}^{-1} \tag{6}$$

in which $Z = 1 + (ha)/k$.

For $Z \gg 1$, i.e. $a \gg k/h$, $(Z^3 - 1)Z^{-3} \approx 1$ and hence

$$t \approx \frac{1}{4\alpha_s S}\left[1 + \left(1 + \frac{4S}{3}\right)^{1/2}\right]\left[a^2 + 2\frac{k}{h}a\right]. \tag{7}$$

This solution can be extended to smaller values of $Z \gg 1$ by taking a binomial expansion of the last term on the right hand side of equation (6). This leads to

$$t = t_A + \frac{1}{3}\left(\frac{k}{h}\right)^2 \frac{1}{\alpha_s}(1 + 4/3S)^{-1/2}\left(1 + \frac{ha}{k}\right)^{-1} \tag{8}$$

in which t_A is the time given by equation (7).

3. Numerical model

There are several different numerical techniques for the solution of moving boundary problems. The most well known of these are the finite difference, finite element and boundary element methods. For this work a fixed mesh finite element method was chosen, because of its ease at handling more complex geometries and boundary conditions.

The solidification of a pure material always produces a sharp phase change boundary, or in metallurgical terms, a planar interface. However most casting materials are composites and have a finite freezing range, which produces a so-called 'mushy-zone'. It was decided that the numerical model should, if possible cope with both these physical situations, although only a pure material is to be modelled in this work. By using a weak solution (enthalpy) method the phase change boundary is not explicitly defined and a single partial differential equation is used to describe the heat transfer throughout the casting. This can be written as

$$\rho\left(\frac{\mathrm{d}H}{\mathrm{d}T}\right)\frac{\partial T}{\partial t} = \nabla \cdot (k(T)\nabla T). \tag{9}$$

By applying the standard Galerkin formulation of the finite element method where the element shape function N is used as the weighting function in the numerical integration, the final system of equations can be written as

$$\mathbf{C}\frac{\mathrm{d}T}{\mathrm{d}t} + \mathbf{k}\boldsymbol{T} = \boldsymbol{f} \tag{10}$$

where

$$C_{ij} = \rho\int_\eta\left(\frac{\mathrm{d}H}{\mathrm{d}T}\right)N_iN_j\,\mathrm{d}w$$

$$K_{ij} = \int_\eta \nabla N_j\cdot(k\,\nabla N_i)\,\mathrm{d}\eta \tag{11}$$

$$f_j = \int_\eta N_jq\,\mathrm{d}\eta - \int_\tau N_jq\cdot n\,\mathrm{d}\tau.$$

Due to the change in phase, solidification problems are strongly nonlinear. Near the phase change region the capacity $\mathrm{d}H/\mathrm{d}T$ changes rapidly and for pure materials tends towards infinity. It is well known [16,17], that the mass matrix $\boldsymbol{C}$ can be diagonalised by assuming the thermal mass is lumped at the nodes similar to the finite difference formulation, rather than being distributed across the element. This is an advantage for phase change problems because it removes the spacial variation of $(\mathrm{d}H/\mathrm{d}T)$ within each element.

Because of the non-linearity of the assembled global matrices of $\mathbf{C}$ and $\mathbf{K}$ a predictor – corrector method is used to ensure that the correct physical properties are used during any timestep. The backward difference equation for (3.1) which is unconditionally stable is

$$\mathbf{C}^*[T^{n+1} - T^n] = -\Delta tn\left[\mathbf{K}(T^*)T^{n+1} - \boldsymbol{f}(T^*)\right] \tag{12}$$

where $*$ denotes mean properties for the time step.

For the predictor step C^* and T^* are calculated from T^n, the temperature at the end of the previous timestep. For the correction steps C^* is calculated from the mean enthalpy gradient given by

$$C_c^* = \frac{\Delta H_p}{T\left(H_p^{n+1}\right) - T^n}, \quad p \text{ from predictor step} \tag{13}$$

where

$$\Delta H_p = C_p\left(T_p^{n+1} - T^n\right), \quad H_p^{n+1} = H^n + \Delta H_p \tag{14}$$

and T^* is calculated from the average enthalpy during the timestep.

$$T^* = \left(H_p^{n+1} + H^n\right)/(2C^*). \tag{15}$$

For a pure material changing phase, C would be infinite, which is of course not possible in a numerical scheme, and therefore a small phase change interval must be defined. With many schemes this interval must be relatively wide and the timestep small, to prevent a node from jumping the phase change peak. With this method, there is no restriction on the minimum size of the phase change interval providing C_{ij} is finite. However if too small an interval is defined, the well known oscillatory behaviour of the enthalpy method [18,19] may occur. This phenomena is particularly a problem when the Stefen number for the material changing phase is much less than unity.

In this work linear quadrilateral elements have been used exclusively, as the geometries studied so far have been fairly simple, and a moderately small number of elements required. Different materials are easily incorporated within a finite element scheme, the only restriction being that each element can contain only one material. This program also allows the specification of fixed temperature or a heat transfer coefficient at the boundaries. In order to simulate the initial heat transfer during filling, these boundary conditions can be specified as a function of time.

4. Experimental

One of the most important aspects of this work was to test the simple analytical results and the numerical model against experimental observations. Experimental work on solidification has proceeded on two different levels. Firstly, a number of people [3,8,13,20,21] have conducted laboratory and industrial tests with metallic castings. The difficulty with such experiments is obtaining accurate and consistent data. Of primary importance to all this work is the exact location of the solid-liquid interface. In experiments with metals this is usually calculated from temperature measurements obtained from thermocouple probes within the cavity and mould. The number of measurements required usually limits the experiment to unidirectional solidification or the results of a two or three dimensional test to a single line through the casting. Another technique frequently employed is to empty out the liquid portion of the casting at specified times after pouring. However the decanting time is often a significant fraction of the solidification time and also the freezing of a residual liquid layer at the interface can introduce additional errors [22]. Finally in some arrangements it is possible to probe the solid surface through the melt but where the interface is irregular the data must be averaged [21].

Table 1. Physical properties of various melts and moulds.

Solid properties Materials		Density ρ kg/m^3	Conduct $k \cdot$ W/ mK	Spec. Heat. C_p jH/ kg °C	Therm. Diff. α m^2/ sec	Latent Heat ΔH_f kJ/kg	α metal/ α mould
Wax ($C_{20}H_{42}$)	(36 °C)	780	0.33	2.04	2.1×10^{-7}	160/80	1.9
Perspex	(20 °C)	1200	0.19	1.38	1.1×10^{-7}		
Tin	(200 °C)	7300	57	0.23	3.4×10^{-5}	60	2.4
Carbond Steel	(100 °C)	7800	52	0.47	1.4×10^{-5}		
Aluminium Bronze	(1000 °C)	7600	42	0.45	1.2×10^{-5}	237	1.5
Grey Iron	(500 °C)	7200	38	0.67	7.9×10^{-6}		
Steel (1.5% C)	(1000 °C)	7500	28	0.90	4.1×10^{-6}	270	3.4
Sand		1300	1.4	0.88	1.2×10^{-6}		

Experiments on solidification are made somewhat easier with the use of non-metals. High temperatures and/or the use of expensive alloys are not required. In addition, solidification occurs more slowly allowing a greater time for taking measurements. Several people [22–25] have used a wax or aqueous mixture to study solidification, because of the ease of observing the interface through the transparent melt. The usual criticism levelled at these experiments is the large difference in the properties of the wax in comparison with those of metals. In this work n-eicosyne wax has been used in solidification experiments within a perspex mould. The thermal properties of the wax and the mould at the solidification temperature are given in Table 1 together with some typical values for metal casting and die systems. Although the absolute values are orders of magnitude different, the ratios of the thermal diffusivities between casting and mould are quite similar. The results will show that the behaviour of the wax/perspex system is not unlike that of a metallic casting.

The well known corner freezing problem was chosen as the geometry for these solidification experiments. It has the advantage over other two dimensional arrangements that away from the corner the interface is parallel with the sides and hence can be considered as a one dimensional problem. The corner freezing problem has been studied analytically [26] and several others have obtained numerical results [17,18], but to the best of the authors' knowledge nobody has completed any experimental work with this geometry. In addition all previous studies have assumed constant temperature boundary conditions, which simplify calculations, but are extremely difficult to obtain in practice.

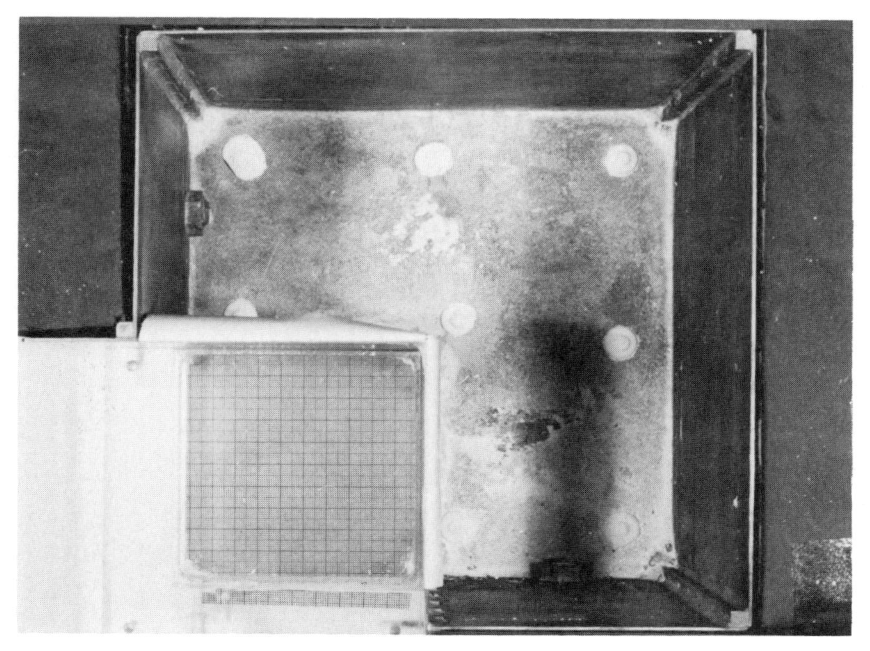

Fig. 1. Brass cooling chamber and perspex wax container (viewed from above).

The wax mould was milled from a 45mm thick perspex sheet. The square cavity had a side length of 100mm with the adjacent cooling walls having an average thickness of 2 mm, and the insulating sides a thickness of 20mm. The radius in each of the horizontal corners approx was 4mm. This wax container was screwed up against a brass cooling chamber. A thin coating of silicon grease was used to ensure good contact between surfaces.

Much care was taken to ensure that heat flow was only in the two horizontal directions. The bottom surface of the perspex mould was made as thin as possible to minimize the heat conducted by this surface. In addition the mould sat upon two cavities, of which the bottom one was evacuated, thereby insulating the bottom of the mould in the vertical direction. A perspex lid was made for the mould, again making the surface in contact with the wax as thin as possible. The lid had two compartments both of which were evacuated. Finally, as there is an increase in density during solidification, a small secondary reservoir was connected near the insulated corner, to ensure that the volume remained constant through out the experiment. Figure 1 shows a plan view of the perspex mould and cooling chamber.

As the melting point of the wax (36.5 °C) was above the normal room temperature, it was necessary to perform the experiment in a temperature-controlled environment. This was achieved using an incubating oven, which operates by a positive circulation of warm air. The layout of the oven is shown

Fig. 2. Layout of the apparatus within the constant temperature cabinet.

in Fig. 2. A mechanical stirrer is mounted centrally above the cooling chamber. The camera is supported from a frame directly above the wax container. The mould assembly sits on an A3 size light box, which was switched on just before to each photograph, to illuminate the molten wax.

N-eicosyne was chosen as the solidifying medium. It is a straight chain alkane of chemical composition $C_{20}H_{42}$. Its main attribute is that being a wax it is transparent when liquid but opaque as a solid. This meant that the solidification front could be easily located and recorded by photography. The wax used in the experiments was 99% pure. A differential scanning calorimeter (D.S.C.) was used to determine the melting and freezing range and specific heat data. The trace on minimum heating and cooling rate is shown in Fig. 3. The cooling curve shows a second peak which is attributed to a solid-solid phase transition which does not occur during melting. This phenomenon is not referred to in the extensive work carried out by NASA [27] on paraffin waxes. The latent heat of n-eicosyne on cooling measured in the D.S.C. was only 160 kJ/kg which is substantially different from the literature value of 245 kJ/kg [27]. The heat given up in the solid phase transition was about 80 kJ/kg giving a total enthalpy change of about 240 kJ/kg. The thermal conductivity of solid n-eicosyne was determined separately by Griggs [28], in further work for NASA on energy storage devices.

Two different types of experiments were performed with this apparatus.

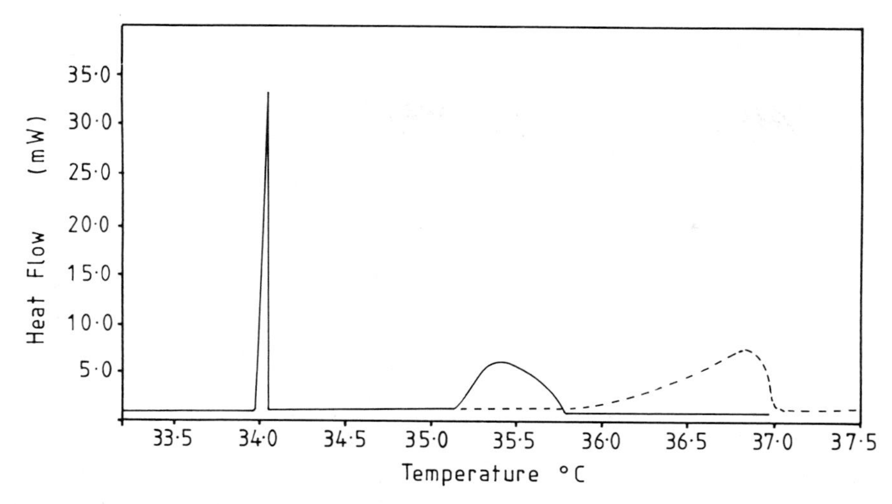

Fig. 3. D.S.C. trace for N-eicosyne at 0.1° C/min. ······ heating curve; ——— cooling curve.

The first was with the molten wax already in the mould prior to cooling. In this experiment the mould is initially warm and the molten wax is stagnant and a few degrees above the melting temperature. The second type of experiments were with an initially cold empty mould. The molten wax, at about the same temperature, would be poured into the mould and the mould lid replaced. In these experiments the wax level took approximately 30 seconds to stabilise and turbulence due to pouring also persisted for some time.

During the course of any experiment the solidification interface was photographed at regular intervals. Data such as ice-water temperature, cabinet temperature and approximate interface position were also recorded. Each experiment ws run for approximately three hours. In this time, approximately two thirds of the wax solidified. The repeatability of the experimental procedure is illustrated in Table 2.

Table 2. A comparison of the solidification distance for different experimental runs of the initially warm mould experiment.

Run No.	Time (min)	Dist from side (mm)	Time (min)	Dist from Side (mm)
4	20	8.0	60	17.5
10	20	7.9	60	16.9
11	20	7.5	60	17.0
12	20	8.0	60	16.8
18	20	7.8	60	17.7

5. Results

The results of experiments and calculations are plotted as t^*/a^* versus a^*, where t^* and a^* are dimensionless time and solidification distance respectively:

$$a^* = ah/k_s$$

$$t^* = t\alpha_s h^2/k_s^2. \tag{16}$$

Here h is the overall heat transfer coefficient through the perspex mould to the melt, k_s the thermal conductivity and α_s the thermal diffusivity of solid n-eicosyne.

Figure 4 shows results for solidification progress along the diagonal, when the mould and melt are at the same temperature initially. The only unknown parameter is the heat transfer coefficient on the external surface of the mould, which is chosen to be $h = 400$ W/m K in order to give best agreement between calculation and experiment. There is seen to be good agreement between theory and data; the experimental data have measurement errors of around ± 0.5 mm, shown as an error bar in the corner of each figure. The results are not too sensitive to the value of h, as is indicated in Fig. 5, where calculations of the motion of the solidification front in a 1-D geometry for

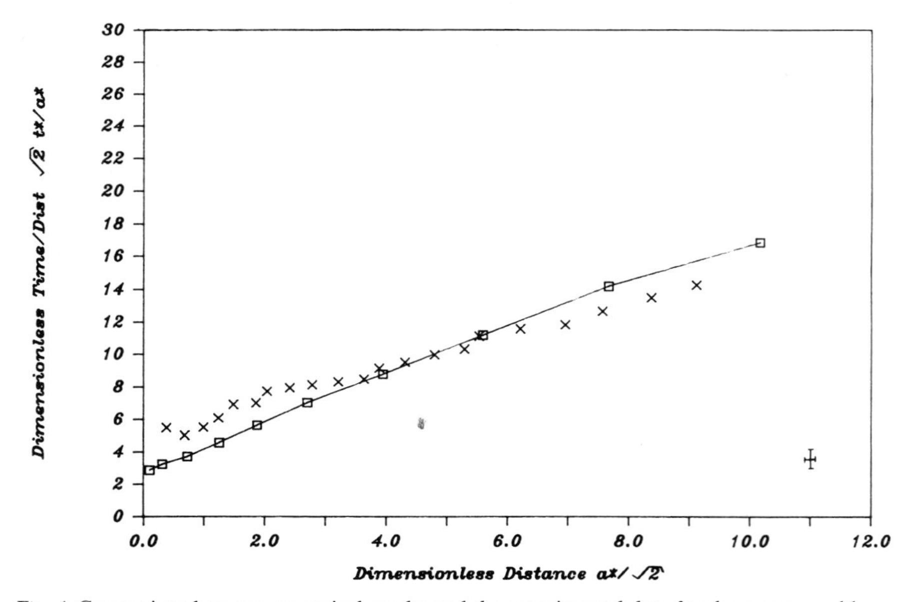

Fig. 4. Comparison between numerical results and the experimental data for the warm mould case, along the diagonal. Data $T_{\text{initial}} = 39\,°\text{C}$; $T_{\text{boundary}} = 0.1\,°\text{C}$; $h_{\text{boundary}} = 400$ W/m^{2}K; X experimental data; □2D numerical results.

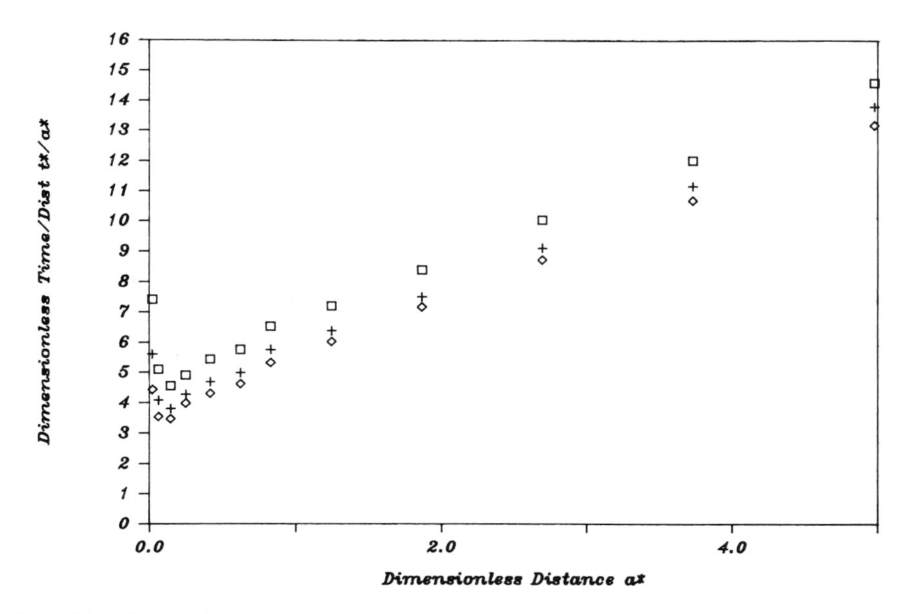

Fig. 5. The effect of the external heat transfer coefficient on the 1D numerical simulation. Data: T initial = 37 °C; T boundary = 0.1 °C; ◇h = 800 W/m^{2}K; + h = 400 W/m^{2}K; □h = 200 W/m^{2}K.

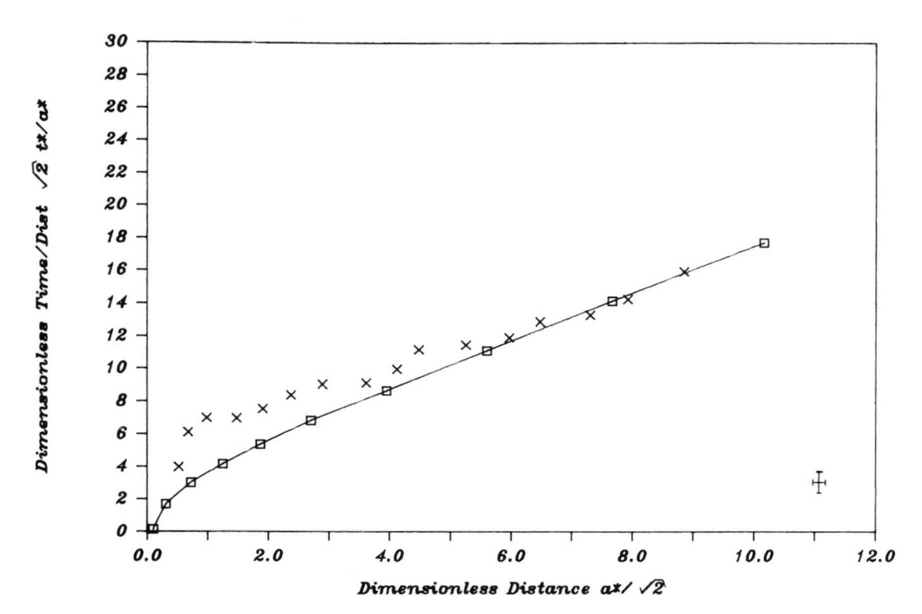

Fig. 6. Comparison between numerical results and the experimental data for the cold mould case, along the diagonal. Data: T_{poured} = 39 °C; $T_{boundary}$ = 0.1 °C; $h_{boundary}$ = 400 W/m^{2}K; X experimental data; □2D numerical results.

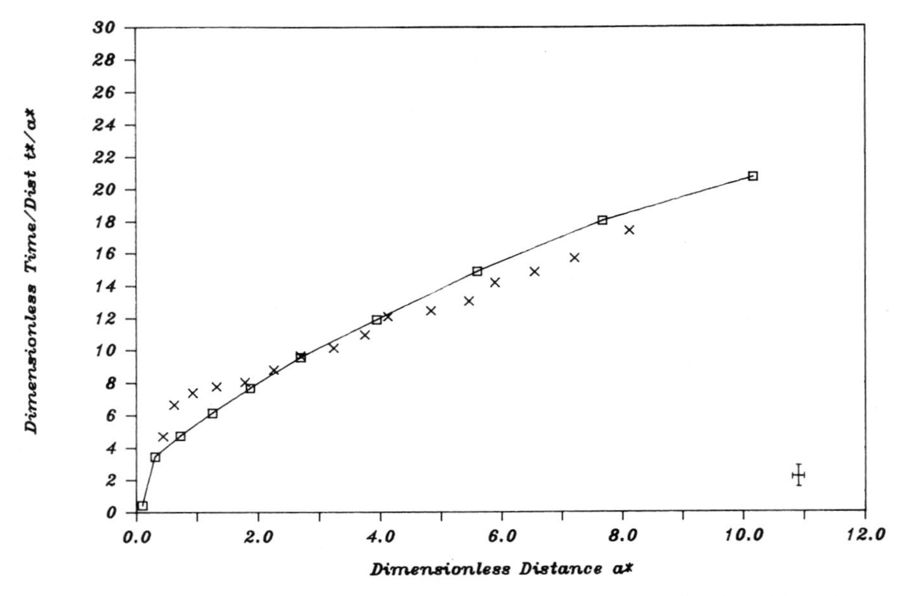

Fig. 7. Comparison between numerical results and the experimental data for the cold mould case, perpendicular to one edge. Data: $T_{poured} = 39°$ C; $T_{boundary} = 0.1°$ C; $h_{boundary} = 400$ W/m^{2}K; X experimental data; □2D numerical results.

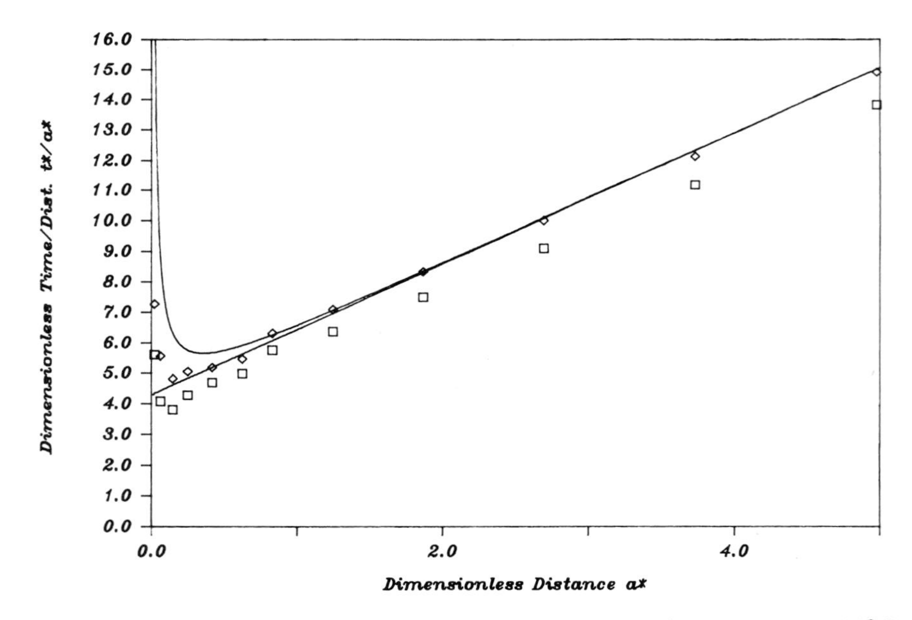

Fig. 8. Adams analytical results compared with the 1D numerical results. Data: $T_{initial} = 37°$ C; $T_{boundary} = 0.1°$ C; $h_{boundary} = 400$ W/m^{2}K; Upper line is modified Adams method. Lower line is Adams Method. ◇1D numerical results for single C_p peak at fusion temperature; □1D numerical results for double C_p peak at fusion and solid state transition.

three different values of h are presented. A change in h of a factor 4 from 200 to 800 W/m K produces at most a 30% change in the result.

Figure 6 shows results corresponding to Fig. 4, but with the mould initially cold and the melt poured in afterwards. Initially there is some disagreement between theory and data, which may be caused by the mixing due to pouring, but for later times the disagreement disappears.

Solidification along a line parallel to one of the rides of the cavity at a distance 0.065 m from the cold face is shown next in Fig. 7. The mould was initially cold. Again there is good agreement between theory and data.

In Fig. 8 theoretical results for 1-D solidification are shown. The analytic approximations of Equations (7) and (8) are compared to two numerical calculations: in the first all the latent heat of fusion is contained in the single peak, and in the second the experimentally measured double peak structure is used. At large times the analytic approximations and the numerical result with one peak in the heating curve agree well, but differ from the result obtained from the correct heating curve. At short times there is scatter between results from the different schemes.

Finally in Fig. 9, the data are compared to the numerical result, and to the modified Adams Equation (8). The deviation of the analytic approximation from data and numerical result is clear.

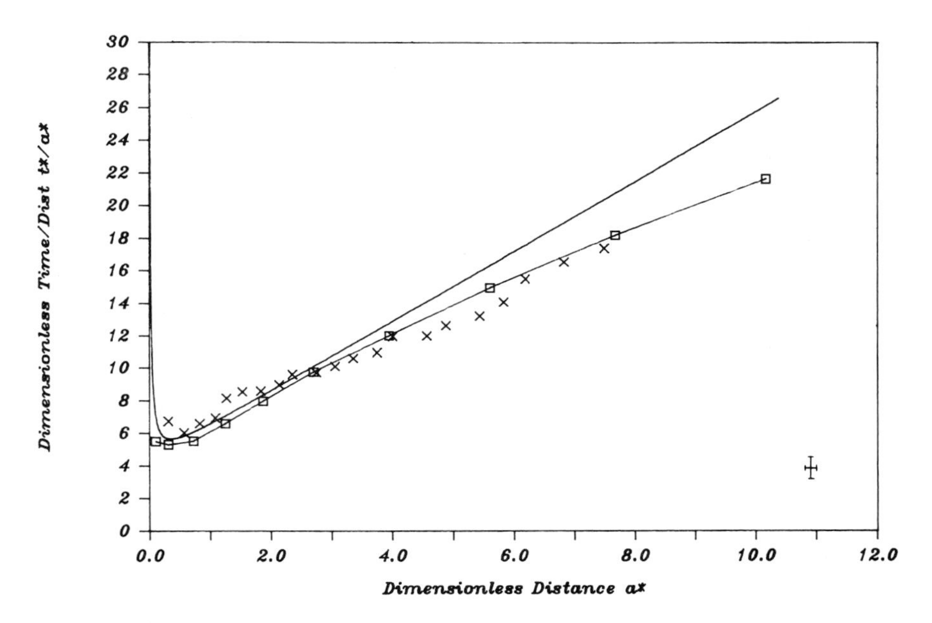

Fig. 9. Comparison between Adams, numerical results and experimental data for the warm mould case, perpendicular to one edge. Data: $T_{\text{initial}} = 39\,°\text{C}$; $T_{\text{boundary}} = 0.1\,°\text{C}$; $h_{\text{boundary}} = 400$ W/m^{2}K; X experimental data; □2D Numerical results; ——— modified Adam's method.

6. Discussion

Experiments on the solidification of n-eicosyne wax in a 2-D geometry have been performed. It is possible to carry out a numerical simulation of the solidification which agrees well with the experiment. This however requires information on the heat transfer coefficient between the mould and the melt, and on the heating curve for the melt; neither of these may be known accurately.

The standard 1-D analytic approximation (Adams) for solidification has been shown to agree quite well with the numerical result for the case where the melt's heating curve has a single peak and heat flow is unidirectional. For the case of a material such as n-eicosyne where the heating curve has two peaks and particularly in the vicinity of the diagonal, the analytic approximation gives a less accurate answer. The most marked scatter between the different theoretical schemes appears at short times; this may be important for thin walled castings where the dimensionless solidification distance a^* are less than unity.

Acknowledgements

We are grateful to the SERC, the Wolfson Foundation and St. John's College, Cambridge for financial support, and to Professor J.F. Davidson for advice and discussions.

References

1. W.C. Erickson: Computer Simulation of Solidification. *AFS International Cast Metals Journal* 5 (1980) 30–40.
2. P.R. Sahm: Applications – properties – microstructure – solidification technologies. In: P.R. Sahm and P.N. Hansen (eds) *Solidification Processes – Computer Simulation and Modelling*. CIATF-Cairo (1983) pp. 2–18.
3. M. Samonds, R.W. Lewis, K. Morgan and R. Symberlist: Efficient three dimensional finite element metal casting simulation employing a quadratic conductivity approximation. In: R.W. Lewis and K. Morgan (eds) *Numerical Methods in Thermal Problems*. Pineridge Press (1985) pp. 64–77.
4. J.T. Berry and R.D. Pehlke: Computer-aided design system for castings. In: *Solidification technology in the foundry and casthouse*. The Metals Society, London (1983) pp. 432–440.
5. J.G. Hartley and D. Babock: Thermal properties of mold materials. In: H.G. Brody and D. Apelian (eds) *Modelling of Casting and Welding Processes*. The Metallurgical Soc. of AIME (1981) pp. 83–92.
6. Kai Ho and R.D. Pehlke: Metal-Mold Interfacial Heat Transfer. *Met. Trans. B* 16B (1985) 585–594.
7. T.W. Clyne: The use of heat flow modelling to explore solidification phenomena. *Met. Trans. B* 13B (1982) 471–478.
8. R.S. Dutton: Thermal design of a blister casting wheel. In: R.G. Rice (ed) *Proceedings of the Second National Chemical Engineering Conference*. The Institution of Chemical Engineers, Australia, (1974) pp. 115–125.

9. F. Weinberg, J. Lait and R. Pugh: Solidification of high-carbon steel ingots. In: *Solidification and Casting of Metals*. The Metals Society, London (1979) pp. 334–339.
10. G.S. Cole: Transport processes and fluid flow in solidification. In: *Solidification*. American Society for Metals, Cleveland, Ohio, (1971) pp. 201–274.
11. G.S. Cole and G.F. Bolling: The importance of natural convection in casting. *Trans. of the Metallurgical Soc. of AIME* 233 (1965) 1568–1572.
12. J.P. Birat, J. Foussal, Mr. Larrecq, C. Saguez and M. Warin: Influence of convective heat transfer on solidification in the mould during continuous casting of steel. In: *Solidification Technology in the Foundry and Casthouse*. The Metals Society, London, (1983) pp. 536–543.
13. R. Harnar and F. Hernandez: Numerical simulation in precision castings In: R.W. Lewis and K. Morgan (eds) *Numerical Methods in Thermal Problems*. Pineridge Press (1983) pp. 1342–1352.
14. H.S. Carslaw and J.C. Jaeger: *Conduction of Heat in Solids*, OUP (1986).
15. C.M. Adams: Thermal considerations in freezing. In: *Liquid Metals and Solidification*. American Society for Metals, Cleveland, Ohio, (1958) pp. 187–217.
16. O.C. Zienkiewicz: *The Finite Element Method*. 3rd edn. McGraw-Hill, New York (1977).
17. Q.T. Pham: The use of lumped capacitance in the finite element solution of heat conduction problems with phase change. *Int. J. Heat Mass Transfer* 29 (1986) 285–291.
18. V. Voller and M. Cross: Accurate solutions of moving boundary problems using the enthalpy method. *Int. J. Heat Mass Transfer* 24 (1981) 545–556.
19. G.E. Bell: On the performance of the enthalpy method. *Int. J. Heat Mass Transfer* 25 (1982) 587–589.
20. A.W.D. Hills, S.L. Malhotra and M.R. Moore: The solidification of pure materials (and eutectics) under uni-directional heat flow conditions. *Met. Trans. B* 6B (1975) 131–142.
21. C. Gau and R. Viskanta: Effect of crystal anisotropy on heat transfer during melting and solidification of a metal. *Trans. ASME J. Heat Transfer* 107 (1985) 706–708.
22. M.E. Glicksman: Direct observations of solidification. In: *Solidification*. American Society for Metals, Cleveland, Ohio, (1971) pp. 155–200.
23. E.M. Sparrow, E.D. Larson and J.W. Ramsey: Freezing on a finned tube for either conduction-controlled or natural convection-controlled heat transfer. *Int. J. Heat Mass Transfer* 24 (1981) 273–284.
24. N. Ramanchandran, J.P. Gupta and Y. Jaluria: Experiments on solidification with natural convection in a rectangular enclosed region. *Int. J. Heat Mass Transfer* 25 (1982) 595–596.
25. H.E. Huppert and M.G. Worster: Dynamic solidification of a binary melt. *Nature* 314 (1985) 703–707.
26. H. Budhia and F. Kreith: Heat transfer with melting or freezing in a wedge. *Int. J. Heat Mass Transfer* 16 (1973) 195–211.
27. W.R. Humphries and E.I. Griggs: A design handbook for phase change thermal control and energy storage devices. NASA technical paper 1074 (1977).
28. E.I. Griggs and D.W. Yarbrough: Thermal conductivity of solid unbranched alkanes from n-hexadecane to n-eiosane. In: *Proceedings of the 14th South Eastern Seminar on Thermal Sciences* (1978) pp. 256–267.

Applied Scientific Research 44: 111–137 (1987)

Thermal rhythms in composite ladle walls

C. REES [1], G. POOTS [1] & V.J. SMALL [2]
[1] *Centre for Industrial Applied Mathematics, University of Hull, HU6 7RX, England;*
[2] *Process Mathematics, British Steel Corporation, Scunthorpe, England*

1. Introduction

The ladle is essential to the operation of a modern integrated iron and steel plant. It is a refractory lined vessel, typically holding 300 tonnes of liquid metal, which is used to transport iron or steel in molten form with a minimum heat loss and also to pour steel into moulds prior to solidification and further processing.

This paper is concerned with the thermal response of a composite ladle wall during day-to-day operations in a steel making plant. Typically a ladle will be filled with molten metal at, for example, 1600 °C. The system will lose heat by conduction through the top slag surface and the ladle wall and then by convection and radiative transfer into the environment. Internally the molten metal will thermally stratify by a process of conduction and convective motion.

The ladle standing period normally lasts for one hour and during that time the molten metal/inner wall local temperature is observed to decrease slowly. For simplicity it is here taken to be 1600 °C. During the next hour the liquid metal is poured from the ladle either into the mould of a concast machine or into moulds for producing ingots. The convective motion and heat transfer within the liquid is now more complex than in the standing period, see for example the mathematical model proposed in Egerton et al. [2,3] where viscous flow effects, buoyancy force and heat capacitance of the walls were ignored. Here the heat loss at the inner ladle wall is approximated by an average constant heat flux which is found experimentally. Of course this local condition at the inner wall depends on whether it is in contact with molten metal or if it is radiating heat back into the empty portion of the vessel. This work is concerned with the former situation.

The total time from filling to re-filling the ladle is typically four hours and this process is repeated continuously in day-to-day operations. After a long time when initial thermal responses have died away, the conduction profile in the ladle wall becomes periodic in time. These transient responses (evolving to a state which is periodic in time) are herein called the thermal rhythms of the ladle wall during cycles of standing and pouring.

The ladle wall is constructed of high Al_2O_3 refractory material and in an attempt to control the heat loss from the ladle wall it is proposed to insert a layer of insulating material into the wall, see Fig. 1. Assuming that the heat conduction across the composite wall is locally at any point one-dimensional then in a steady state situation the heat loss from the outer ladle surface is independent of the location of the inserted insulated layer; it is dependent on the relative thickness of the insulating layer and on the thermal properties of the refractory lining and the insulating layer. However, as previously discussed, the day-to-day operations involving a ladle is a time dependent process and in the case of thermal rhythms in the composite wall the heat loss at the outer ladle surface will also depend on the thermal capacitance of the composite wall and on the location of the insulating layer.

It is the objective of this paper to construct a one-dimensional mathematical model to simulate the thermal response of a composite ladle wall, to various cycles of heating and cooling as would occur in day-to-day operations. There are three main questions to be answered:

(a) If the composite ladle wall starts out as "cold" how many cycles (heating followed by cooling) must take place so that the thermal rhythm becomes periodic?
(b) What is the effect of preheating the composite ladle wall on the heat loss from the outer ladle surface during the first few plant operations? and
(c) For specified thickness and properties of the insulating layer is there an optimal location for it, so as to achieve minimum heat loss from the outer ladle surface?

Another interesting aspect of this investigation would be to use the transient thermal profiles to compute the thermal stresses in the ladle wall. Such results could be useful in ladle design technology.

In Section 2 the transient heat conduction problem is formulated for the nonlinear boundary conditions simulating day-to-day operations. Section 3 deals with simplified linear models in which either the ladle wall temperature or heat flux is periodic. In Section 4 a fully implicit Crank-Nicolson method is used to solve the full nonlinear equations. Analytical and numerical results are discussed in Section 5. Conclusions are given in Section 6.

2. Formulation

A typical composite ladle wall is shown in Fig. 1. Let D denote the thickness of the composite wall and λD $(\lambda < 1)$ that of the insulating layer which is located at distance γD from the inner ladle wall, so that $(\lambda + \gamma) \leqslant 1$; measure z along the normal to the inner ladle surface, $z \in [0, D]$.

The general case will consist of three heat conduction regions. The governing equations and boundary conditions for the related thermal fields T_i $(i = 1, 2, 3)$ are as follows:

Region 1: For $z \in [0, \gamma D]$, $t > 0$

$$k_1 \, \partial^2 T_1/\partial z^2 = \partial T_1/\partial t, \tag{2.1}$$

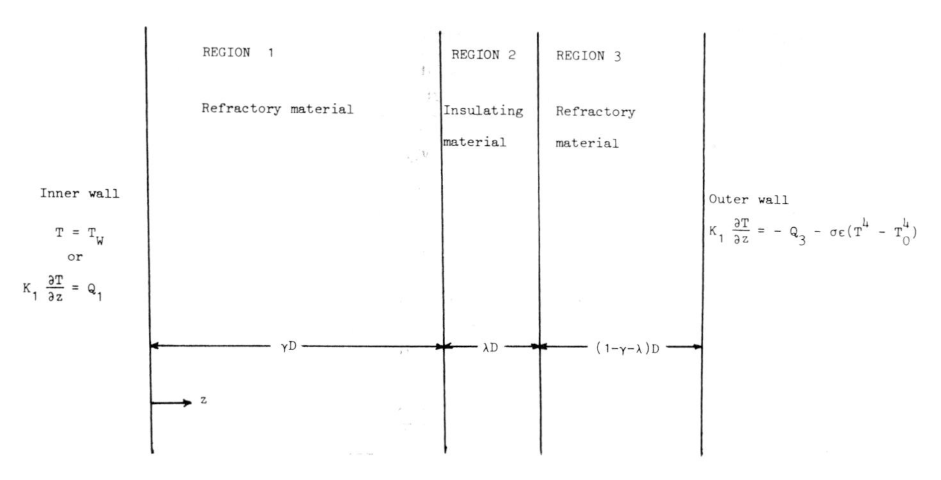

Fig. 1. Composite ladle wall.

subject to the boundary conditions: at $z = 0$, $T_1 = T_W$ for $(2n-2)t_0 < t < (2n-1)t_0$

$$k_1\ \partial T_1/\partial z = Q_1 \quad \text{for } (2n-1)t_0 < t < 2nt_0 \tag{2.2}$$

where $n = 1, 2, 3, 4\ldots$ denotes the number of heating and cooling cycles of period $2t_0$;

$$\text{at } z = \lambda D, \quad T_1 = T_2, \quad K_1\ \partial T_1/\partial z = K_2\ \partial T_2/\partial z. \tag{2.3}$$

Region 2: For $z \in [\gamma D, (\gamma+\lambda)D]$, $t > 0$

$$k_2\ \partial^2 T_2/\partial z^2 = \partial T_2/\partial t, \tag{2.4}$$

subject to the boundary conditions:

$$\text{at } z = (\gamma+\lambda)D, \quad T_2 = T_3, \quad K_2\ \partial T_2/\partial z = K_1\ \partial T_3/\partial z. \tag{2.5}$$

Region 3: For $z \in [(\gamma+\lambda)D, D]$, $t > 0$

$$k_1\ \partial^2 T_3/\partial z^2 = \partial T_3/\partial t, \tag{2.6}$$

subject to the boundary condition:

$$\text{at } z = D, \quad K_1\ \partial T_3/\partial z = -Q_3 - \sigma\epsilon\left(T_3^4 - T_0^4\right). \tag{2.7}$$

The initial condition is taken as

$$T_1 = T_2 = T_3 = T_0 \quad \text{at } t = 0,\ z \in [0, D], \tag{2.8}$$

where T_0 is the environmental temperature. In the case of a preheated ladle wall the non uniform temperatures $T_1(z)$, $T_2(z)$ and $T_3(z)$ will be prescribed at $t = 0$ and this will be discussed later.

Table 1. Physical properties and dimensions of a typical ladle wall.

Property (units)	Symbol used	Material 1 (refractory mat.)	Material 2 (insulating mat.)
Conductivity (W/mk)	K	1.5	0.51
Density (Kg/m^3)	ρ	2200.0	1071.0
Specific heat (J/Kg/k)	c	1072.0	1130.0
Thermal diffusivity (m^2/S)	k	6.36×10^{-7}	4.22×10^{-7}
Wall thickness (m)	D	0.386	
Insulating thickness (m)	λD	0.076	
Inner wall temperature (K)	T_W	1873.0	
Air temperature (K)	T_0	293.0	
Inner wall heat flux (W/m^2)	Q_1	15.0	
Outer wall heat flux (W/m^2)	Q_3	8.0	
Boltzmann constant	σ	5.6697×10^{-8}	
Emissivity	ϵ	0.7	

In the above K denotes the thermal conductivity, $k = K/\rho c$ the thermal diffusivity, Q is constant heat flux, σ the Boltzmann constant, ϵ the emissivity of the outer wall, ρ the density and c the specific heat. The physical properties and dimensions for the composite ladle wall are given in Table 1.

Obviously there are two special cases to be considered when the composite wall consists of only two regions

(a) $\gamma = 0$ and the insulating layer is located at the inner ladle wall, and
(b) $\gamma = 1 - \lambda$ and the insulating layer is located at the outer ladle wall.

The governing equations and boundary conditions for these follow directly from (2.1)–(2.8).

The above system of equations are nonlinear because of (2.7) and must be solved numerically. However, as will be verified later, it is a weakly nonlinear system and a great deal of information on thermal response in a composite ladle wall can be obtained from analytical solutions using the linear condition

$$K_3 \, \partial T_3/\partial z = -Q_3 - h(T_3 - T_0) \quad \text{at } z = D, \quad t > 0 \tag{2.9}$$

to replace (2.7).

In principle the methods for calculating the periodic temperature in composite slabs either by a matrix algorithm or by Laplace Transform, and for dealing with switch 'on-off' external conditions by Duhamel's theorem are well known, see Carslaw and Jaeger [1]. In Section 3, these methods are implemented, albeit, in a limited way but are shown to exhibit some novel features and yield some useful results. In Section 4 the numerical solution of the full system is briefly discussed. Section 5 deals with a discussion of the numerical solutions. In Section 6 conclusions are given.

3. Analytical solutions

In this section three simplified conduction models are considered: steady state, periodic temperature and switch 'on-off' inner wall temperature. Moreover in the first two of these it is assumed that Q_3 in equation (2.9) is zero. The reason for doing so is that its inclusion just adds a continuous piecewise linear temperature profile to the following solutions and hence would mask the special features that appear.

Model 1: steady state

Prior to constructing transient solutions it is useful to record here the well known and obvious steady state result for the 1-dimensional composite ladle wall, namely that the heat flux within the wall and into the environment is independent of the location of the insulating layer.

The steady state equations are (2.1)–(2.9) with $\partial/\partial t \equiv 0$ and (2.2) and (2.8) replaced by

$$T_1 = T_M \quad \text{at } z = 0, \tag{3.1}$$

and

$$K_1 \; \partial T_3/\partial z = -h(T_3 - T_0) \quad \text{at } z = D, \tag{3.2}$$

respectively. In terms of the dimensionless variables.

$$\begin{aligned} &\theta_j = (T_j - T_0)/(T_M - T_0) \quad \text{for } j = 1, 2, 3, \\ &Z_1 = z/\gamma D, \quad Z_1 \in [0, 1], \\ &Z_2 = (z - \gamma D)/\lambda D, \quad Z_2 \in [0, 1], \\ &Z_3 = (z - (\gamma + \lambda)D)/(1 - \gamma - \lambda)D, \quad Z_3 \in [0, 1], \end{aligned} \tag{3.3}$$

the piecewise linear temperature field is

$$\begin{aligned} &\theta_1(Z_1) = 1 - \gamma Z_1 \Big/ \left(1 + \frac{1}{Bi} + \lambda\left(\frac{K_1}{K_2} - 1\right)\right), \\ &\theta_2(Z_2) = 1 - K_1 \lambda Z_2 \Big/ \left\{K_2\left(1 + \frac{1}{Bi} + \lambda\left(\frac{K_1}{K_2} - 1\right)\right)\right\}, \\ &\theta_3(Z_3) = 1 - (1 - \gamma - \lambda) Z_3 \Big/ \left(1 + \frac{1}{Bi} + \lambda\left(\frac{K_1}{K_2} - 1\right)\right), \end{aligned} \tag{3.4}$$

where the outer wall Biot number is defined by

$$\text{Bi} = hD/K_1. \tag{3.5}$$

The outer wall Nusselt number is defined, for $\gamma \neq 1 - \lambda$, as

$$\mathrm{Nu} = -K_1 \, \partial T_3/\partial z \,|_{z=D} \frac{D}{K_1(T_M - T_0)} = \frac{\mathrm{Bi}}{1 + \mathrm{Bi} + \lambda \, \mathrm{Bi}(K_1/K_2 - 1)}. \tag{3.6}$$

Thus although the thermal profiles are dependent on γ, the location of the insulating layer, the outer wall Nusselt number is not. Moreover with regard to the former the implication is that the state of thermal stress in the composite wall is dependent on γ.

Finally if (3.2) is replaced by the radiative boundary condition:

$$K_1 \, \partial T_3/\partial z = -\sigma\epsilon\left(T_3^4 - T_0^4\right) \quad \text{at } z = D, \tag{3.7}$$

the heat transfer from the outer wall is again independent of γ. Quantitatively Nu is then determined from the transcendental equation:

$$\mathrm{Nu} = \sum_1^4 A_r \left\{1 + \mathrm{Nu}\left(1 + \lambda\left(\frac{K_1}{K_2} - 1\right)\right)\right\}^r, \tag{3.8}$$

where

$$\begin{aligned} A_1 &= 4\sigma\epsilon D T_0^3/K_1, & A_2 &= 6\sigma\epsilon D(T_M - T_0)T_0^2/K_1 \\ A_3 &= 4\sigma\epsilon D(T_M - T_0)^2 T_0/K_1, & A_4 &= \sigma\epsilon D(T_M - T_0)^3 K_1. \end{aligned} \tag{3.9}$$

Some typical results for a composite ladle wall are given in Table 2 for the data of Table 1, $T_M = 1000\,^\circ\mathrm{C}$ and in the case of Newton Cooling Bi = 1. These results are compared with those for a wall consisting of refractory material only and for a wall consisting entirely of insulating material. In the case of Newton Cooling, as expected, the outer wall temperature using the insulating material is lowest. However for a composite wall with radiative cooling there is an improvement in heat retention using the composite wall.

Consider now some simple transient models designed to simulate the inner wall temperature condition (2.2).

Model 2: periodic wall temperature

The theory, based on the analogy between electrical and thermal phenomena, for the calculation of periodic heat flow in a composite wall has been

Table 2. Steady state outer wall temperature for inner wall temperature $T_M = 1000\,^\circ\mathrm{C}$.

	Ladle Wall constructed of		
	Refractory material	Insulating material	Composite material
Newton cooling Bi = 1, eqn. (3.2)	509	267	429
Radiative cooling $Q_3 = 0$, eqn. (2.9)	258	225	159

developed by van Gorcum [6] and Vodika [7]; see also Jaeger [4] and the considerable discussion in Carslaw and Jaeger [1]. These methods are readily applied to composites of two materials but become tedious and cumbersome for more complex regions. Only a brief outline of the analysis will be given.

The system of equations to be solved is (2.1)–(2.6) with (2.2) replaced by:

$$\text{at } z=0,\ T_1 = \tfrac{1}{2}(T_W + T_c) + \tfrac{1}{2}(T_W - T_c)\cos \omega t, \tag{3.10}$$

and once again (3.2) is assumed at $z = D$. There is no initial condition, as it is assumed that (3.10) has applied for a long time. Perhaps it is interesting to note here that one of the earliest solutions for a thermal rhythm initiated by a periodic wall temperature applied at time $t = 0$ was provided by Stokes [5], see Carslaw and Jaeger [1].

The wall temperature (3.10) oscillates, with period $2\pi/\omega$, between T_W and T_c with mean temperature $T_M = \frac{1}{2}(T_W + T_c)$ and amplitude $T_A = \frac{1}{2}(T_W - T_c)$.

Introduce the dimensionless temperatures

$$\boldsymbol{T}(\boldsymbol{Z}, t) = T_0 + (T_M - T_0)\boldsymbol{\theta}(\boldsymbol{Z}) + T_A\boldsymbol{\theta}^*(\boldsymbol{Z}, \tau) \tag{3.11}$$

where

$$\boldsymbol{\theta}(\boldsymbol{Z}) = (\theta_1(Z_1), \theta_2(Z_2), \theta_3(Z_3))^T \tag{3.12}$$

are the steady state solutions as given in (3.4), and

$$\tau = k_1 t/D^2 \tag{3.13}$$

is the dimensionless time. In the following since the $Z_i \in [0, 1]$ for $i = 1, 2, 3$ the suffix i will be dropped for convenience.

The governing system of equations become, for $Z \in [0, 1]$, the following:

Region 1: $\partial^2\theta_1^*/\partial Z^2 = \gamma^2\, \partial\theta_1^*/\partial\tau$,

$$\theta_1^*(0, \tau) = R(\exp \mathrm{i}\Omega\tau),\ \Omega = D^2\omega/k_1, \tag{3.14}$$

$$\theta_1^*(1, \tau) = \theta_2^*(0, \tau),\quad \partial\theta_1^*(1, \tau)/\partial Z = (K_2\gamma/K_1\lambda)\, \partial\theta_2^*(0, \tau)/\partial Z;$$

Region 2: $\partial^2\theta_2^*/\partial Z^2 = \lambda^2(k_1/k_2)\, \partial\theta_2^*/\partial\tau$

$$\theta_2^*(1, \tau) = \theta_3^*(0, \tau),$$

$$\partial\theta_2^*(1, \tau)/\partial Z = (\lambda K_1/(1-\gamma-\lambda)K_2)\, \partial\theta_3^*(0, \tau)/\partial Z; \tag{3.15}$$

Region 3: $\partial^2\theta_3^*/\partial Z^2 = (1-\gamma-\lambda)^2\, \partial\theta_3^*/\partial\tau$,

$$\partial\theta_3^*(1, \tau) = -\mathrm{Bi}(1-\gamma-\lambda)\theta_3^*(1, \tau); \tag{3.16}$$

The solution of this system of equations is

$$\boldsymbol{\theta}^*(\boldsymbol{Z}) = \left[A_j \cosh \mu_j(1+\mathrm{i})Z + B_j \sinh \mu_j(1+\mathrm{i})Z\right] \exp(\mathrm{i}\Omega\tau) \tag{3.17}$$

for $j = 1$, 2 and 3, where

$$\mu_1 = \gamma(\Omega/2)^{1/2}, \quad \mu_2 = \lambda(k_1\Omega/2k_2)^{1/2} \quad \text{and}$$

$$\mu_3 = (1 - \gamma - \lambda)(\Omega/2)^{1/2}. \tag{3.18}$$

$A_1 = 1$ and the remaining constants of integration $\boldsymbol{A}$ and $\boldsymbol{B}$ are all complex and because of the nature of (3.14)–(3.18) the related set of simultaneous equations with complex coefficients has to be solved numerically; this was carried out using a modification of the Choleski method.

Profiles of the thermal rhythm will not be given for this model as similar results for more realistic models will be discussed later. However at this stage it is useful to consider the periodic Nusselt number at the outer ladle wall. This is defined by

$$\mathrm{N\overset{*}{u}} = -K_1 \left.\partial(T_A\theta_3^*)/\partial z\right|_{z=D} \frac{D}{T_A K_1} \tag{3.19}$$

and its amplitude is

$$|\mathrm{N\overset{*}{u}}| = |\partial\theta_3^*(1, \tau)/\partial Z|/(1 - \gamma - \lambda). \tag{3.20}$$

In Fig. 2 the amplitude $|\mathrm{N\overset{*}{u}}|$ is given as a function of γ (the location of the insulating layer) for infinite Biot number and for values of Ω in the range $1 - 100$; $B_i = \infty$ corresponds physically to the case when the outer wall is maintained at the environmental temperature T_0. As $\Omega \downarrow 0$ $|\mathrm{N\overset{*}{u}}| \uparrow 1/\left(1 + \lambda\left(\frac{K_1}{K_2} - 1\right)\right)$, the steady state value which is independent of γ. As Ω is increased a maximum value of $|\mathrm{N\overset{*}{u}}|$ is appearing when the insulating layer is centrally placed in the composite, although its magnitude is decreasing to zero. For large Ω, $|\mathrm{N\overset{*}{u}}|$ varies with γ and achieves minimum values when the insulating layer is located at the inner or outer part of the wall.

In Fig. 3 the amplitude $|\mathrm{N\overset{*}{u}}|$ is given as a function of γ for various Biot numbers when $\Omega = 100$, a typical value encountered in day-to-day operations. For large Biot number $|\mathrm{N\overset{*}{u}}|$ has, as previously stated, minimum values when the insulating layer is placed at either the inner or outer walls, it takes a maximum value at $\gamma = 0.15$ and 0.65 and has a shallow minimum at $\gamma = 0.40$. For $\Omega = 100$ and Bi decreasing $|\mathrm{N\overset{*}{u}}|$ is no longer symmetrical about $\gamma = 0.40$ as clearly indicated by Bi = 10 and 1.

Moreover for large Ω, $|\mathrm{N\overset{*}{u}}| = 0\ (\exp(-\Omega))$ and this is not surprising from a physical viewpoint as the wall temperature is then fluctuating 'rapidly' about its mean value and there is limited thermal penetration into the composite medium.

Finally the total Nusselt number at the outer wall corresponding to the inner wall temperature (3.10), namely

$$T_1 = T_M + T_A \cos \omega t, \tag{3.21}$$

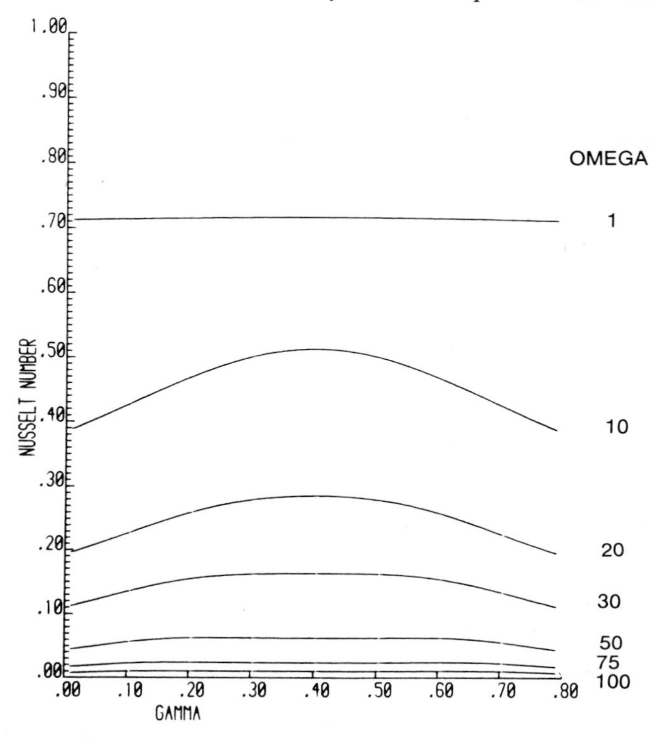

Fig. 2. Outer wall Nusselt Number for various values of omega.

may be defined as

$$\mathrm{Nu}_T = -(\partial T_3/\partial z)_{z=D}\frac{D}{(T_M - T_0)} \tag{3.22}$$

and this will have stationary values

$$\mathrm{Nu}_T(\gamma) = \mathrm{Nu}(\gamma) \pm \frac{1}{(T_M - T_A)}|\mathrm{N\overset{*}{u}}(\gamma)|, \tag{3.23}$$

where $\mathrm{Nu}(\gamma)$ and $\mathrm{N\overset{*}{u}}(\gamma)$ have already been computed in (3.6) and (3.19), respectively. Since the second term of (3.23) is only significant when the wall temperature varies slowly with time it follows that the implication of the above model in regard to plant operations (when Ω is large) is that the heat loss at the outer wall is insensitive to the location of the insulating layer.

The above model partially answers Question (c) as raised in Section 1. The next model is constructed using a switch on-off thermal condition at the inner wall and is thus aimed at finding an analytical solution related to Questions (a) and (b).

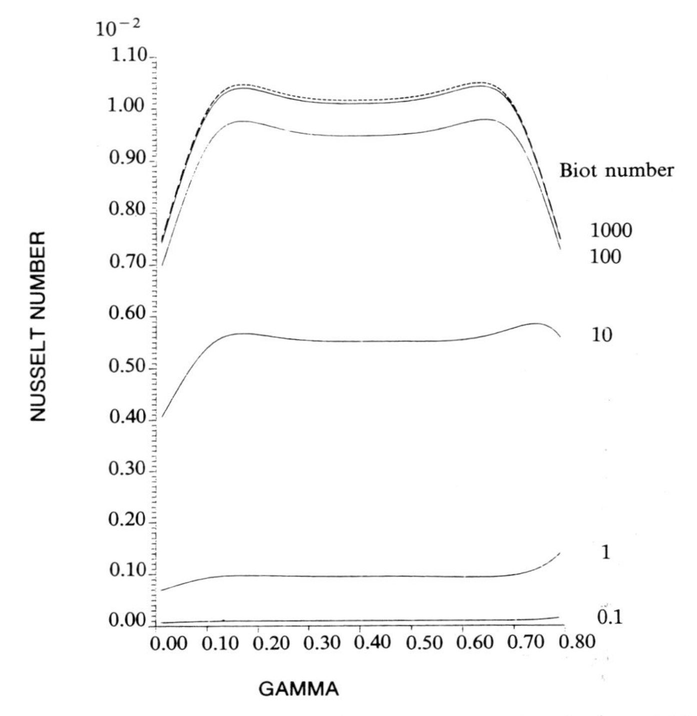

Fig. 3. Outer wall Nusselt Number for Various Biot Numbers (omega = 100).

Model 3: switch on-off wall temperature

To simulate a switch on-off thermal condition at the inner ladle wall consider the system (2.1)–(2.6) with (2.2) replaced by

$$T_1 = T_0 + (T_W - T_0)\phi(t), \tag{3.24}$$

where

$$\begin{aligned}\phi(t) &= 1, \ (2n-2)t_0 < t < (2n-1)t_0 \\ &= 0, \ (2n-1)t_0 < t < 2nt_0\end{aligned} \tag{3.25}$$

for $n = 1, 2, 3 \ldots$. Thus the inner ladle wall is maintained at temperature T_W for time t_0 and then reduced to temperature T_0 for time t_0 and repeated giving a cycle time of $2t_0$. At the outer ladle wall the linear condition (2.9) is employed. The initial condition is given in (2.8).

Introduce the dimensionless temperature fields:

$$\Theta_i = (T_i - T_0)/T_W - T_0) \tag{3.26}$$

for $i = 1, 2$ and 3, together with the dimensionless time τ (given by (3.12)) and the variables $\boldsymbol{Z}$ (given by (3.3)).

An analytical solution to this system of equations can then be obtained using Duhamel's theorem, see Carslaw and Jaeger [1]. This solution is expressed in terms of the eigenfunctions of the related Sturm-Liouville system for the composite domain satisfying the Dirichelet condition $\Theta_1(0) = 0$ and the Neumann condition $\Theta_3'(1) = -\text{Bi}\ \Theta_3(1)$.

The formal solution is as follows:

$$\boldsymbol{\Theta}(\boldsymbol{Z}, \tau) = \boldsymbol{F}(\boldsymbol{Z}) + \sum_1^{\infty} A_n \boldsymbol{\Phi}_n(\boldsymbol{Z}) e^{-\alpha_n^2 \tau} + \sum^{\infty} B_n \alpha_n^2 \boldsymbol{\Phi}_n(\boldsymbol{Z}) e^{-\alpha_n^2 \tau} \int_0^{\tau} \phi(\lambda) e^{\alpha_n^2 \lambda}\, \mathrm{d}\lambda \tag{3.27}$$

Here $\boldsymbol{F}(\boldsymbol{Z})$ is a piecewise differentiable linear function defined by

$$\boldsymbol{F}(\boldsymbol{Z}) = \left(\frac{\gamma q_3 Z}{H\ \text{Bi}}, \frac{q_3\left(\gamma + \dfrac{K_1 \lambda}{K_2} Z\right)}{H\ \text{Bi}}, \frac{q_3\left(\gamma + \dfrac{K_1 \lambda}{K_2} + (1 - \gamma - \lambda) Z\right)}{H\ \text{Bi}} \right)^T, \tag{3.28}$$

where

$$q_3 = \frac{Q_3 D}{K_1 (T_W - T_0)}, \qquad H = 1 + \frac{1}{\text{Bi}} + \lambda\left(\frac{K_1}{K_2} - 1\right). \tag{3.29}$$

Consider now the eigenfunctions $\boldsymbol{\Phi}_n(\boldsymbol{Z})$ and eigenvalues α_n appearing in (3.27). For convenience the suffix n will be dropped. The eigenfunction is represented by

$$\boldsymbol{\Phi}(\boldsymbol{Z}) = (\Phi_1(Z), \Phi_2(Z), \Phi_3(Z))^T. \tag{3.30}$$

It is piecewise differentiable in each of the three regions $Z \in [0, 1]$ and is continuous at the surfaces of the insulating region. The Sturm-Liouville system for the composite region is as follows:

Region 1: $Z \in [0, 1]$

$$\Phi_1'' + \alpha^2 \gamma^2 \Phi_1 = 0, \tag{3.31}$$

$$\Phi_1(0) = 0, \quad \Phi_1(1) = \Phi_2(0), \quad \Phi_1'(1) = \frac{K_2 \gamma}{K_1 \lambda} \Phi_2(0), \tag{3.32}$$

Region 2: $Z \in [0, 1]$

$$\Phi_2'' + \alpha^2 \frac{k_1}{k_2} \lambda^2 \Phi_2 = 0, \tag{3.33}$$

$$\Phi_2(1) = \Phi_3(0), \quad \Phi_2'(1) = \frac{K_1 \lambda}{K_2 (1 - \gamma - \lambda)} \Phi_3'(0); \tag{3.34}$$

Region 3: $Z \in [0, 1]$

$$\Phi_3'' + \alpha^2(1-\gamma-\lambda)^2\Phi_3 = 0, \tag{3.35}$$

$$\Phi_3'(1) = -\mathrm{Bi}(1-\gamma-\lambda)\Phi_3(1). \tag{3.36}$$

The eigenfunctions are:

$$\Phi_1(Z) = \sin \gamma\alpha Z, \tag{3.37}$$

$$\Phi_2(Z) = \frac{K_1}{K_2}\sqrt{\frac{k_1}{k_2}} \cos \gamma\alpha \sin \sqrt{\frac{k_1}{k_2}}\lambda\alpha Z + \sin \gamma\alpha \cos \sqrt{\frac{k_1}{k_2}}\lambda\alpha Z, \tag{3.38}$$

$$\begin{aligned}\Phi_3(Z) = {} & \left\{\cos \gamma\alpha \cos \sqrt{\frac{k_1}{k_2}}\lambda\alpha - \frac{K_2}{K_1}\sqrt{\frac{k_1}{k_2}} \sin \gamma\alpha \sin \sqrt{\frac{k_1}{k_2}}\lambda\alpha\right\} \\ & \times \sin(1-\gamma-\lambda)\alpha Z \\ & + \left\{\frac{K_1}{K_2}\sqrt{\frac{k_1}{k_2}} \cos \gamma\alpha \sin \sqrt{\frac{k_1}{k_2}}\lambda\alpha + \sin \gamma\alpha \cos \sqrt{\frac{k_1}{k_2}}\lambda\alpha\right\} \\ & \times \cos(1-\gamma-\lambda)\alpha Z.\end{aligned} \tag{3.39}$$

It can be shown that all the eigenvalues α are real and positive. The eigenvalues α are determined from the transcendental equation:

$$\begin{aligned}& \sqrt{\frac{k_1}{k_2}}\lambda\Bigg\{\alpha\left[\sin \gamma\alpha \cos \sqrt{\frac{k_1}{k_2}}\lambda\alpha \sin(1-\gamma-\lambda)\alpha - \cos \gamma\alpha \cos \sqrt{\frac{k_1}{k_2}}\lambda\alpha\right. \\ & \left.\times \cos(1-\gamma-\lambda)\alpha\right] \\ & - \mathrm{Bi}\left[\sin \gamma\alpha \cos \sqrt{\frac{k_1}{k_2}}\lambda\alpha \cos(1-\gamma-\lambda)\alpha + \cos \gamma\alpha \cos \sqrt{\frac{k_1}{k_2}}\lambda\alpha\right. \\ & \left.\times \sin(1-\gamma-\lambda)\alpha\right]\Bigg\} \\ & + \frac{k_1}{k_2}\frac{K_2}{K_1}\lambda\left\{\alpha \sin \gamma\alpha \sin\sqrt{\frac{k_1}{k_2}}\gamma\alpha \cos(1-\gamma-\lambda)\alpha\right. \\ & \left. + \mathrm{Bi} \sin \gamma\alpha \sin \sqrt{\frac{k_1}{k_2}}\lambda\alpha \sin(1-\gamma-\lambda)\alpha\right\} \\ & + \frac{K_1}{K_2}\lambda\left\{\alpha \cos \gamma\alpha \sin\sqrt{\frac{k_1}{k_2}}\lambda\alpha \sin(1-\gamma-\lambda)\alpha\right. \\ & \left. - \mathrm{Bi} \cos \gamma\alpha \sin \sqrt{\frac{k_1}{k_2}}\lambda\alpha \cos(1-\gamma-\lambda)\alpha\right\} = 0.\end{aligned} \tag{3.40}$$

Moreover, it is readily shown from the Strum-Liouville system (3.31)–(3.36) that the eigenfunctions are orthogonal. The orthogonality condition is written in terms of the inner product:

$$\langle \boldsymbol{\Phi}_m(\boldsymbol{Z}), \boldsymbol{\Phi}_n^T(\boldsymbol{Z})\rangle_\omega = \gamma \int_0^1 \Phi_{1m}\Phi_{1n}\, dZ + \frac{k_1}{k_2}\frac{K_2}{K_1}\lambda \int_0^1 \Phi_{2m}\Phi_{2n}\, dZ$$
$$+ (1-\gamma-\lambda)\int_0^1 \Phi_{3m}\Phi_{3n}\, dZ$$
$$= 0 \text{ if } m \neq n, \tag{3.41}$$

A novel feature of this condition is the appearance of the weight function.

Returning now to (3.27) the coefficients A_n and B_n are readily evaluated using (3.40). They are as follows:

$$A_n = -\langle \boldsymbol{F}(\boldsymbol{Z}), \boldsymbol{\Phi}_n^T(\boldsymbol{Z})\rangle_\omega / \langle \boldsymbol{\Phi}_n(Z), \boldsymbol{\Phi}_n^T(Z)\rangle_\omega \tag{3.42}$$

and

$$B_n = -\langle \boldsymbol{R}(\boldsymbol{Z}), \boldsymbol{\Phi}_n^T(\boldsymbol{Z})\rangle_\omega / \langle \boldsymbol{\Phi}_n(\boldsymbol{Z}), \boldsymbol{\Phi}_n^T(\boldsymbol{Z})\rangle_\omega, \tag{3.43}$$

where $\boldsymbol{R}(\boldsymbol{Z})$ is a piecewise differentiable linear function given by

$$\boldsymbol{R}(\boldsymbol{Z}) = \left(1 - \frac{\gamma Z}{H}, 1 - \frac{\gamma + \frac{K_1}{K_2}\lambda Z}{H}, 1 - \frac{\gamma + \frac{K_1}{K_2}\lambda + (1-\gamma-\lambda)Z}{H}\right). \tag{3.44}$$

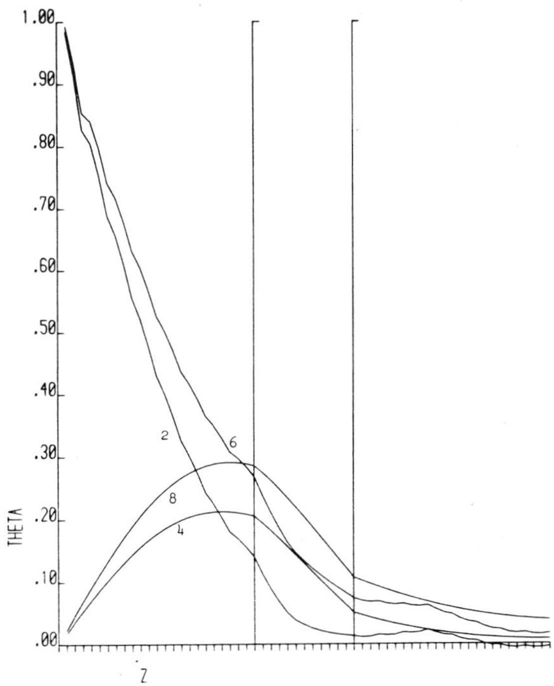

Fig. 4. Temperature profiles after 2, 4, 6 and 8 hours.

The above analytical solution (3.27) is exact. It is complicated algebraically but never the less it is a relatively simple matter to write a program to compute $\Theta(Z, \tau)$. The eigenvalues α are computed from (3.40) using the Newton-Raphson method; also the compactness of the eigenfunction expansion was validated on expanding a known function of Z in terms of the functions (3.30) and checking the result numerically across the composite region.

In Fig. 4 is displayed in the dimensionless temperature within the ladle wall after 2, 4, 6 and 8 hours for the switch on-off condition (3.24) taking $t_0 = 2$ hrs. The insulating layer is placed centrally at $\gamma = 0.4$. In this case, as will be discussed later, 12 complete cycles are required to establish the final periodic rhythm. In Fig. 5 a comparison is given between the analytical solution and the full numerical solution after 2 hrs, i.e. the end of the first hot period. The analytical results were computed using 500 eigenfunctions and the computing time required was found to be a small fraction of that required in the full numerical solution which will be discussed in the next section.

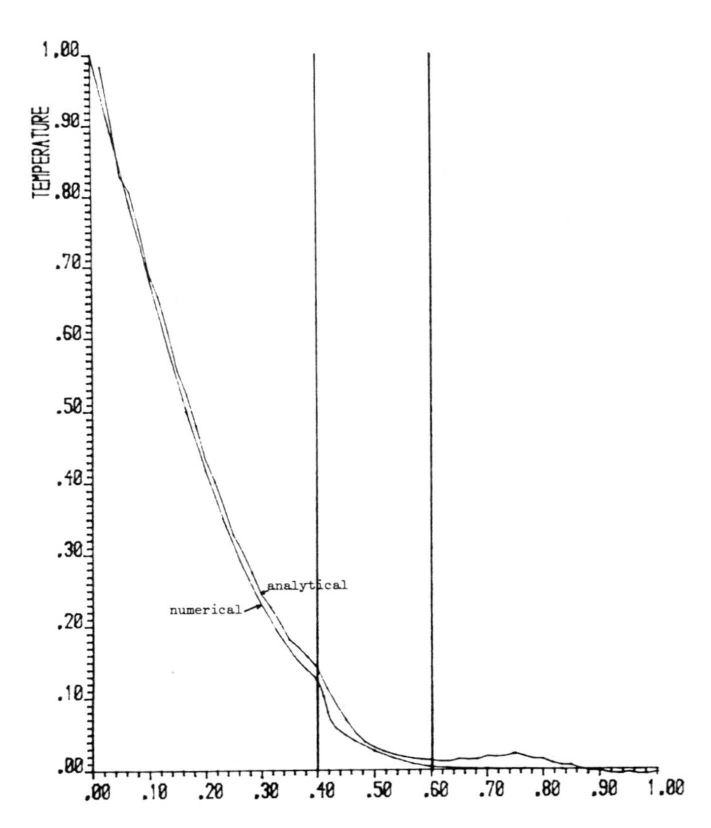

Fig. 5. Comparison of the 2 hour temperature profiles obtained using analytical and numerical methods (Gamma = 0.4).

Finally it should be mentioned that this model can be extended to cope with the actual condition (2.2) applied at the inner ladle wall. All that is necessary is to investigate the Sturm-Liouville system (3.31)–(3.36) with the Dirichelet condition $\Phi_1(0) = 0$ replaced by the Neumann condition $\Phi_1'(0) = 0$. The two (compact) sets of eigenfunctions can then be used to construct eigenfunctions expansions giving $\Theta(Z, \tau)$ for the heating and cooling cycles.

4. Numerical solution

Here a brief account is given of the numerical method employed to solve the full nonlinear system (2.1)–(2.8) which simulates thermal rhythms in a ladle wall in day-to-day operations.

In terms of the dimensionless temperature, time and spatial variables:

$$\Theta_i = \frac{(T_i - T_0)}{(T_W - T_0)} \quad (i = 1, 2, 3), \quad \tau = k_1 t/D_2 \quad \text{and } Z = z/D, \tag{4.1}$$

respectively, the required boundary condition for Region 1 at $Z = 0$ becomes

$$\begin{aligned} &\Theta_1(0, \tau) = 1, \quad (2n-2)\tau_0 < \tau < (2n-1)\tau_0. \\ &\partial\Theta_1(0, \tau)/\partial Z = q_1, \quad (2n-1)\tau_0 < \tau < 2n\tau_0, \end{aligned} \tag{4.2}$$

where

$$q_1 = Q_1 D/(K_1(T_W - T_0)). \tag{4.3}$$

The nonlinear boundary condition (2.7) for Region 3 at $z = 1$ becomes

$$\partial\Theta_3(1, \tau)/\partial Z = -q_3 - \sum_1^4 A_n \Theta_3^n(1, \tau) \tag{4.4}$$

where the coefficients A_n are as already given in (3.9) with T_M replaced by T_W. The heat conduction equation for each region is conveniently written as

$$\partial\Theta/\partial Z^2 = \alpha\, \partial\Theta/\partial\tau, \tag{4.5}$$

where for

Region 1: $\Theta = \Theta_1, \ \alpha = 1, \ Z \in [0, \gamma]$,

Region 2: $\Theta = \Theta_2, \ \alpha = k_1/k_2, \ Z \in [\gamma, \gamma + \lambda]$ and (4.6)

Region 3: $\Theta = \Theta_3, \ \alpha = 1, \ Z \in [\gamma + \lambda, 1]$.

In each region the fully implicit Crank-Nicolson method yields, in the usual notation, the finite difference equation:

$$\begin{aligned} &-r\Theta_{i-1,j+1} + 2(r+\alpha)\Theta_{i,j+1} - r\Theta_{i+1,j+1} \\ &\quad = r\Theta_{i-1,j} + 2(\alpha - r)\Theta_{i,j} + r\Theta_{i+1,j}, \end{aligned} \tag{4.7}$$

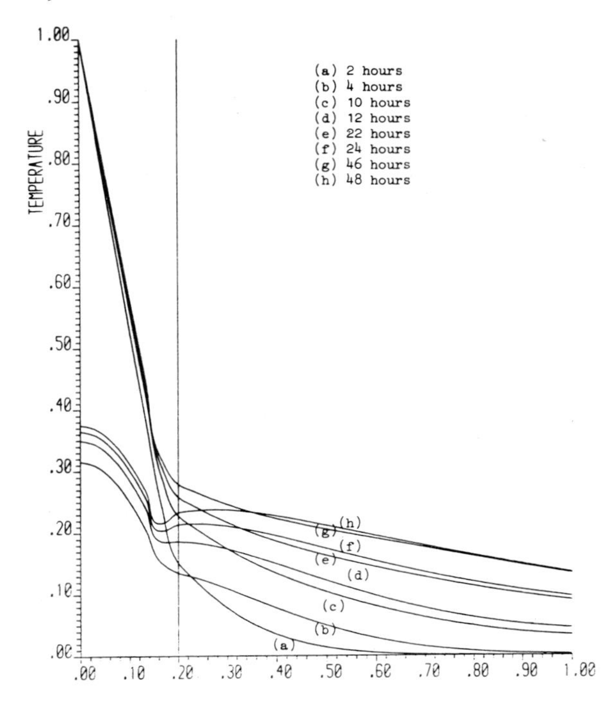

Fig. 6. Temperature profiles when insulating layer forms inner wall (Gamma = 0.0).

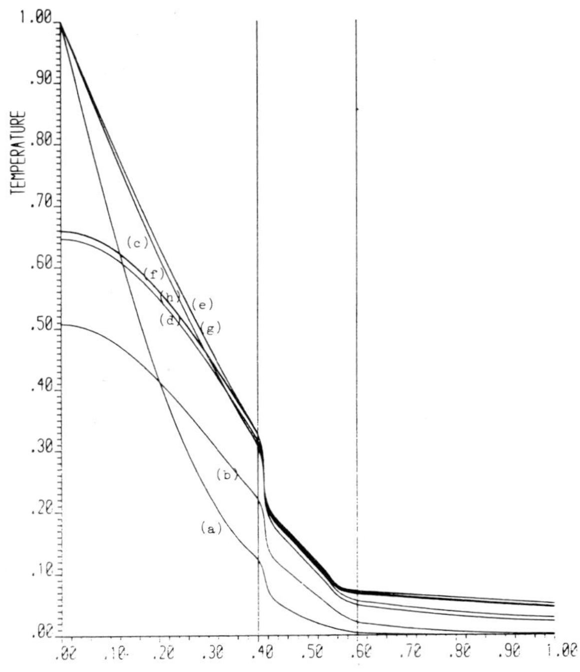

Fig. 7. Temperature profiles when insulating layer placed centrally (Gamma = 0.4).

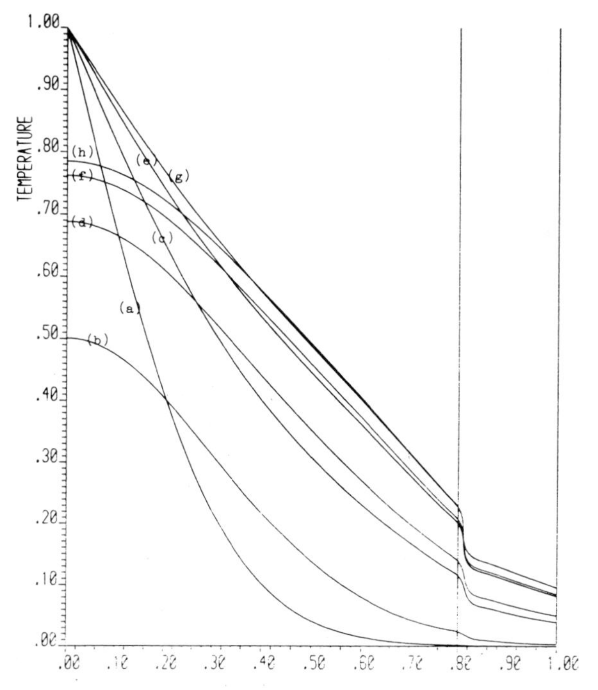

Fig. 8. Temperature profiles when insulating layer forms outer wall (Gamma = 0.8).

where $r = k/h^2$ and k and h denote the time and spatial mesh lengths. In the usual fashion (4.4) is quasi-linearized using a perturbation from the current value of $\Theta_3(1, \tau)$. The tri-diagonal form of the matrix equation for $\Theta_{i,j}$ in each region is maintained on introducing fictitious nodal values at surfaces

Table 3.

Dimensionless location of insulating region (γ)	Total time taken to achieve a period thermal rhythm (hrs)	Temperature of outer ladle wall (°C)
0.0	112	251
0.1	104	114
0.2	68	118
0.3	68	107
0.4	48	100
0.5	48	95
0.6	48	91
0.7	60	89
0.8	60	173
No insulating layer	144	770

when calculating derivative conditions. For example when $Z=\gamma+\lambda$ and $i=I$ then elimination of these fictitious nodes gives

$$-\frac{K_2}{K_1}r\Theta_{I-1,j+1}+\left\{1+\frac{k_1K_2}{k_2K_1}+r\left(\frac{K_2}{K_1}+1\right)\right\}\Theta_{I,j+1}-r\Theta_{I+1,j+1}$$

$$=\frac{K_2}{K_1}r\Theta_{I-1,j}+\left\{1+\frac{k_1K_2}{k_2K_1}-r\left(\frac{K_2}{K_1}+1\right)\right\}\Theta_{I,j}+r\Theta_{I+1,j}, \tag{4.8}$$

with similar derivative conditions at $Z=\gamma$ and at $Z=0$ and 1; note that in (4.8) $\Theta_{I-1,j}$ is in Region 2, $\Theta_{I,j}$ is on the interface between Regions 2 and 3 and $\Theta_{I+1,j}$ is in Region 3.

The matrix equation for the $\Theta_{i,j}$ is solved using the Choleski method with typically $h=\frac{1}{30}$ and $k=0.0012$. Solutions are computed until the thermal rhythm becomes periodic i.e., the temperature profile across the slab at any time from the start of the cycle is identical to the profile at the same time in the preceding cycle.

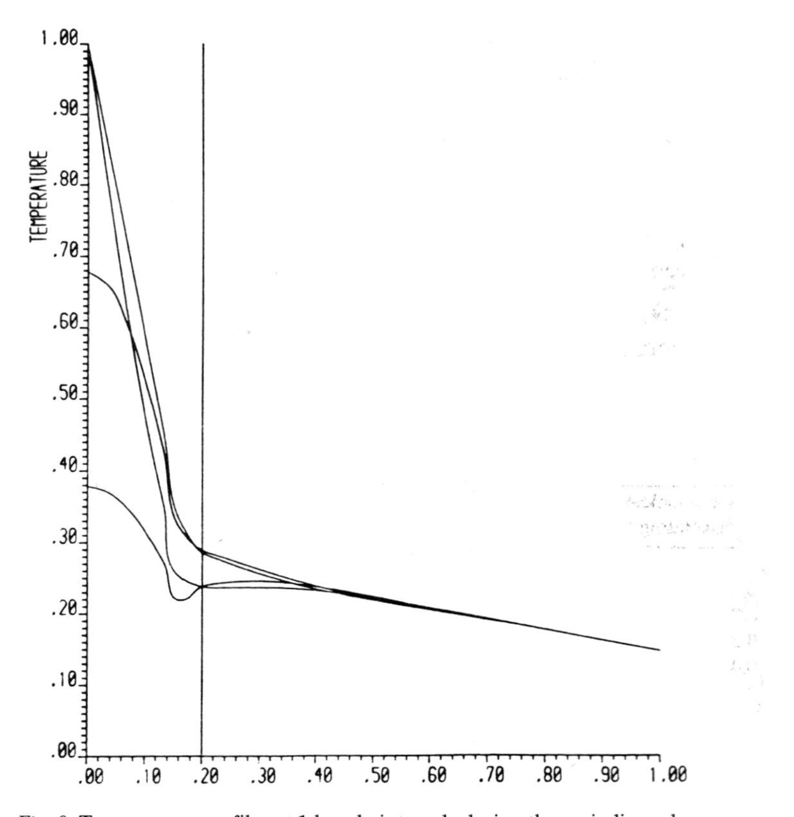

Fig. 9. Temperature profiles at 1 hourly intervals during the periodic cycle.

5. Discussion of results

Results will now be given on the thermal response of the composite ladle wall for typical heating and cooling cycles for various locations of the insulating region.

Figure 6 shows the thermal profile across the composite region, with the insulating layer located at the inner wall $\gamma = 0$, after

(a) the first heating up period (2 hrs),
(b) the end of the first cooling down period or the first complete cycle (4 hrs),
(c) the third heating up period (10 hrs),
(d) the third complete cycle (12 hrs),
(e) the sixth heating up period (22 hrs),
(f) the sixth complete cycle (24 hrs),
(g) the twelfth heating up period (45 hrs), and
(h) the twelfth complete cycle (48 hrs).

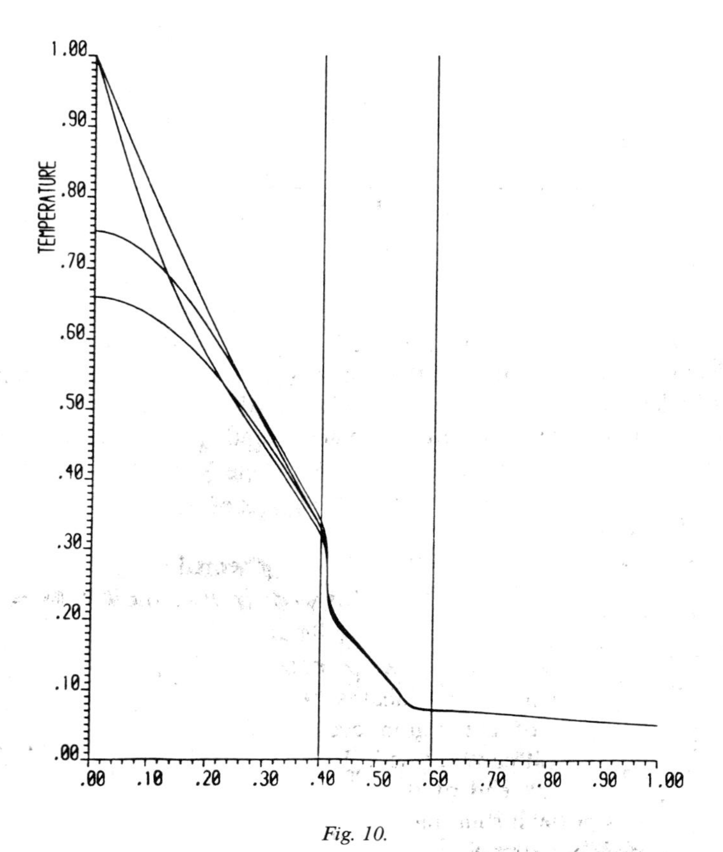

Fig. 10.

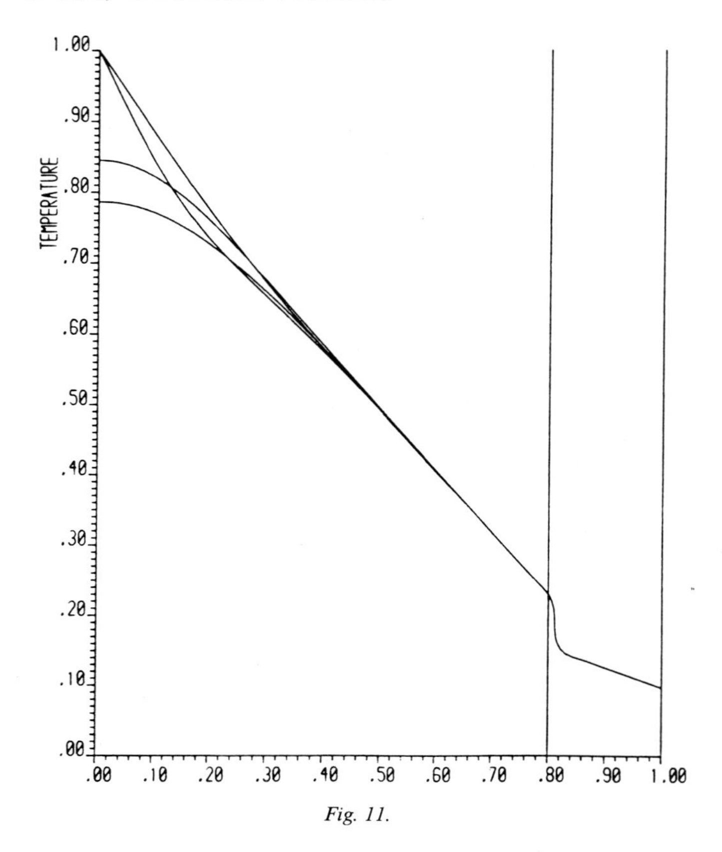

Fig. 11.

Comparison of these results with those in Figs. 7 and 8, when the insulating layer is placed at the centre of the region ($\gamma = 0.4$) and at the outside wall ($\gamma = 0.8$), respectively, show that a periodic thermal rhythm is achieved sooner when the insulating layer is centrally placed and that the temperature of the outside wall is then at its lowest value. In Table 3 the time taken to reach a periodic thermal rhythm and the temperature of the outside ladle wall are displayed for various γ.

Figures 9–11 shows the evolution of the thermal response when the final period thermal rhythm has been established; in these the insulating layer is located at the inner wall, centrally and at the outer wall. It is of particular interest to note in Fig. 9 the fall in temperature when the insulating layer is placed at the inner wall. This indicates that heat is still being lost from the insulator to the rest of the region even though the input of heat to the insulating layer has already ceased. Figure 12 displays the final periodic thermal rhythms at the end of the heating up and cooling down phase for various locations of the insulating layer.

As in the previous models it is useful to record results for the outer wall

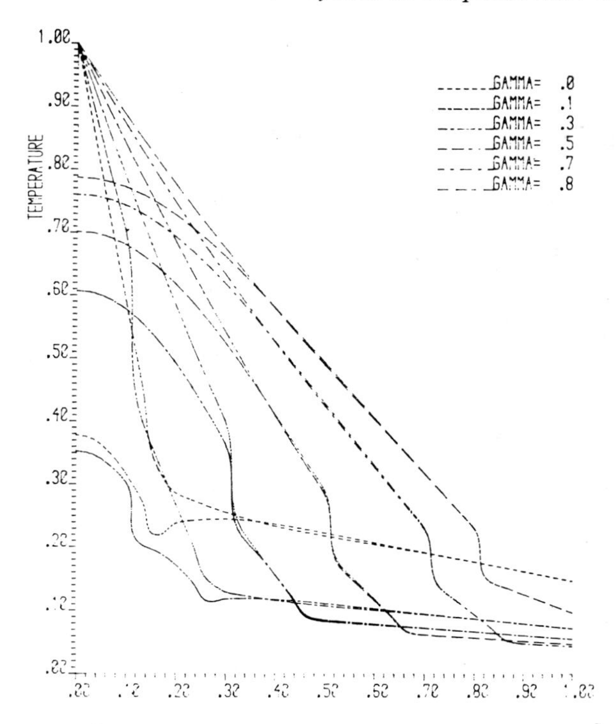

Fig. 12. Temperature profiles after heating and cooling phases of the equilibrium cycle, for various values of gamma.

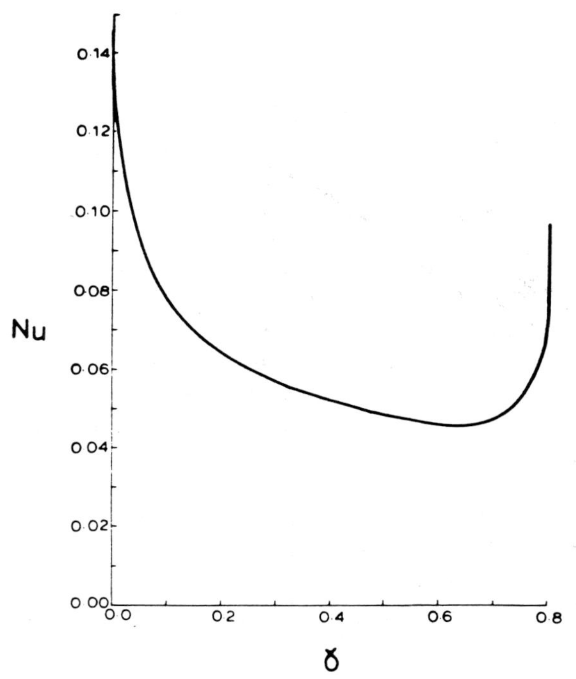

Fig. 13. The Nusselt Number at the outer wall.

Nusselt number Nu defined for $\gamma \neq 1 - \lambda$, as

$$\mathrm{Nu} = -K_1 \, \partial T_3/\partial z|_{z=D} \frac{D}{K_1(T_W - T_0)} \tag{5.1}$$

These are displayed graphically in Fig. 13 as a function of γ.

From the viewpoint of day-to-day operations in an integrated steel plant it is important to know how much heat is lost from the steel whilst it is in the ladle. Ideally it would be desirable to minimize the heat loss either by inserting an insulating layer within the ladle wall or by preheating the ladle wall. Let the temperature field at the start of the nth cycle be $T(z, \tau_{n-1})$. The heat gained during the nth cycle will be

$$Q_n = \int_0^D \rho c \{T(z, \tau_n) - T(z, \tau_{n-1})\} \, \mathrm{d}z, \tag{5.2}$$

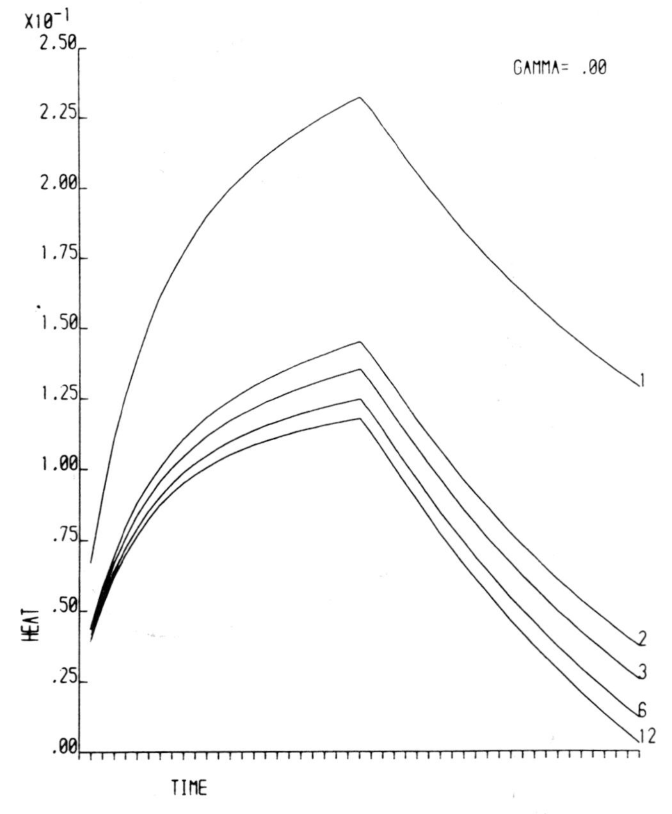

Fig. 14. Heat gained by the ladle wall during various cycles.

where the integral is evaluated with appropriate properties of the composite region. In dimensionable form (5.2) may be written as

$$q_n^* = \frac{Q_n}{D(T_W - T_0)\rho_1 c_1} = \gamma \int_0^\gamma (\Theta_1(Z, \tau_n) - \Theta_1(Z, \tau_{n-1}))\, \mathrm{d}Z$$
$$+ \lambda \frac{\rho_2 c_2}{\rho_1 c_1} \int_\gamma^{\gamma+\lambda} (\Theta_2(Z, \tau_n) - \Theta_2(Z, \tau_{n-1}))\, \mathrm{d}Z$$
$$+ (1-\gamma-\lambda) \int_{\gamma+\lambda}^1 \Theta_3(Z, \tau_n) - \Theta_3(z, \tau_{n-1})\, \mathrm{d}Z. \tag{5.3}$$

q_n^* has been computed from the numerical solutions for $n = 1, 2, 3, \ldots, 12$ and displayed graphically for cycles $n = 1, 2, 3, 6$ and 12 for various positions of the insulating layer in Figs. 14–16. Each case displays the characteristic rise in q_n^* when the steel is in the ladle followed by the fall in q_n^* when the ladle

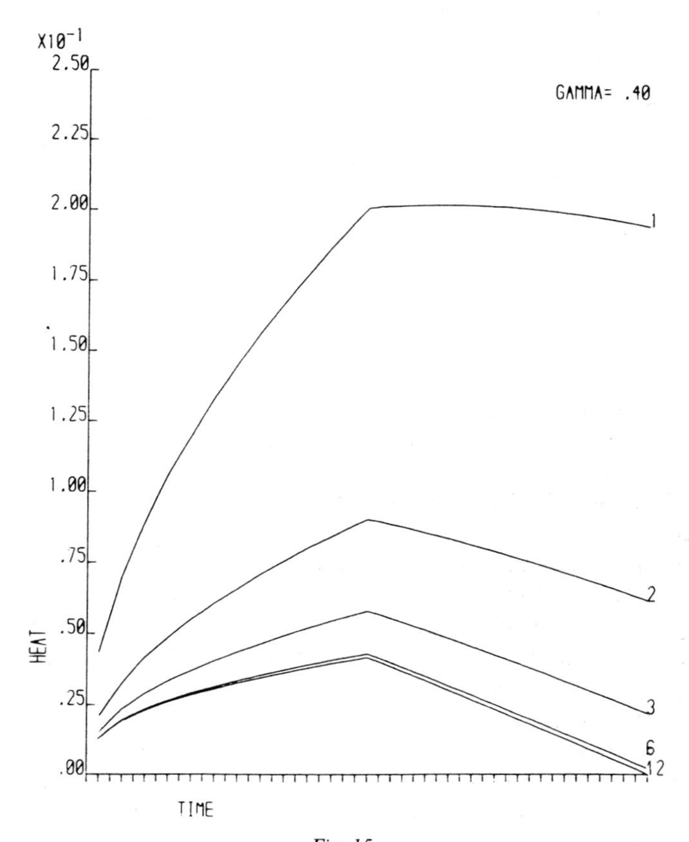

Fig. 15.

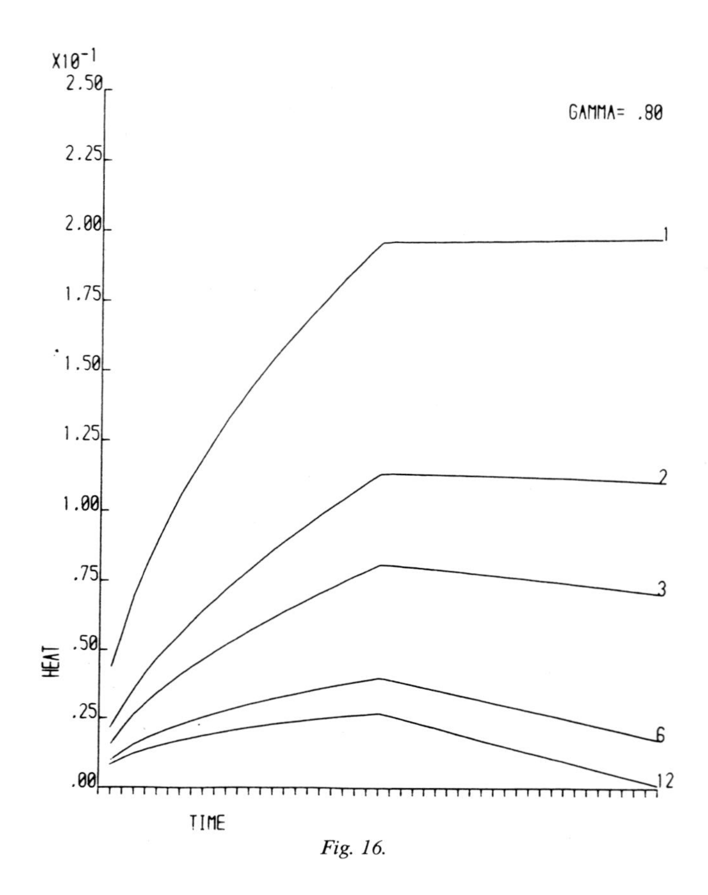

Fig. 16.

stands empty. It is the early part of the cycle which is of practical interest and in particular the value of q_n^* at the end of 2 hrs i.e., the total heat lost from the ladle during that cycle.

Figure 14 shows the heat loss from the steel for $\gamma = 0$ when the insulating layer is at the inner ladle well. Clearly as the number of cycles progress the heat gained per cycle decreases. After 12 cycles (48 hrs) the composite wall has not achieved its periodic cycle. Figure 15 displays the results when the insulating layer is centrally placed. Then the ladle wall does not lose as much heat in the second half of the cycle and the periodic rhythm is achieved at the 12th cycle. When the insulating layer is at the outer wall the ladle loses even less heat in the second half of the cycle, see Fig. 16.

Figure 17 displays the total heat gained for various cycles as a function of γ. From these results it is seen that during the early cycles the least heat loss from the steel is achieved when the insulating layer is quite close to the inner wall. However as the number of cycles increase the minimum heat loss occurs

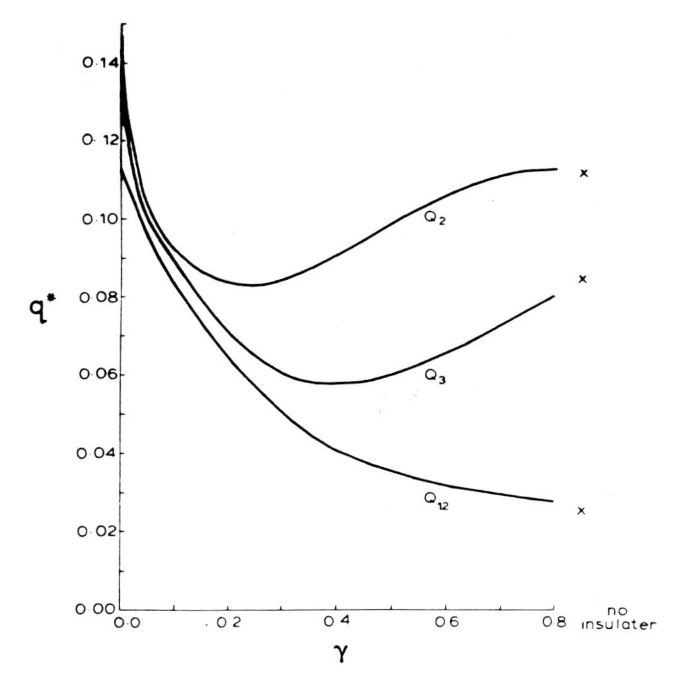

Fig. 17. Heat lost from the steel during 2nd, 3rd and 12th cycles.

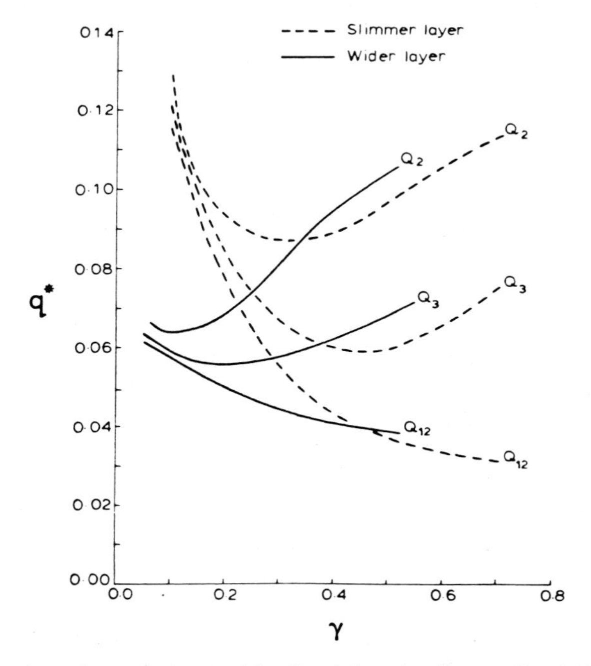

Fig. 18. Heat lost from the steel when a wider ($\lambda = 0.4$) and a slimmer ($\lambda = 0.13$) insulating layer is used.

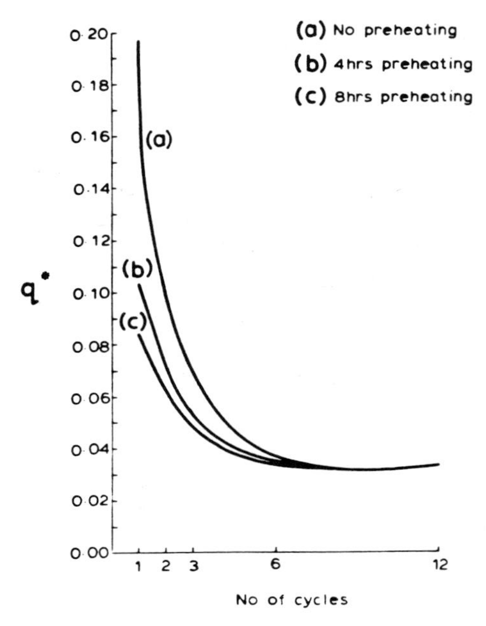

Fig. 19. The effects of preheating the ladle (Gamma = 0.6).

when the insulating layer is closer to the outer wall, or even when it is left out altogether. Note that the results in Fig. 17 have also been computed using Newton Cooling at the outer wall with $\mathrm{Bi} = A_1$. The agreement for all practical purposes was satisfactory.

In Fig. 18 a wider and a slimmer insulating layer was employed. As can be seen the width of the insulating layer has little effect on the optimal location for minimum heat loss.

Finally in Fig. 19 the effects of preheating the ladle are displayed for $\gamma = 0.6$. Clearly preheating the ladle for 4 hrs virtually halves the heat loss during the first cycle.

6. Conclusions

Analytical and numerical solutions have been obtained for the propagation of thermal rhythms in a composite ladle wall. In particular an analytical model simulating switch 'on-off' thermal conditions at the inner ladle wall with Newton Cooling at the outer ladle wall is found to be in exact agreement with the full numerical solution for the composite wall. This analytical solution is seen to possess novel mathematical features.

From a practical viewpoint, to obtain minimum heat loss from the ladle of molten steel it is clear that in early operations the insulating region is most effective if it were located close to the inner wall. As the thermal rhythms approach periodicity it is seen that minimum heat loss would be achieved if the insulating region were placed close to the outer ladle wall. Moreover in the later operational stages it was shown that the minimum heat loss is actually obtained if the insulating layer was omitted altogether. Since it is also shown that the time taken to achieve a periodic thermal rhythm is always least when the insulating region is located centrally in the composite ladle wall it can be argued that the evolution of the thermal rhythm is controlled by its inclusion. Inclusion of the insulating layer yields a marked reduction in the outside wall temperature and this could have considerable bearing on the life of the ladle in an integrated steel works.

Preheating the ladle walls was found to reduce the ladle heat loss by more than 50 percent during the initial cycle.

Finally it is hoped that the above theoretical results will be of considerable value in the understanding of thermal stratification in ladles of molten steel during standing and pouring.

References

1. H.S. Carslaw and J.C. Jaeger: *Conduction of Heat in Solids.* Oxford University Press (1959).
2. P. Egerton, J. Howarth, G. Poots and S. Taylor-Reed: *Int. J. Heat and Mass Trans.* 22 (1979) 1525–1532.
3. P. Egerton, J. Howarth and G. Poots: *Int. J. Heat and Mass Trans.* 24 (1981) 557–562.
4. J.C. Jaeger: *Quart. Appl. Math.* 8 (1950) 187–198.
5. G.G. Stokes: *Trans. Camb. Phil. Soc.* 9 part 2 (1851) 8–106.
6. A.H. van Gorcum: *Appl. Sci. Res. A* 2 (1959) 272–280.
7. V. Vodika: *App. Sci. Res. A* 5 (1955) 108–114.

Applied Scientific Research 44: 139–160 (1987)

The integration of geometric modelling with finite element analysis for the computer-aided design of castings

T.J. SMITH [1] & D.B. WELBOURN [2]

[1] *GST Professional Services Limited, 8 Green Street, Willingham, Cambridge, CB4 5JA, UK*
[2] *DUCT Development Unit, 20 Trumpington Street, Cambridge, UK;*

Abstract. This paper considers the problem of interfacing geometric modelling with finite element analysis in the numerical simulation of the casting process. It is shown that geometrical modellers are capable of describing uniquely the shape of the most complex cast object and that finite element meshes may be constructed within the component. However, such a mesh is defined by the geometry of the object and, hence, may not be the ideal for subsequent simulations. This difficulty is overcome by the use of an adaptive mesh created during the finite element analysis on the basis of information provided by the current solution. The fundamental background to adaptive mesh generation is presented together with the details of its implementation. The method is then applied to the calculation of the flow into a mould cavity from which the accuracy and efficiency of the method is demonstrated. It is concluded that the technique is viable for use with computer-aided design tools for the foundry industry.

1. Introduction

In principal, casting is a simple, straightforward metal forming process. In practice, it may be extremely difficult to ensure a final product that is within the required dimensional tolerances and free from defects, a number of which may be invisible to the naked eye. Many of the difficulties arise from the volume changes in the metal during solidification and cooling. The vast majority of commercially cast alloys contract during liquid cooling and during solidification although there are important exceptions. For example, ductile and grey irons undergo a period of expansion after an initial contraction and, more significantly, the pattern of the volume change is not constant but depends upon a number of parameters including the cooling rate.

For these reasons, most castings require a feeding system to provide molten metal to those areas undergoing contraction. If this feed metal is not provided then significant dimensional changes can occur together with the formation of shrinkage porosity. In addition to the feeding system, a gating system must also be provided, this being the means by which the molten metal is introduced into the mould cavity. Often the desirable gating and feeding system to ensure a high quality casting leads to a low yield and hence, greater cost, negating the advantages of casting over other forming processes. Therefore, efficient foundry practice requires design tools which allow a component to be designed not only for use but also for manufacture.

The success of casting, particularly sand casting, depends on a large number of parameters, many of which are only controlled to loose tolerances within many foundries. This has led to casting being something of a 'black-art' rather than a science, with experience being the most desirable possession. However, the foundry industry has come under increased commercial pressure from other metal forming processes such that a more scientific approach to casting design is being introduced into many foundries. This has also coincided with the availability of relatively cheap, high-capacity, high-speed mini and microcomputers which enable the relatively complex calculations of scientifically based casting design to be carried out at an acceptable cost.

The main features of the casting process which contribute to the quality of the final product include:

i) The flow of molten metal into the mould cavity.
ii) The solidification of the metal.
iii) The feed of liquid metal to the solidification front to counteract shrinkage.
iv) The generation of residual thermal stresses due to postsolidification cooling.

In principle, numerical simulations of each of these processes can be applied to any castings using currently available models, some of which are described in this volume [6,13,17] and others elsewhere [5,12,18,21]. In practice, such models are severely limited in their application by the difficulties of describing unambiguously the geometry of the casting, together with its associated gating and feeding system and of discretising that geometry for the purpose of numerical calculations. These problems were recently highlighted in two surveys of the use of computer-aided casting design, that by Berry and Pehlke [2] covering experience in the United States and Europe while that by Aizawa and Umeda [1] covers the state of the art in Japan.

Examples of the type of complex geometry involved are shown in Figs. 1 and 2. The aluminium die-cast gearbox casing shown in Fig. 1 is from a well known automobile whilst Fig. 2 shows a typical exhaust manifold with the tooling necessary to make it. At the top of the Figure is the pattern while at the sides are the core boxes. Geometrical descriptions of objects as complex as these can be provided only by a surface modeller [e.g. 22] and not a solid modeller [e.g. 20]. The fundamental difference between the two types of modeller is that a solid modeller is capable of distinguishing between solid and space whereas a surface modeller is incapable of doing so, knowing only the details of surfaces and nothing about the intervening volumes. However, at present, solid modellers are restricted to geometries which can be assembled from combinations of various building blocks each of which has an analytically simply geometry such as a cuboid, cylinder, core, sphere or torus. Such restrictions do not apply to surface modellers which, as a result, are capable of providing the exact intersection between two entirely general surfaces. This facility is of considerable use in determining parting lines during design.

Fig. 1. Aluminium die-cast gear-case; tooling designed to bottom tolerances and made using the DUCT system.

This paper discusses the interfacing of geometrical modelling with numerical simulation of the casting process to provide an integrated design tool for the foundry. Particular emphasis is placed on mesh generation and the

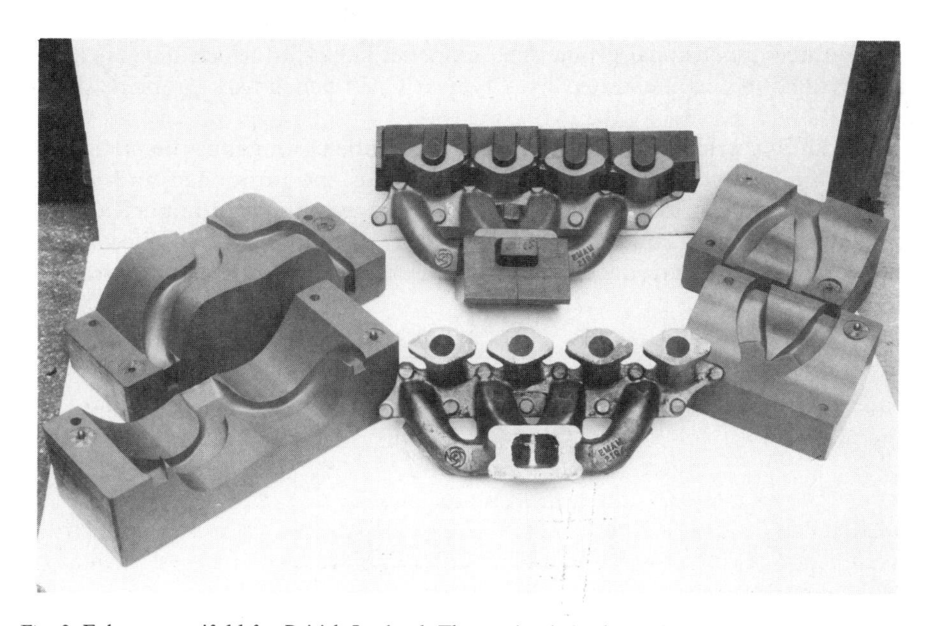

Fig. 2. Exhaust manifold for British Leyland. The casting is in the middle with the pattern at the top and the core boxes at either side. Pattern and core boxes were designed using the DUCT system.

potential for adaptive refinement. Also, the often conflicting requirements of design for use and design for manufacture are considered and it is argued that such differences are more easily resolved through the use of computer-aided casting design.

2. Requirements of computer-aided design

The traditional sequence of events in the design of a component to be cast starts with the designer setting down the geometric shape of the object. In general, this shape is determined primarily by the use to which the component will be put, with only secondary consideration given to its manufacture. The drawings so produced are then translated by the patternmaker who introduces parting lines so that cope ('top') and drag ('bottom') may be found for the mould and who also introduces draft angles so that the pattern may be removed from the sand mould or a die-casting from the die. Frequently this introduces major changes in the detailed geometry which can be highly significant. One automobile manufacturer found that a finite element based stress analysis of a connecting rod was valueless because of the alteration in radii produced by the toolmaker at the manufacturing stage. In fact, the presence of ghost lines (Fig. 3) as typical features in conventional drawings means that a draughtsman does not know what he has designed until a patternmaker has made a model of it. This difficulty is overcome by the use of a fully three-dimensional geometrical modeller [23,24] in which blends can be defined uniquely as illustrated by a typical CAD benchmark problem shown in Fig. 4.

In addition, prior to the pattern being made, the casting must be methoded; that is the gates, feeders and feeder necks must be positioned and dimensioned. These are heavily dependent on casting geometry and in particular on the development of solidification within the casting. Many detailed design changes are often introduced by the patternmaker in order to control the

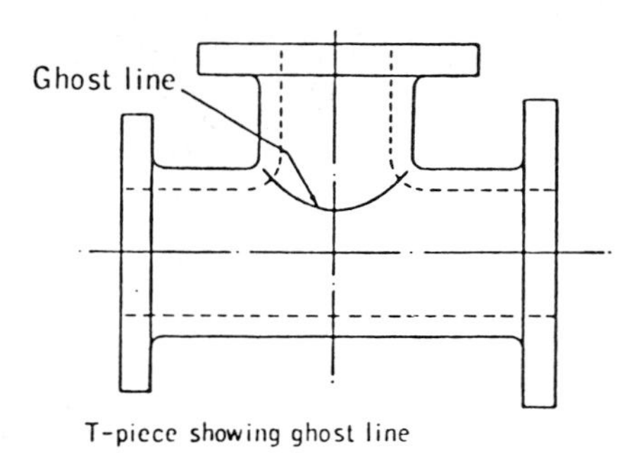

Fig. 3. Typical drawing of valve body showing ghost line to indicate blending of two cylinders.

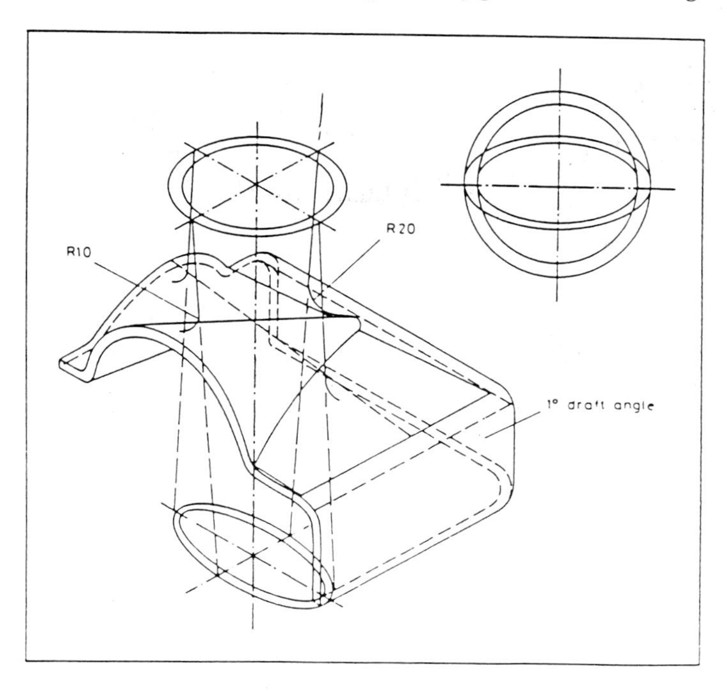

Fig. 4. Typical CADCAM benchmark test piece. (a) The working drawing, (b) The finished product.

progress of solidification through the casting and the efficiency of the feeding system. At present, such changes are based as much on guesswork and experience as on detailed analysis. Thus, given any arbitrary geometry, it is desirable to be able to predict solidification development within that geometry. Models for the thermal analysis of cast components are now available and allow predictions to be made [e.g. 13]. However, care is required in their application due to their sensitivity to certain physical parameters required as input [17].

Care is also required in prescribing the initial temperature distribution in the melt resulting from the flow of metal into the mould. Methods do exist for estimating heat transfer during pouring although the complexity of the problem is such that gross simplifications of the physics governing the problem have to be introduced. Shrinkage effects should also and can be included in such analyses to indicate regions where shrinkage induced defects or high residual stress levels are likely to occur [12].

The results of such calculations can then be used to redesign the methoding system or even the geometric design of the component. Any such redesign may then be re-analysed for performance in use, ultimately to produce a product which not only satisfies its design specification but which can be easily made for the minimum cost. This interactive nature of the design process also has the added advantage of reducing the lead time for a new product as well as improving productivity and quality.

The ideal requirements of a computer-aided design system for foundry use are straightforward and comprise the following modules:

i) A geometric modeller
ii) A mesh generator
iii) A computerised foundry methoding system
iv) Thermal and stress analysis simulators
v) A database.

All these modules should be fully integrated to allow unrestricted feedback during the iterative design process.

Examples of each module currently exist, some of which are discussed in this volume. However, as yet all five have not been brought together in a single package for computer-aided casting design. It is unlikely that such a situation will persist for much longer although a number of problems have still to be overcome. One such is the generation of an optimum computational mesh for thermal and/or stress analysis of a given geometrical shape. Such a mesh will depend on the geometry, physical parameters and initial conditions and may vary with time as solidification proceeds. It is also likely that the optimum computational mesh will be different for a thermal analysis from that for a stress analysis which leads to the proposition that adaptive mesh schemes may be useful.

The remainder of this paper discusses the various available options for geometric modelling of cast components and their interface with suitable mesh generators. Some new ideas on the implementation of adaptive mesh generation are also introduced.

3. Geometric modelling

Geometric modelling packages are generally divided into three categories: wire frame modellers, surface modellers and solid modellers. Wire frame modellers are inherently unsuited to casting design while the other two categories have both advantages and disadvantages. Here, we will concentrate on the capabilities of surface modellers as these have received least attention in the past few years and offer a number of features not generally available with solid modellers.

Designers and draughtsmen seem to think in three different ways about a design. They conceive of objects as being surfaces bounded by lines (as in a wire frame model), of assemblies of standard solids (as in a solid model) and of objects defined by reference lines with sections normal to them. These sections have to be faired with a surface to turn them into solids. In addition, they think of blending both surfaces and solids into one another and of points rather than vectors for defining lines and surfaces. However, it should be noted that a few points together with appropriate angles are a far better way of defining surfaces than are a large number of points.

Patternmakers have yet another way of looking at objects, since the mould is the female of the male object and the holes in the object are produced by cores which themselves are made in moulds. Therefore, an object can be thought of as what is left when the holes have been defined!

One surface modelling program which is based on the concept of reference lines (spines) in three-dimensional space and sections normal to them is the Duct package developed at the University of Cambridge. The sections may be open or closed while the necessary means is provided for joining these together to give tangent and surface normal continuity at the edges. For the foundry, higher order continuity is unnecessary although in other industries great importance is laid on second order continuity because of surface reflections. Various examples of the different types of surface together with the means of intersecting and blending these surfaces are shown in Fig. 5. The patches used in the DUCT program have third order Bezier curves as boundaries but sixth order internally [19] allowing efficient intersection and blending. Further details of the DUCT system were given by Welbourn [22].

The crucial feature of all castings is that they must have parting lines with draft angles on them. The unique feature of DUCT is that provision has been made for simulating the patternmakers working practice for a complicated object for a three-dimensional parting line. With DUCT a draughtsman can readily examine the horizon lines of an object he has designed and then modify the object, at the same time introducing draft angles so that it is castable. Figure 6 shows a simple object as originally designed in which the horizon line on one side represents a possible parting line for a mould. Unfortunately, the other horizon line (marked 2) is unsuitable as it becomes vertical which would give an undesirable parting surface on the mould. Therefore, this face would have to be modified but with DUCT the designer can decide how, rather than leave it to the pattern maker. A further simple example is that of an uncastable geometry shown in Fig. 6. This component incorporates the basic concept behind many inlet and exhaust ports and is

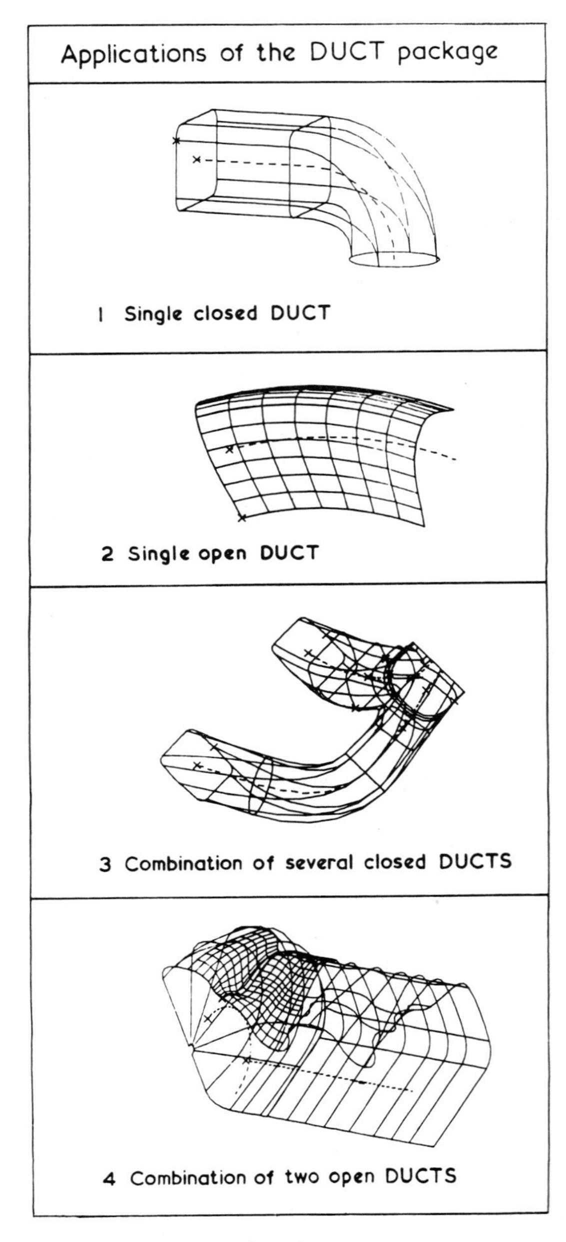

Fig. 5(a). Examples of closed and open DUCT. The spines or reference lines are shown dotted.

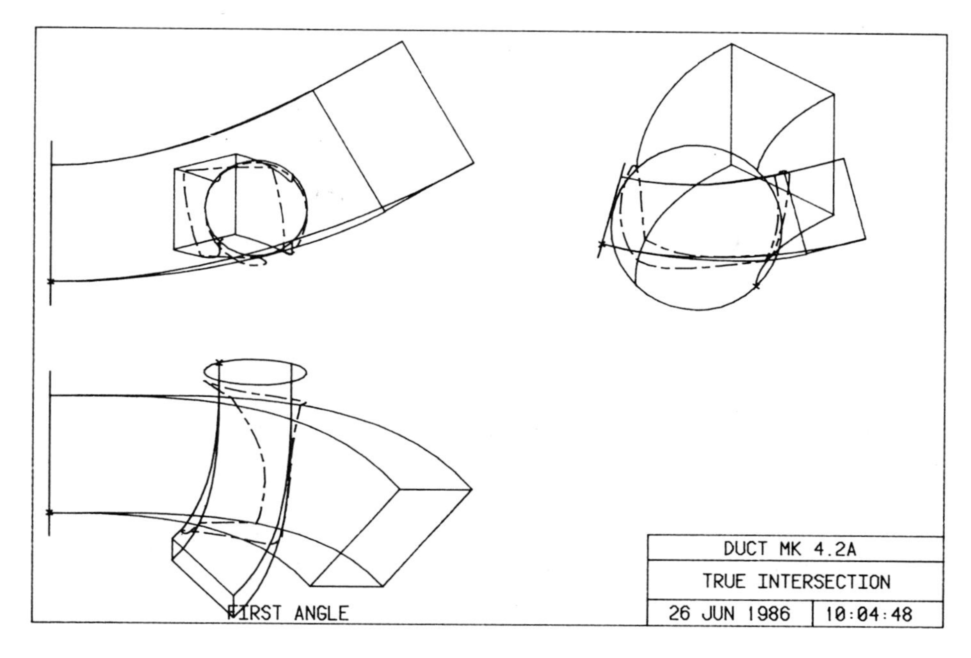

Fig. 5(b). Example of a true intersection. The large DUCT has been shrunk, rotated and moved to give the smaller one intersecting it. The line of intersection is shown dashed.

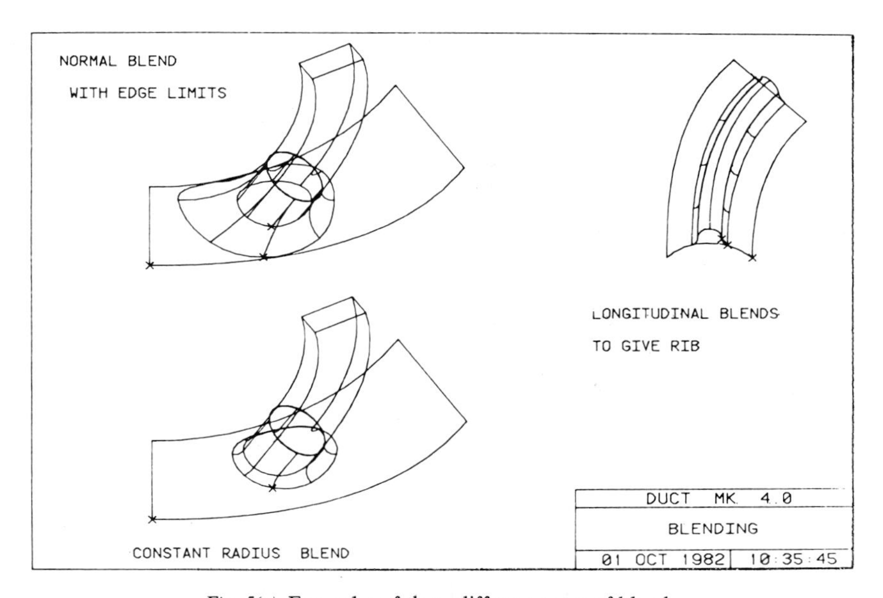

Fig. 5(c). Examples of three different types of blend.

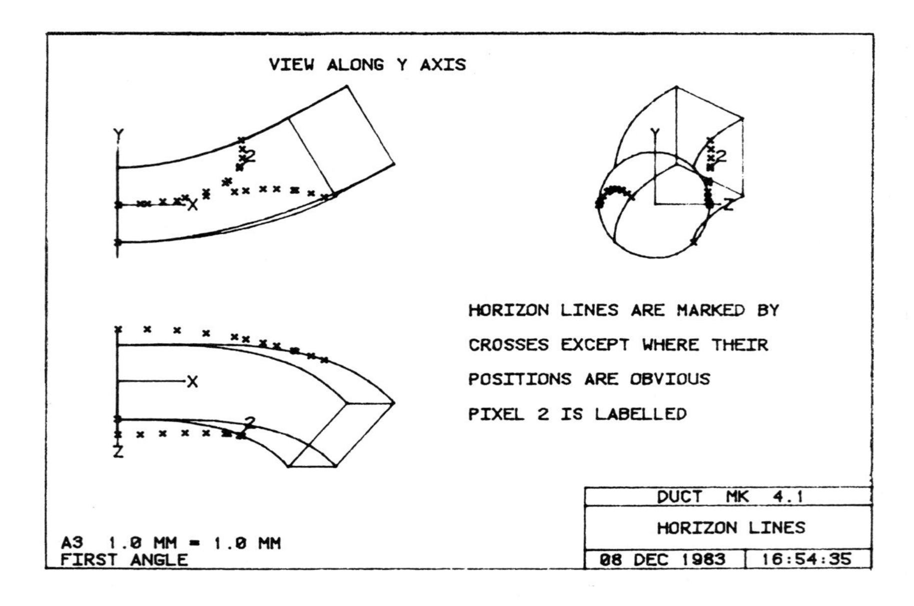

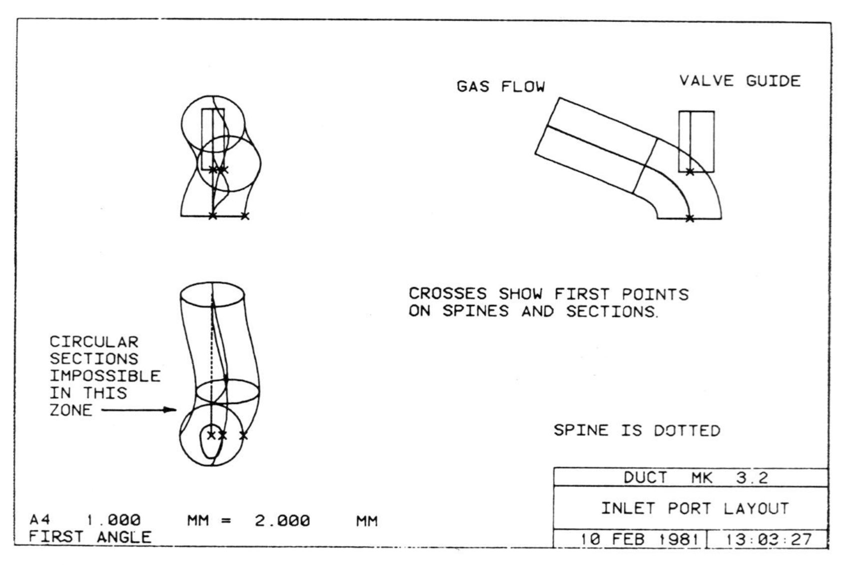

Fig. 6(a). Horizon lines for a simple object. One of these becomes vertical at the point marked 2 and therefore is unusable for a satisfactory mould. The design must be modified if it is to be split in this manner. (b). Example of an uncastable geometry.

representative of the fundamental inconsistencies which are all too frequent in engineering drawings. With a geometrical modelling package, these inconsistencies would be brought to light much earlier in the design process and, therefore considerably reducing costs.

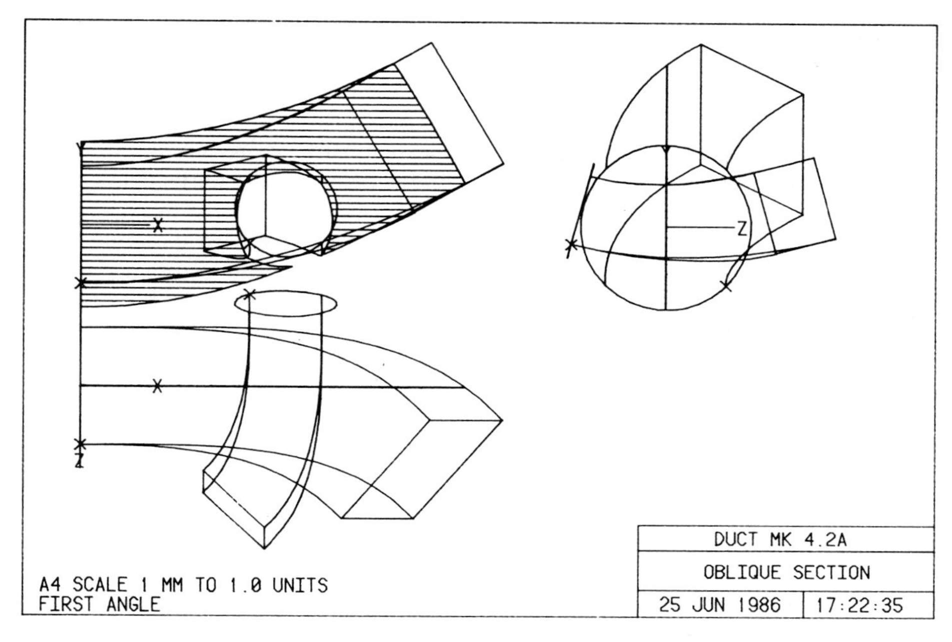

Fig. 7. Example of an oblique section through the intersection shown in Fig. 5b.

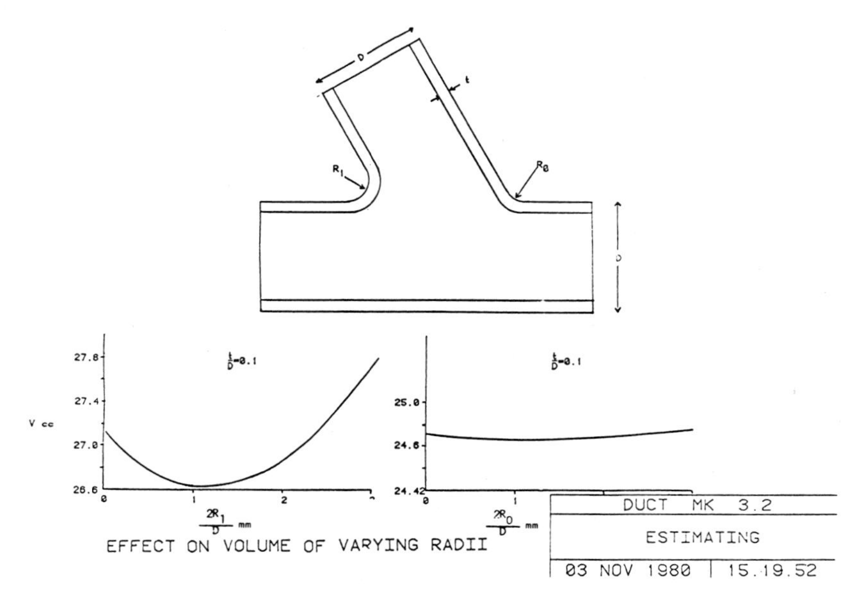

Fig. 8. Effect of the blend radii on the total volume of two intersecting tubes whose centre lines are in a plane.

A further requirement of a geometric modeller for the foundry industry is that allowance must be introduced into the pattern to provide for contraction (or expansion) of metal in the mould during solidification. With DUCT, the design can be scaled automatically by the necessary amount differentially along all three axes. This also affects the volume, surface areas and cross-sectional areas including oblique sections, these being easily derived (Fig. 7). Volumes for simple components are easy to obtain accurately using Gauss' theorem but, currently, those for complex components can sometimes be difficult to obtain to an accuracy greater than $\pm 1\%$. Volume estimates are of great importance in the foundry since the price of castings for tendering purposes is largely based on weight. In addition, the foundry does not want to melt unnecessary metal.

The ability to describe objects uniquely and then to analyse them can lead to surprising results. Few people are aware of the results shown in Fig. 8 where blend radii are varied. The effect of this variation on solidification will be significant such that there is little or no penalty in designing this component for manufacture rather than for minimum weight in use.

4. Adaptive finite elements models

Having derived a geometrical model, it is often necessary to define a finite element model based on the geometrical model. Here a finite element model is taken to be the assemblage of finite elements which represent the geometrical model. Note that in geometrical terms the two are not identical even before the finite element model is reduced to take account of axes of symmetry and other simplifying features.

Finite element models of a surface can be generated very simply using plate elements as illustrated in Fig. 9 which shows a finite element model of the exterior surface of the manifold for the new Ford 2.5 litre diesel engine [3]. By describing a number of internal surfaces for which finite element models are easily defined it is straightforward to build up a finite element model of a solid component as illustrated by the twin ports and pipe elbow in Fig. 10.

However, such models are defined by the geometry of the object being modelled and not by the requirements of the finite element analysis with which they are to be used. Simple tests can be made on the suitability of the finite element model for finite element analysis. For example, the shape of each element can be examined and highly distorted elements replaced or subdivided either automatically or manually. Again, this revision is often based on geometrical rather than physical criteria.

Unfortunately, such finite element models are not necessarily the optimum for the subsequent finite element analysis with which they are to be used. Indeed, if a number of different finite element analyses, for example, thermal and stress, are to be carried out, the optimum finite element model may be

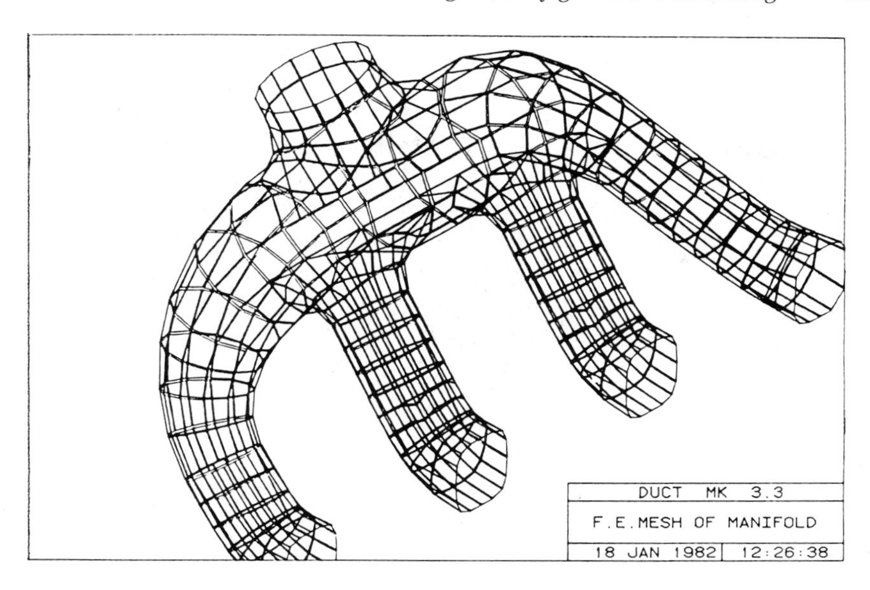

Fig. 9. Finite element shell mesh for a Ford manifold.

different for each analysis. Further, the choice of the finite element model is also often a compromise between spatial resolution and computational cost. This compromise is achieved before the solution of the finite element analysis is known even though the choice of mesh influences the solution obtained.

One way of overcoming these difficulties is to use an adaptive scheme which couples the finite element model with the finite element analysis. In such a scheme the finite element model is continually and automatically modified during the analysis on the basis of information provided by the current solution. In this way, an optimum finite element model is achieved at all times. It also has the advantage of considerably reducing the time required to establish the initial finite element model as this does not need to be particularly refined. Unfortunately, while a self-adaptive scheme avoids the disadvantages of conventional finite element models, it does introduce additional difficulties, chiefly the estimation of errors in the existing solution and the assessment of whether these errors can be reduced by further adjustment of the finite element model. Both these requirements together with the necessity to include mesh adjustment facilities can lead to a significant increase in software complexity. Also, the theory underlying adaptive mesh generation is still under development although significant advances have been made recently [7,9] particularly with regard to the *a posteriori* estimation of errors and the use of correction indicators to assess the advantages of further refinement. While these theories have been developed for linear problems only, they have been successfully extended empirically and applied to non-linear

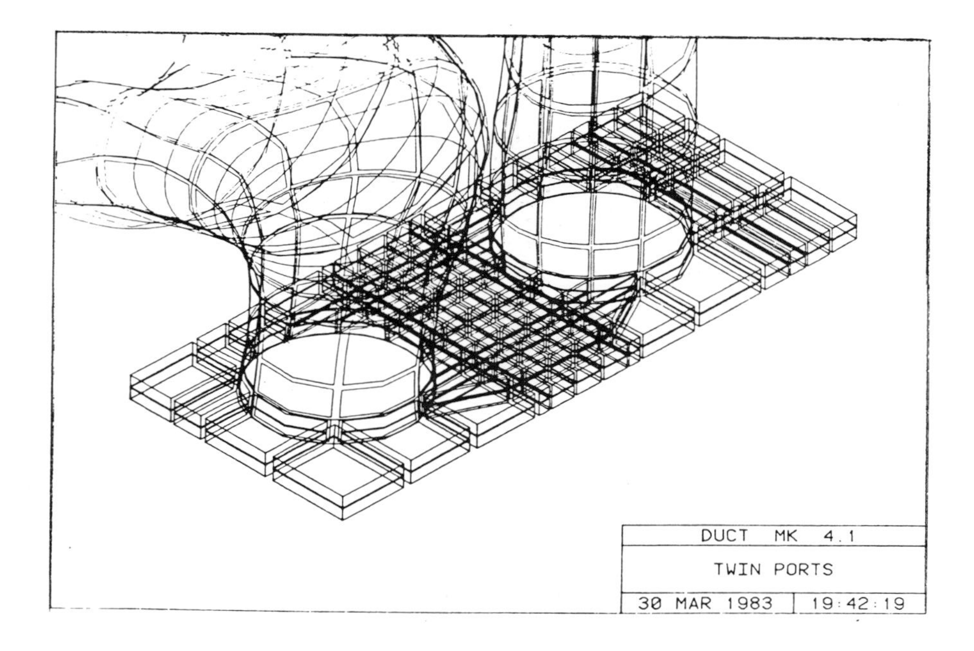

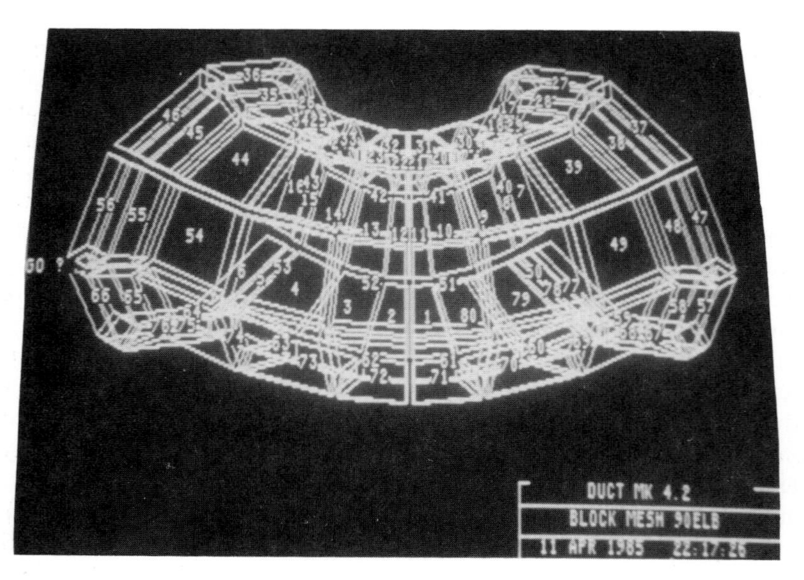

Fig. 10(a). Part of a three-dimensional finite element mesh for a diesel engine cylinder head. (b). A three-dimensional finite element mesh for a malleable iron pipe fitting.

problems [10,16]. These applications have also demonstrated that significant benefits are obtained in terms of the computational time required to achieve a solution of any required degree of accuracy [10].

5. Implementation of mesh adaptation

Consider $\Omega \subset \mathbb{R}^3$ to be a domain bounded by Γ representing the geometric model and let Ω_d with boundary Γ_d be a discretisation of Ω into disjoint finite elements representing the finite element model. Let E_k ($k \in \mathbb{N} \geqslant 0$) denote the set of elements e_{ik} and N_k denote the corresponding set of nodes n_{ik}.

A particular problem might be to solve

$$\frac{d\theta}{dt} + \mathscr{D}(\theta) = 0 \quad \text{on } \Omega,\ t \geqslant 0 \tag{1a}$$

with

$$\theta = 0 \quad \text{on } \Gamma_0, \quad t \geqslant 0 \tag{1b}$$

$$\mathscr{L}(\theta) = 0 \quad \text{on } \Gamma_L, \quad t > 0 \tag{1c}$$

$$\theta = \theta_0 \quad \text{on } \Omega, \quad t = 0 \tag{1d}$$

in which $\Gamma = \Gamma_0 \cup \Gamma_L$, $\mathscr{D}$ is a non-linear differential operator and $\mathscr{L}$ is a linear differential operator. The application of variational principles to the suitably normalised equations (1a, b, c) on Ω_d is a set of initial value problems of the form

$$\boldsymbol{K}\boldsymbol{\Theta} + \boldsymbol{C}\frac{d}{dt}\boldsymbol{\Theta} = \boldsymbol{q} \quad t \in [0, t] \tag{2}$$

in which T is a finite but possibly large dimensionless time scale, $\boldsymbol{\Theta}$ denotes the vector of nodal values of the normalised θ while $\boldsymbol{K}$, $\boldsymbol{C}$ and $\boldsymbol{q}$ are the global stiffness and mass matrices and force vector respectively. The forms of $\boldsymbol{K}$, $\boldsymbol{C}$ and $\boldsymbol{q}$ will reflect the choice of discretisation scheme but are also dependent on the choice of the computational mesh, E_k.

Now let $\boldsymbol{\Theta}^{(j)}(E_k, N_k)$ denote the solution of equation (2) on mesh E_k with nodes N_k at the discrete time t_j ($j = 0, 1, \ldots, J$) and assume that an algorithm exists to compute $\boldsymbol{\Theta}^{(j+1)}(E_k, N_k)$ from $\boldsymbol{\Theta}^{(j)}(E_k, N_k)$ for any E_k belonging to a specific set of meshes G.

The purpose of the adaptive mesh algorithms is to determine a mesh $E_k \in G$ on which a solution is to be obtained at t_{j+1} such that the error in the solution, estimated *a posteriori* under a suitable norm is less than a specified value everywhere within Ω_d.

The simplest implementation is to let the set of meshes G consist of a base mesh E_0 and a number of refined meshes E_k obtained by equally subdividing some or all of the elements within E_0. A further condition is also imposed on all meshes within G, namely

$$N_0 \subset N_k$$

as this simplifies the construction of E_k and hence, the resulting algorithm. However, one problem with this strategy is achieving compatibility of the coarse and fine parts of the mesh along their common boundaries. The main difficulty concerns those fine mesh nodes lying on those boundaries which do not form part of the coarse mesh. These are termed irregular nodes. This difficulty is most efficiently overcome by extending the influence of the coarse mesh into the first elements of the refined mesh by imposing constraints on the irregular nodes so that their values are consistent with the degree of interpolation used on the coarse mesh [16]. These constraints can then be introduced directly into the discretised form of the set of initial value problems (2) given by

$$\boldsymbol{A}\Theta^{(j+1)} = \boldsymbol{f}. \tag{3}$$

Initially, all coefficients of $\boldsymbol{A}$ are computed using a conventional discretisation scheme on E_k. The matrix coefficients $a_{ik}^{(k)}$ related to the irregular nodes are then distributed among the coefficients $a_{ik}^{(0)}$ related to the adjacent nodes on E_0 using weighting coefficients obtained by polynomial interpolation of the same degree as that implicit in the discretisation scheme being used. In general, the symmetry of the modified matrix $\boldsymbol{A}$ is apparently upset by this procedure. However, for most discretisation schemes symmetry can be restored by simple rearrangement.

The application of the above algorithm appears to be computationally more efficient than the elimination of irregular nodes by the direct imposition of constraints during the assembly of $\boldsymbol{A}$ [4] which is particularly important in transient problems. Also while this algorithm was initially developed for use with the finite element method, it is independent of the discretisation scheme used to construct $\boldsymbol{A}$. Consequently, it is equally applicable to finite difference based solution schemes.

The final requirement for the implementation of an adaptive mesh scheme is the provision of an error estimate and a correction indicator. The error estimate should give both global and local errors and is used to define those regions within Ω_d in which mesh adaption is required. Thus, any error estimate should be an overestimate of the true error and tend to zero as mesh size tends to zero. In practice, error estimates based on an energy norm have been found to be applicable to wide variety of problems [7,9]. Thus, if h_{ik}^2 is the area of element e_{ik} and h is the length of the common boundary (if any), Γ_{ik}, between e_{ik} and Γ_d then the local error ϵ_{ik} within e_{ik} is given by

$$\epsilon_{ik}^2 = \frac{h_{ik}^2}{24p^2}\int_{e_{ik}} r^2 \, d\Omega + \frac{h}{24p^2}\int_{\Gamma_{ik}} J^2 \, d\Gamma \tag{4}$$

in which r and J are the residuals over the domain and boundary respectively. The global error on an energy norm $\| e \|_E^2$ is then given by

$$\| e \|_E^2 = \sum_{\forall e_{ik} \in E_k} \epsilon_{ik}^2. \tag{5}$$

Equations (4) and (5) provide an *a posteriori* estimate of the error on a given mesh. However, they give no indication of by how much the error will be reduced if the mesh is refined further. If the error is orthogonal to the mesh then further refinement will not reduce the error. Consequently, a correction indicator is required to indicate what the likely reduction in error will be if the mesh is further refined. This saves considerable computational expense.

The simplest correction is provided by equation (4). Let the error on a particular element of size h_0 be $\epsilon_0^2 \sim ch_0^x$. If the element is subdivided into elements of size h_0/m then the error is $\epsilon_m^2 \sim c(h_0/m)^x$. Provided these two errors are available, further subdivision by a factor of n can be estimated to lead to an error of the order of $\epsilon_n^2 \sim [(\epsilon_m^4)/(\epsilon_0^2)](m/n)^x$. As the exponent $x \in [1, 2]$ and we require an overestimate of the error, then x is taken to be 2 for $m > n$ and 1 for $m < n$.

6. Application to flow into a mould

As a simple example of the application of the above adaptive mesh scheme consider the casting of a square plate whose thickness, H, is much less than the length of its side, L. The metal flows into the horizontal mould cavity through an ingate located at one corner of the plate while a riser with a neck of identical dimension to the ingate is located at the diagonally opposite corner. This geometry is shown schematically in Fig. 11. As the plate is thin, $H < L$ the variation of velocity and pressure across the thickness of the plate is of little interest and only the average velocity, $\boldsymbol{u}'$, and pressure, p', across the plate thickness need be calculated. Hence, the plate $\Omega \subset \mathbb{R}^2$ and by suitable normalisation $\Omega \in [0, 1]^2$. The boundary of Ω, denoted by Γ, is divided into three disjoint zones, $\Gamma = \Gamma_0 \cup \Gamma_I \cup \Gamma_R$. Γ_I denotes the ingate, Γ_R denotes the riser neck and Γ_0 denotes the impervious part of the boundary.

For each of illustration, it is assumed that heat losses to the mould are negligible during the filling of the cavity such that the thermal field need not be considered. However, thermal effects can be easily included, the details of which may be found in Morgan et al. [15]. It is also assumed that the Reynolds

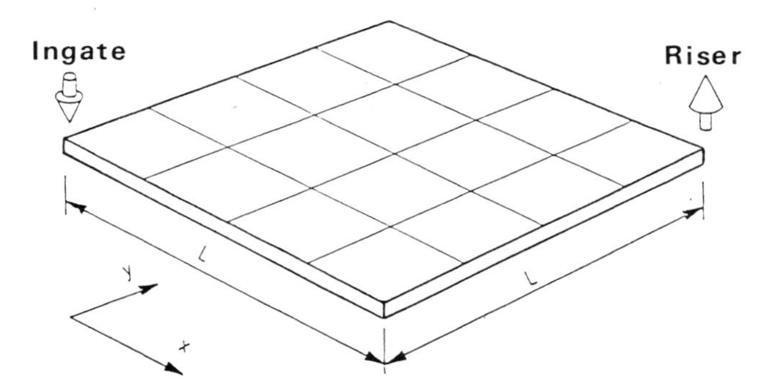

Fig. 11. Schematic diagram of the casting geometry used to illustrate the application of adaptive mesh generation.

number of the flow is sufficiently small that the velocity vector, $\boldsymbol{u}'$, within the mould is given by

$$\boldsymbol{u}' = -\frac{8H}{\mu}\nabla p' \tag{6}$$

in which μ denotes the dynamic viscosity of the metal taken to be very much greater than that of the gas initially filling the cavity. If it is further assumed that the capilliary pressure is negligible then the pressure is continuous across the gas-metal interface. Mass conservation then requires

$$\nabla^2 p = 0 \quad \text{on } \Omega \in [0, 1]^2 \tag{7}$$

with p being the suitably normalised pressure. The boundary conditions on the pressure are

$$p = 1 \quad \text{on } \Gamma_I \tag{8a}$$

$$\nabla p \cdot \boldsymbol{n} = 0 \quad \text{on } \Gamma_0 \tag{8b}$$

and

$$\nabla p = Q(\boldsymbol{i} + \boldsymbol{j}) \quad \text{on } \Gamma_R \tag{8c}$$

in which Q is the normalised flow rate through the ingate, $\boldsymbol{n}$ is the outward facing normal to Γ_0 and $\boldsymbol{i}$ and $\boldsymbol{j}$ are the unit vectors in the x and y directions respectively. The position of the metal free surface as it propagates through the mould cavity can be obtained in terms of a conservative 'saturation' function S, defined to be unity in the metal and zero elsewhere. Conservation of S requires [11],

$$\frac{\partial S}{\partial t} + \nabla \cdot (\boldsymbol{u}S) = 0 \quad (\boldsymbol{x}; t) \in \Omega \times [0, T] \tag{9}$$

in which t denotes time and T is a finite but possibly large dimensionless time scale. Boundary and initial conditions for (9) are

$$S(\boldsymbol{x}, t) = 1 \quad (\boldsymbol{x}; t) \in \Gamma_I \times [0, T] \tag{10a}$$

$$\nabla S(\boldsymbol{x}; t) \cdot \boldsymbol{n} = 0 \quad (\boldsymbol{x}; t) \in \Gamma_0 \times [0, T] \tag{10b}$$

$$S(\boldsymbol{x}; 0) = 0 \quad \boldsymbol{x} \in \Omega. \tag{10c}$$

On $\boldsymbol{x} \in \Gamma_R$ special provision has to be made to allow the free surface to pass out of the plate into the riser neck. This requires

$$Q_{\Gamma_R}(\boldsymbol{x}; t) = QS(\boldsymbol{x}; t) \quad (\boldsymbol{x}; t) \in \Gamma_R \times [0, T] \tag{10d}$$

Equation (9) is the two-dimensional kinematic wave equation and is hyperbolic. It is valid over the whole of Ω and hence, a front tracking algorithm is not essential to obtain a solution. In practice, accurate numerical solutions can only be obtained with very high spatial resolution in the vicinity of the interface [10,11].

As this interfacial region comprises only a small proportion of Ω, the use of a uniform high resolution mesh is computationally very inefficient. Hence, an adaptive mesh scheme is ideal for this type of problem. Here, numerical solutions to equations (7) to (10) are presented on four different computational meshes each of which was based on bilinear quadrilateral elements. These correspond to:

i) A uniform, non-adaptive mesh in the square domain $\Omega \in [0, 1]^2$
ii) An adaptive mesh in the same square domain
iii) A non-uniform, non-adaptive mesh in the triangular domain,

$$\Omega \in \{(x, y): [0, 1], [0, 1], y \leqslant x\}$$

iv) An adaptive mesh in the same triangular domain.

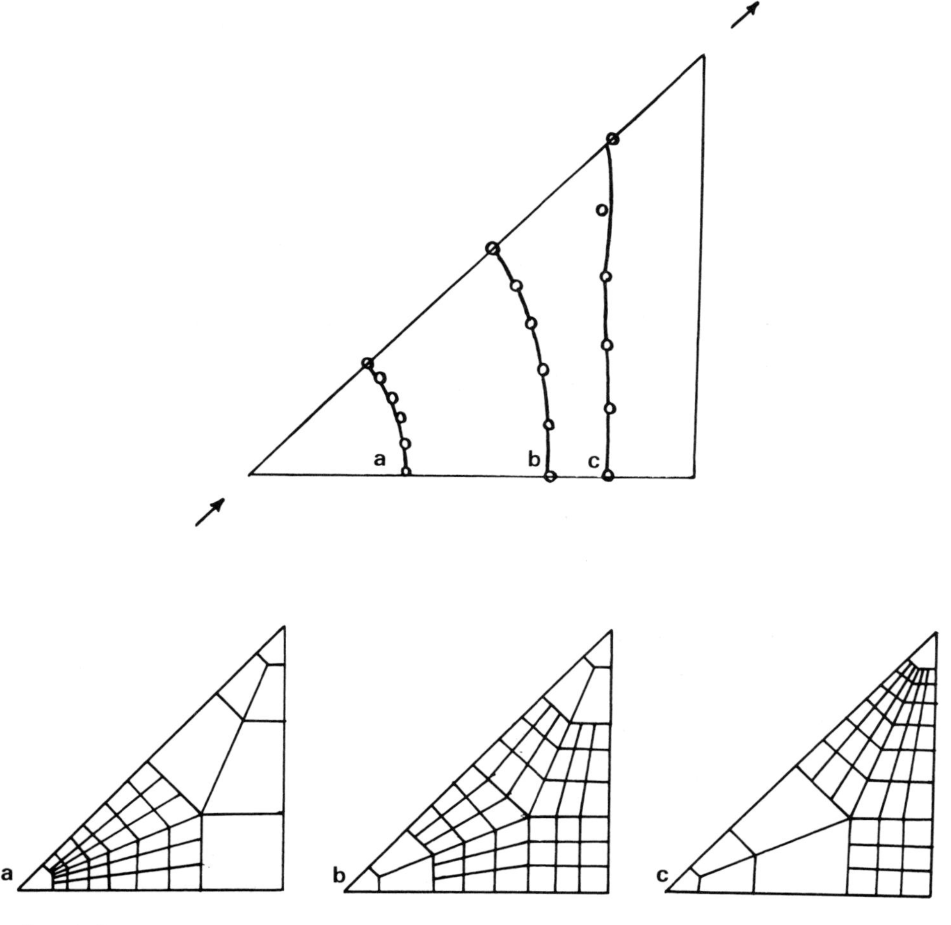

Fig. 12. Propagation of the molten metal interface through the mould cavity. Use has been made of the diagonal axis of symmetry to minimise computational cost. The solid line denotes the numerical solution while the open circles denotes the exact analytical solution. The three curves show the position of the interface at (a) 0.1 (b) 0.4 and (c) 0.65 of the time required to inject one cavity volume. The computational meshes corresponding to these times are also shown.

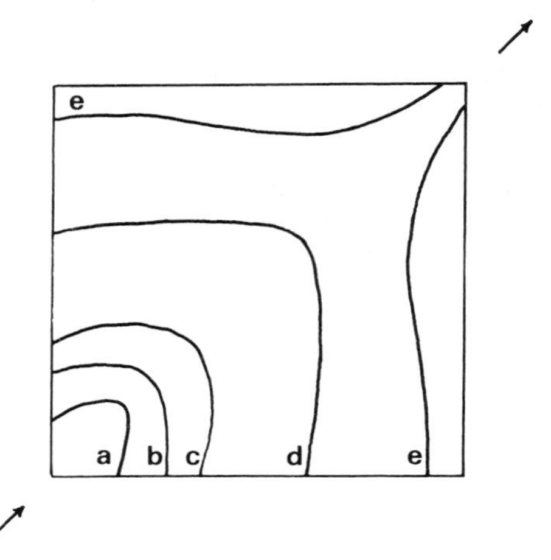

Fig. 13. Adaptive mesh solution on the square domain of the molten metal front propagating through the mould cavity. The lines correspond to the position of the front at (a) 0.04 (b) 0.075 (c) 0.1 (d) 0.36 and (e) 0.8 of the time required to inject one cavity volume.

The last two cases make use of the symmetry either side of the diagonal joining the ingate with the riser neck. This use of symmetry eliminates unnecessary calculation at the expense of having a more difficult geometry to discretise.

Figure 12 shows the progress of the free surface through the mould cavity as predicted by the adaptive mesh scheme in the triangular domain. Also shown in this Figure is the analytical solution to the problem [14]. There is good agreement between the two solutions. The adaptive solution on the square mesh is shown in Fig. 13 which demonstrates the breakthrough of metal into the riser before the mould cavity is filled. The close agreement between the two adaptive solutions demonstrates the elimination of mesh induced errors common in problems such as these. The non-adaptive solutions in the square and triangular domains showed considerable disparity, demonstrating the dependence of the solution on the *a priori* chosen mesh and the magnitude of the errors which can be present if an inappropriate mesh is chosen. The use of an adaptive mesh minimises these difficulties.

7. Concluding remarks

In the preceding sections it has been shown that adaptive finite element models are capable of providing mesh independent solutions to finite element analysis problems. This is a significant step forward toward a fully integrated computer-aided design system. However, its benefits will only be realised fully if there is little or not additional computational expense associated with such a system. This aspect must be investigated further in some detail.

Fortunately, existing evidence shows that to achieve a desired accuracy, adaptive finite element solutions are computationally cheaper than those using a fixed set of elements [10]. This is because the additional software complexity is offset by the fewer number of elements required to achieve the desired solution. Further work is now required to exploit these advantages to the full in computer-aided design systems for the foundry industry.

Acknowledgements

The authors are grateful to Deltacam Systems Limited and the Wolfson Foundation for their partial support of this work. The authors would also like to thank Professor R.W. Lewis and Dr K. Morgan of University College, Swansea, with whom some of the adaptive mesh algorithms discussed in this paper were derived.

References

1. T. Aizawa and T. Umeda: CAD/CAE for casting: State of the art in Japan. In: *CADCAM in the Foundry*, Proceedings 52nd Congress, CIATF (1985).
2. J.T. Berry and R.D. Pehlke: CADCAM and the simulation of solidification. In: *CADCAM in the Foundry*, Proceedings 52nd Congress, CIATF (1985).
3. G.L. Bird: The Ford 2.5 litre direct injection, naturally aspirated diesel engine. *Proc. Inst. Mech. Engrs.* 199 (1985) 122–133.
4. J.C. Cavendish: Local mesh refinement using rectangular blended finite elements. *J. Comp. Phys.* 19 (1975) 211–228.
5. G.S. Cole: Transport processes and fluid flow in solidification. In: *Liquid Metals and Solidification*. American Society for Metals, Cleveland, Ohio (1958) pp. 201–274.
6. S.C. Flood and J.D. Hunt: A model of a casting. *Appl. Sci. Res.* 44 (1987) 27–42.
7. J. Gago, D.W. Kelly, O.C. Zienkiewicz and I. Babuska: A-posteriori error analysis and adaptive processes in the finite element method. Part II – Adaptive mesh refinement. *Int. J. Numer. Methods Engng.* 19 (1983) 1621–1656.
8. I. Imafuku and K. Chijiiwa: A mathematical model for shrinkage cavity prediction in steel castings. *Trans. Am. Foundrymen's Soc.* 91 (1983) 527–540.
9. D.W. Kelly, J. Gago, O.C. Zienkiewicz and I. Babuska: A-posteriori error analysis and adaptive processes in the finite element method. Part I – Error analysis. *Int. J. Numer. Methods Engng.* 19 (1983) 1593–1619.
10. R.W. Lewis, K. Morgan, R. Pietlicki and T.J. Smith: The application of adaptive mesh methods to petroleum reservoir simulation. *Rev. de l'Inst. Français Pet.* 38 (1983) 751–761.
11. R.W. Lewis, K. Morgan, E.D.L. Pugh and T.J. Smith: A note on discontinuous numerical solutions of the kinematic wave equation. *Int. J. Numer. Methods Engng.* 20 (1984) 555–563.
12. R.W. Lewis, K. Morgan and P.M. Roberts: Determination of thermal stresses in solidification problems. In: J.F.T. Pitman, O.C. Zienkiewicz, R.D. Wood and J.M. Alexander (eds) *Numerical Analysis of Forming Processes.* Chichester: John Wiley and Sons (1984) pp. 405–431.
13. R.W. Lewis and P.M. Roberts: Finite element simulation of solidification problems. *Appl. Sci. Res.* 44 (1987) 61–92.
14. H.J. Morel-Seytoux: Unit mobility ratio displacement calculations for pattern floods in a homogeneous medium. *Soc. Petrol. Engrs. J.* 6 (1966) 217–227.

15. K. Morgan, T.J. Smith, R. Pietlicki, E.D.L. Pugh and R.W. Lewis: The simulation of thermal fronts in petroleum reservoirs during enhanced oil recovery operations. In: R.W. Lewis, J.A. Johnson and W.R. Smith (eds) *Numerical Methods in Thermal Problems*, Volume III. Swansea: Pineridge Press (1983) pp. 1154–1170.
16. R. Pietlicki, T.J. Smith, R.W. Lewis and K. Morgan: Adaptive computational grids for EOR simulations. University College of Swansea, Institute for Numerical Methods in Engineering, Preprint (1986) 8 pp.
17. T.J. Smith, A.F.A. Hoadley and D.M. Scott: On the sensitivity of numerical simulations of solidification to the physical properties of the melt and the mould. *Appl. Sci. Res.* 44 (1987) 93–109.
18. R.A. Stoehr and W.-S. Hwang: Modelling the flow of molten metal having a free surface during entry into moulds. In: *Modelling of Casting and Welding Processes II.* New York: AIME (1985) pp. 47–58.
19. D.P. Sturge: Improvements to parametric bicubic surface patches. In: J.A. Gregory (ed.) *The Mathematics of Surfaces*. Oxford: Oxford University Press (1986) pp. 47–58.
20. P.A. Walsham and W.A. Knight: Further developments of a geometric modelling system for the computer-aided manufacture of dies and moulds. *Annals CIRP* 32 (1983) 339–344.
21. C. Wei, J.T. Berry and P.H. Franklin: Riser design using edge functions. In: *Modelling of Casting and Welding Processes II.* New York: AIME (1983) pp. 237–242.
22. D.B. Welbourn: Computer-aided engineering (CADCAM/DUCT) in the foundry. In: *Development for Future Foundry Prosperity*. Alvechurch: BCIRA (1984) pp. 23-1 to 23-19.
23. D.B. Welbourn: Full three-dimensional CADCAM. *Comput. Aided Engng. J.* 1 (1984) 54–60.
24. D.B. Welbourn: Full three-dimensional CADCAM – Part II. *Comput. Aided Engng. J.* 1 (1984) 189–192.

Applied Scientific Research 44: 161–174 (1987)

Heat line formation in a roll caster

M.J. BAGSHAW [1], J.D. HUNT [1] & R.M. JORDAN [2]
[1] *Dept. of Metallurgy and Science of Materials, University of Oxford, Parks Road, Oxford OX1 3PH, UK;* [2] *Alcan International Ltd., Banbury, Oxon OX16 7SP, UK*

Abstract. A steady state two dimensional finite difference model has been developed to describe the roll casting process. The model explains the formation of the speed limiting defects called 'Heat Lines'.

The results of subsequent casting speed trials are compared with the predictions made by the numerical model.

Nomenclature

Symbol	Meaning
A_x	area of control volume face perpendicular to the x direction
A_y	area of control volume face perpendicular to the y direction
C	specific heat capacity
D	thermal diffusivity
F	accumulation of heat in each volume
g_L	fraction liquid
H	enthalpy
H_1	water/roll heat transfer coefficient
H_2	strip/roll heat transfer coefficient
k	distribution coefficient
K	thermal conductivity
L	number of columns of nodes
L_H	latent heat per unit volume
N	number of nodes
SJ	number of rows of nodes in roll
SK	column number at end of contact length
SL	column number at start of contact length
T	temperature
T_A	alloy input temperature
T_C	cooling water temperature
T_L	liquidus temperature
T_M	melting point of pure metal
T_R	roll input temperature
V	casting speed
W	number of rows of nodes
x	distance along length of slab
y	distance across thickness of slab
δT_O	change in central box temperature

Subscript	Meaning
A	alloy
e	east (face)
E	east (grid point)
i	alloy or roll
n	north (face)
N	north (grid point)
O	central (grid point)
R	roll
s	south (face)
S	south (grid point)
w	west (face)
W	west (grid point)

1. Introduction

The traditional method of producing 6 mm-thick metal sheet or strip involves two discrete processes: d.c. casting or extrusion of a slab followed by hot rolling to produce the final strip (involving several passes through the rolls to reach the desired thickness).

The twin-roll casting process [8,11] produces coilable strip directly from the melt by combining the casting and hot rolling into a single process. Although the idea of casting directly between cooled rotating rolls was first proposed by Bessemer [1], it has only in recent years been put into practice. The twin-roll casting process has now assumed importance in the aluminium industry for the production of stock for the further processing into a wide variety of sheet and foil products.

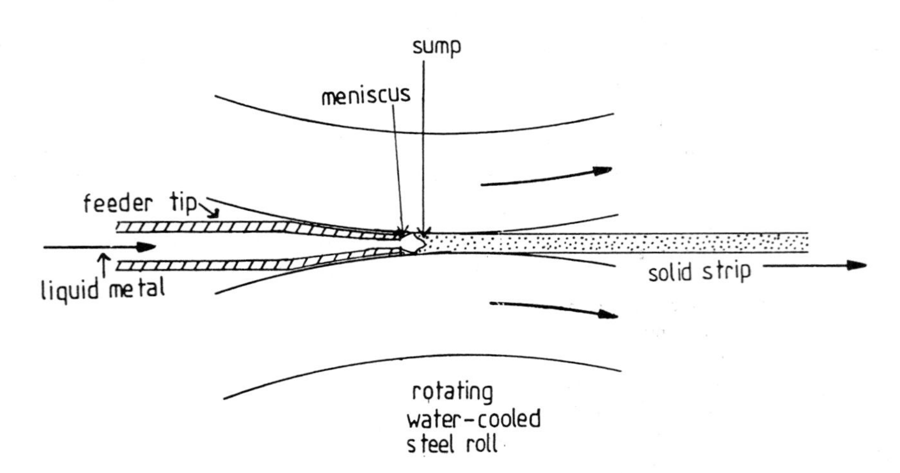

Fig. 1. Twin-Roll strip caster (schematic).

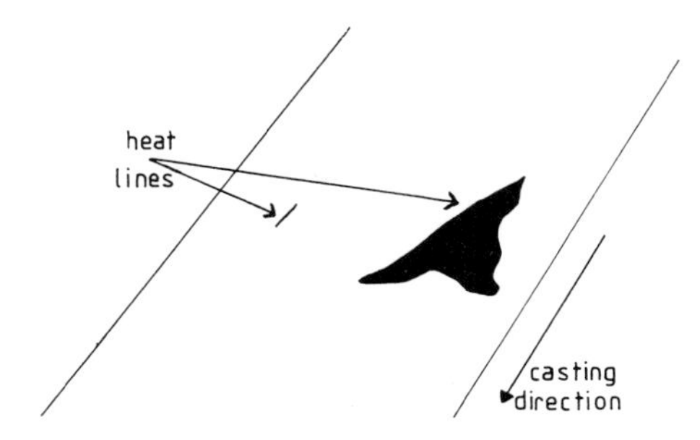

Fig. 2. Heat line in a cast strip (schematic).

The principles of the twin-roll casting process are shown in Fig. 1. The melt is maintained at a constant level in a headbox from which it flows by gravity feed through a transfer system into a ceramic tip assembly. The molten metal flows in a controlled manner from the orifice of the feed-tip into the gap between two rotating water-cooled cylindrical steel rolls. The metal solidifies just before reaching the bite of the rolls and is then hot rolled as it passes through the rolls, emerging as a fine-grained sheet approximately 6 mm thick.

There are obvious economic advantages in producing strip from the melt and eliminating the hot rolling operations of the traditional method [2,4]. There are also potential metallurgical advantages as the high freezing rate which occurs during roll casting can result in unique microstructures.

One of the major problems which is encountered in the roll casting process is the formation of heat lines [9]. Heat lines are semi-continuous longitudinal defects, usually only a few centimetres in width, where the material leaves the roll bite still partially molten. This is illustrated schematically in Fig. 2. These defects occur at relatively low casting speeds when the mean strip exit temperature is several hundred degrees below the solidus temperature.

Figure 3 shows typical strip exit temperatures for a range of commercial aluminium alloys as a function of casting speed for a cast strip thickness of approximately 6 mm. The highest recorded speed for each alloy is very close to that at which heat line formation first occurs even though the mean strip exit temperatures are well below the solidus temperatures of the alloys. If casting stability could be maintained (as shown by the extrapolation in Fig. 3), considerably higher casting speeds and therefore greater productivity of the roll caster could be achieved.

The present work was an attempt to gain a better understanding of the relationship between casting speed, strip exit temperture and the formation of heat lines by developing a mathematical model of the process and deriving conclusions from it. Results of subsequent experimental studies are also given to verify the predictions of the model.

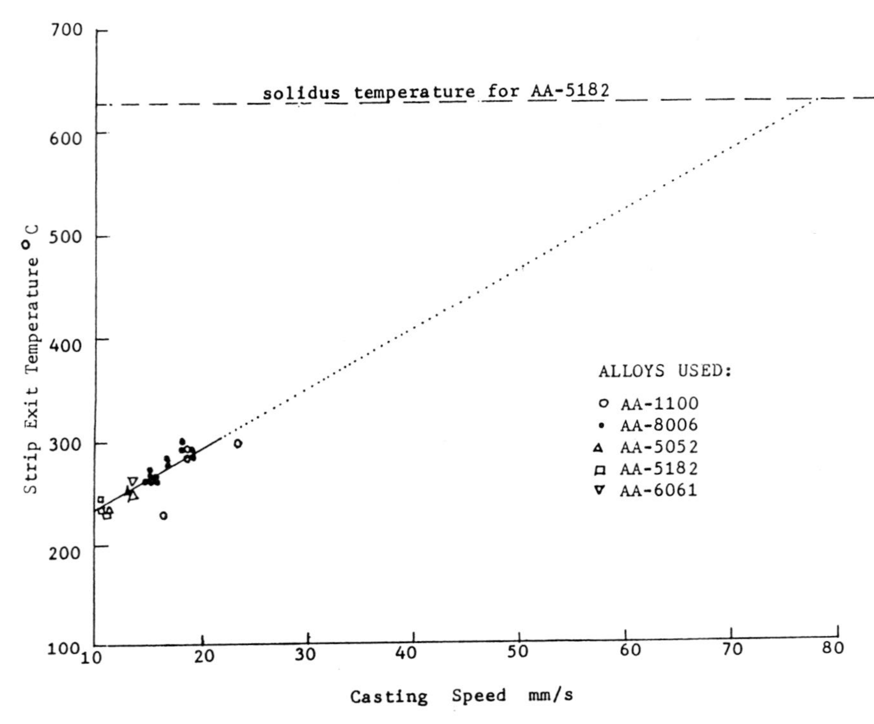

Fig. 3. Experimental strip exit temperatures [9].

2. The numerical model

2.1. Physical basis

During roll casting the heat flow in the roll and alloy strip can be described by the two dimensional finite difference model shown in Fig. 4. In this model positive x is in the casting direction and positive y towards the centre of the strip. Molten alloy at temperature T_A is fed to the rolls at $x = 0$.

The principal assumptions made in the model are:

1. The process is at steady state.
2. The latent heat is liberated over a temperature range: the fraction of the latent heat evolved at a given temperature being determined by the volume fraction liquid.
3. The volume fraction liquid is assumed to be given by the Scheil equation (equation 3) [5,7].
4. A variable strip/roll heat transfer coefficient is assumed through the roll bite. In the current model three regions of heat transfer coefficient are postulated along the strip/roll contact arc (Fig. 5):

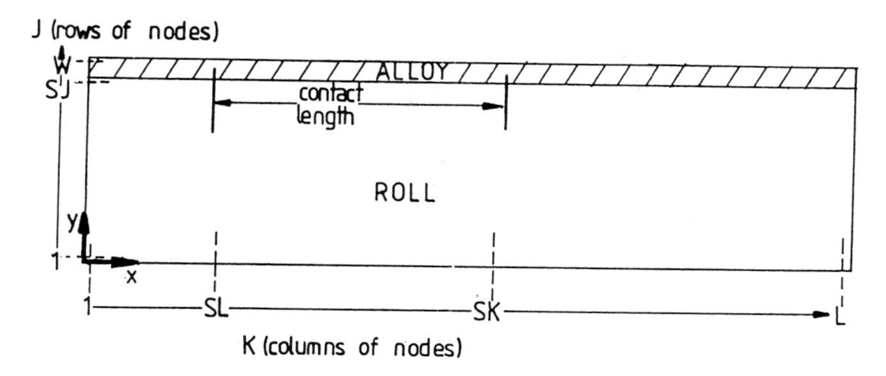

BOUNDARY CONDITIONS:

K=1: $T_W = T_A$ (Alloy input temperature) for $J > SJ$
$T_W = T_R$ (Roll input temperature) for $J \leq SJ$

K=L: The temperature gradients at the east and west box faces are made equal: $\left(\frac{\partial T}{\partial x}\right)_e = \left(\frac{\partial T}{\partial x}\right)_w$

J=1: Heat is removed at a rate characterised by the water/roll heat transfer coefficient, H_1
$T_S = T_C$ (Cooling water temperature)

J=SJ: Heat is accumulated at a rate characterised by the strip/roll heat transfer coefficient, H_2 for $K \geq SL$ and $K \leq SK$

J=SJ+1: Heat is removed at a rate characterised by the strip/roll heat transfer coefficient, H_2 for $K \geq SL$ and $K \leq SK$

J=W: The temperature gradient across the centre of the strip is zero: $\left(\frac{\partial T}{\partial y}\right)_n = 0$

[e,w,n refer to east, west, north box faces respectively
W, S refer to west, south nodal points respectively]

Fig. 4. The physical model.

a) an initial region of relatively high heat transfer where there is intimate contact between the molten metal and roll;
b) an intermediate region of low heat transfer coefficient as the strip buckles away from the roll and an oxide film forms; and

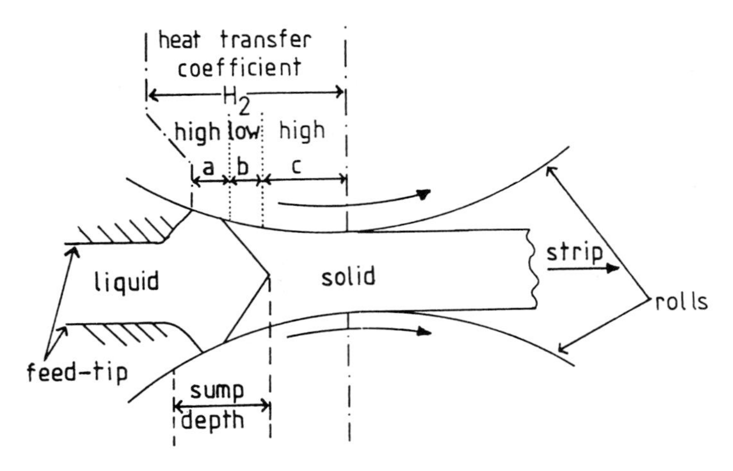

Fig. 5. Schematic variation of the strip/roll heat transfer coefficient, H_2, through the roll bite.

c) a high heat transfer coefficient once again as the solidifying alloy gains in strength in the centre and so a greater pressure is exerted on the strip by the rolls thus achieving intimate contact between the roll and strip.

The length of the final high heat transfer coefficient (region c) is controlled by the fraction solid at the centre of the strip.

The boundary conditions for the model are given in Fig. 4. Values of the casting and physical parameters used in he computation for two alloys are given in Table 1.

2.2. *Mathematical formulation*

The steady state heat equation has to be solved, subject to the relevant boundary conditions, at every point.

A control volume method is used to formulate the difference equations. Thus the region to be modelled is subdivided into a finite number of small sections by setting up a rectangular array of N nodes and placing box walls halfway between adjacent nodes. The current model employs 486 nodes in the roll and 702 nodes in the strip. The problem is a steady state one so that the accumulation of heat in each volume must be zero. The heat equation (equation 1) consists of a conductive flux across the four faces of a box together with a velocity term for the enthalpy crossing the walls which are perpendicular to the x direction.

$$A_x\left\{K_i\left(\frac{\partial T}{\partial x}\right) - VH_i\right\}_e - A_x\left\{K_i\left(\frac{\partial T}{\partial x}\right) - VH_i\right\}_w + A_y K_i\left(\frac{\partial T}{\partial y}\right)_n$$

$$-A_y K_i\left(\frac{\partial T}{\partial y}\right)_s = 0. \tag{1}$$

Table 1. Values of casting and physical parameters used in computation for Al-2% Cu and Al-1% Cu.

Parameter	Al-2% CU	Al-1% Cu
Strip thickness (cm)	0.635	0.635
Strip/roll contact length (cm)	5.5	5.5
Strip/roll heat transfer coefficient, H_2(J/cm^{2}s°C):		
region *a*	3.53	3.53
region *b*	0.1015	0.1015
region *c*	20.0	20.0
Length of region *a* (cm)	0.925	0.525
Fraction solid (at centre of strip) which governs start of region *c*	0.9763	0.9875
Water/roll heat transfer coefficient (J/cm^{2}s°C)	3.4	3.4
Roll shell thickness (cm)	3.66	3.66
Distribution coefficient	0.18	0.18
Specific heat of alloy (J/cm^3°C)	3.4	3.4
Thermal conductivity of alloy (J/cm°Cs)	0.98	0.98
Liquidus temperature of alloy (°C)	654.8	657.4
Latent heat of alloy (J/cm^3)	1020	1020
Thermal conductivity of roll (J/cm°Cs)	0.44	0.44
Specific heat of roll (J/cm^3°C)	3.7	3.7
Alloy input temperature (°C)	700	700
Water temperature (°C)	23	23

The enthalpy for the roll is given as:

$$H_R = C_R T \tag{2a}$$

and for the alloy:

$$H_A = C_A T + g_L L_H. \tag{2b}$$

The volume fraction liquid is assumed to be given by the Scheil equation [6,10]:

$$g_L = \left\{ \frac{T_M - T}{T_M - T_L} \right\}^{1/(k-1)}. \tag{3}$$

The tempertures and temperature gradients at the faces perpendicular to the y direction are calculated by a linear interpolation of adjacent points. For the box faces perpendicular to the x direction the linear interpolation is a good approximation provided that the box size is small compared with D/V, (thermal diffusivity/casting speed). If the box size is of the order of D/V, a better approximation is given by an exponential interpolation of the form:

$$T = A + B \exp(Vx/D). \tag{4}$$

This equation is valid as it is the solution to the steady state heat flow equation. The temperatures and temperature gradients at the box faces perpendicular to the x direction are calculated using this exponential interpolation. A and B are functions of the temperature at nodal points either side of the face, the casting speed, the thermal diffusivity and the box length. This exponential interpolation only makes a difference at the start and end of the strip where the box size is large and not in the vicinity of the front where smaller box sizes have been used.

The heat equation which has to be obeyed for each of the N volumes is of the final form:
Accumulation of heat,

$$F = A_1(T_E - T_O) + A_2(T_W - T_O) + B\{(g_L)_w - (g_L)_E\} + C_1(T_N - T_O) + C_2(T_S - T_O) = 0 \tag{5}$$

A_1 and A_2 depend on the box size, casting speed and relevant specific heat; whereas C_1 and C_2 depend on the box size and relevant thermal conductivity. For boxes in the alloy B depends on the latent heat, casting speed and box size. B is zero for boxes in the roll.

Initially a Gauss-Seidel technique [12] was used to obtain the final solution for the temperature distribution in the roll and strip. This method of solution was inherently unstable because of the non-linearity of the latent heat term with temperature when the Scheil equation is used to determine the fraction liquid. At each iteration the latent heat term was calculated using the value of the central temperature of the box calculated in the previous iteration. Thus the value of the latent heat term used at each step was not consistent with the central box temperature calculated for that iteration. In order to obtain a stable solution an underrelaxation parameter had to be employed, and so this method of solution required a large amount of computation time to converge.

Because of this problem with instability a Newton-Raphson method [3] is employed in the current model. The same iterative technique as before is used to obtain the final solution but at each iteration the central temperature of a box is calculated using a Newton-Raphson method. This ensures that the central temperature of a box used to calculate the latent heat term is consistent with the central temperature of the box calculated at the end of that iteration. The Newton-Raphson method uses the derivative of the heat accumulation in a box to calculate the change in temperature at the centre of each box (equation 6).

$$\delta T_O = -F \Big/ \left(\frac{\partial F}{\partial T_O} \right). \tag{6}$$

The latent heat term is recalculated for each new central temperature and a final value of the central temperature is obtained when δT_O tends to zero.

2.3. Results from the model

The model predicts that the strip exit temperature is decreased by decreasing the alloy input temperature or casting speed or by increasing the strip/roll heat transfer coefficient or roll thermal conductivity. These trends are as expected and as observed in practice.

Theoretical plots of strip exit temperature against casting speed produce a hysteresis loop (continuous line on Figs. 6 and 7). The two possible strip exit temperatures obtained for a given casting speed in the two solution region correspond to the strip initially exiting the roll bite as 'solid' (lower curve) or as 'liquid' (upper curve). This plot explains why alloys cannot be cast at the higher casting speeds that might be expected from the extrapolation of the previously obtained results (Fig. 3). Although all the experimental data in Fig. 3 appear to lie on a straight line, as one would expect assuming a single uniform value for the strip/roll heat transfer coefficient, the model here, which inputs a more logical variable heat transfer coefficient predicts a catastrophic increase in strip exit temperature at a critical velocity. This is the velocity above which heat lines can now be predicted to form. The catastrophic increase in the strip exit temperature is caused by the sudden decrease in the

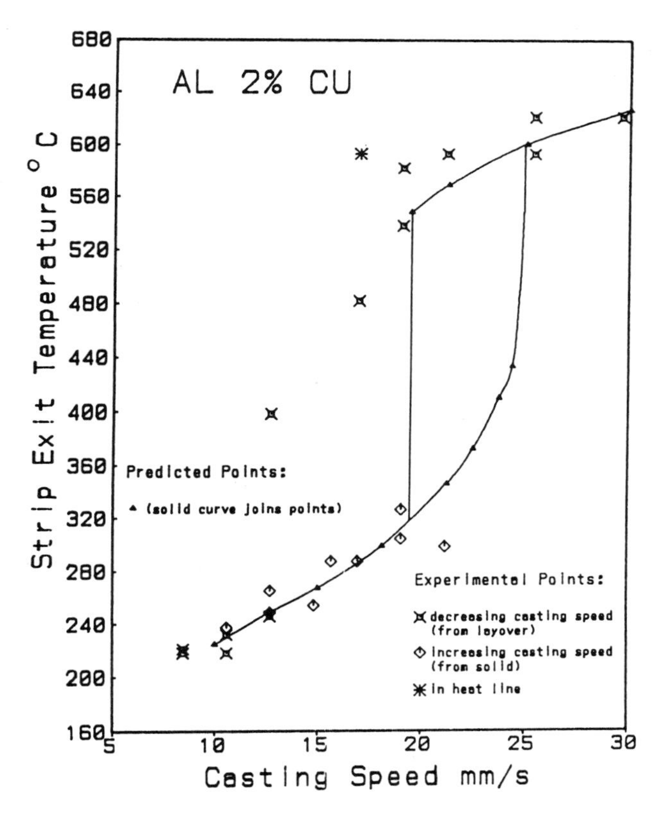

Fig. 6. Experimental and predicted plots of strip exit temperature against casting speed for Al-2%Cu.

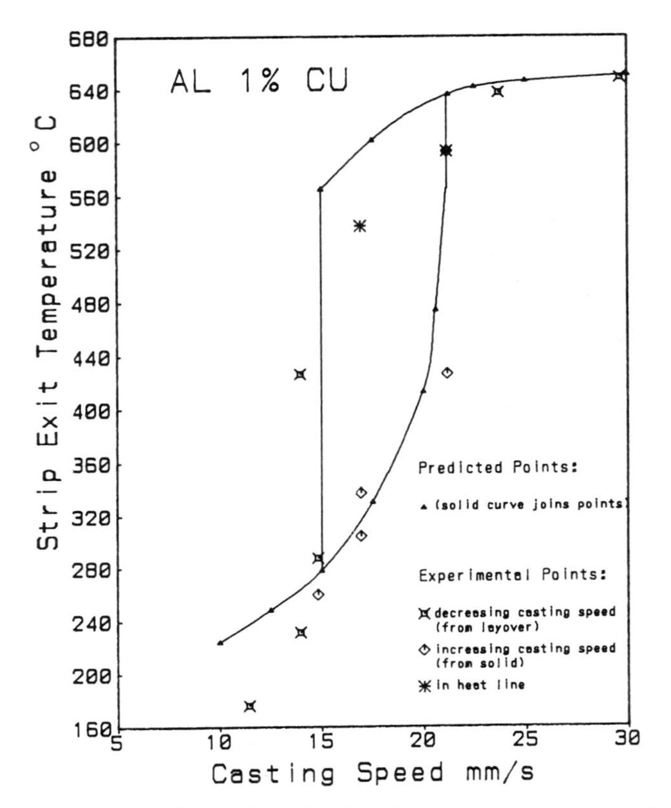

Fig. 7. Experimental and predicted plots of strip exit temperature against casting speed for Al-1% Cu.

length of the final high strip/roll heat transfer coefficient part of the contact length (region *c*). The model also predicts a different curve for each alloy composition giving different critical casting speeds, as found in practice.

In fact in commercial production it is not the casting speed that changes but it is a variaton in local operating conditions which causes heat line formation. At each casting speed there is a critical value of the heat transfer coefficient along the contact length below which heat lines form. A localised variation in the heat transfer coefficient bringing about a low enough value to produce heat lines locally might easily occur because of gas or oxide formation, for example. Another parameter which can vary locally is the alloy input temperature. Metal temperatures across the width of the tip can vary quite significantly. The sensitivity of the critical casting speed to the temperature distribution in the feed-tip is shown in Fig. 8 which was obtained by varying the alloy input temperature in the model. The lower and more uniform the input temperature across the width of the tip the greater the speed at which the alloy can be cast safely. Currently there is a lot of work being carried out

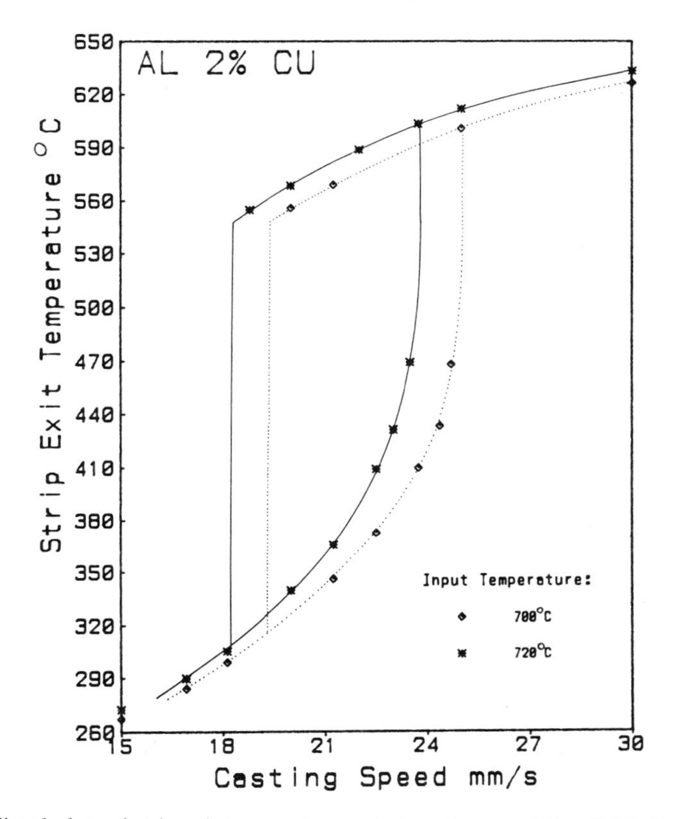

Fig. 8. Predicted plots of strip exit temperature against casting speed for Al-2% Cu at different alloy input temperatures.

on improving the feed-tip design in order to obtain a more uniform flow and temperature distribution of the alloy across the tip.

The model shows that heat lines can be deterred from forming by using a roll material with a high thermal diffusivity. Figure 9 compares plots of strip exit temperature against casting speed for a steel roll ($D_{steel} = 0.138$ cm^2/s) with that of a copper roll ($D_{Cu} = 1.147$ cm^2/s). The critical casting speed is significantly increased by using a copper roll instead of a steel one.

3. Experimental studies

3.1. Procedure

Practical roll casting trials were carried out on a Hunter twin-roll casting machine. Two binary Al-Cu alloys containing 1 and 2% Cu were used and the

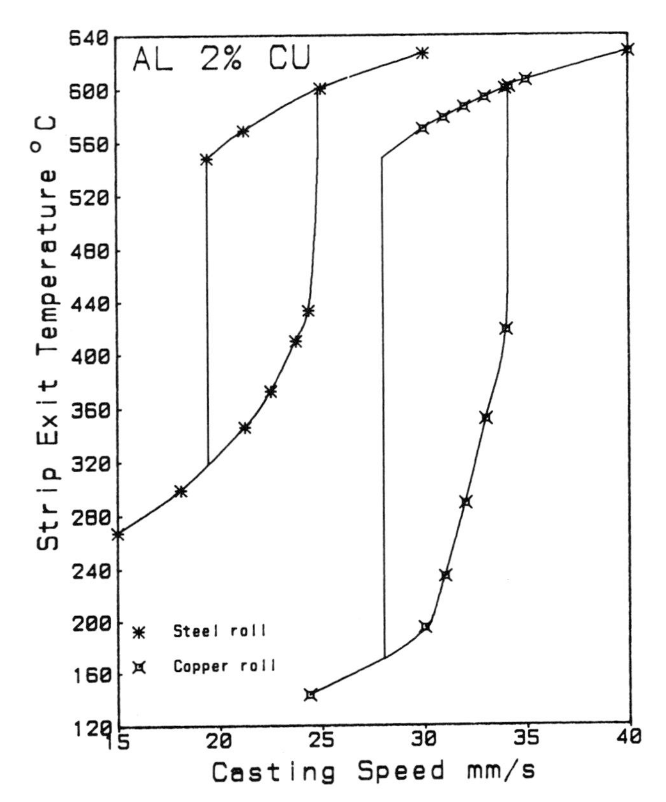

Fig. 9. Comparison of predicted plots of strip exit temperature against casting speed for Al-2% Cu using a steel roll with plots for Al-2% Cu using a copper roll.

machine was deliberately run at rates in excess of the heat line formation critical speed as well as at known stable casting speeds.

The strip thickness, strip/roll contact length, roll shell thickness, alloy input temperature and water temperature all corresponded to the values used in the model. The strip exit temperature was measured by prod thermocouples. During casting a constant metal level in the head box was maintained throughout.

The actual sequence in the casting trials was as follows:

a) The caster was run at a higher than normal rate; the molten alloy was then introduced from the head box and feed-tip into the roll gap. With this operational speed, a semi-solid strip (layover) was produced.
b) The caster speed was gradually reduced, measuring the strip exit temperature corresponding to various speeds. When the speed was reduced to the normal operating speed, a solid strip was obtained across the whole width.

c) The caster speed was increased, measuring the strip exit temperature, until heat lines were produced.

3.2. Experimental results

The results of trials on Al-2% Cu and Al-1% Cu are given in Figs. 6 and 7 respectively. In addition to the experimental points the values predicted by the mathematical model are given in the figures.

The experimental results clearly show the hysteresis loop predicted by the model and, despite the experimental scatter, there is good agreement between the actual results and the predicted curves.

4. Conclusion

Previous work [9] relating casting speed and exit temperature of the strip from a roll caster (Fig. 3) showed that, although strip exit temperatures increased with increase in roll caster speed, the temperatures were well below the solidus temperature. However experience showed that the linear relationship could not be extrapolated to high casting speeds because of the formation of heat lines.

The results from the mathematical model show why the line in Fig. 3 cannot be extrapolated. The model predicts a hysteresis loop (continuous line on Figs. 6 and 7) in the casting speed/strip exit temperature relationship and practical experimental results from two alloy systems confirm this hysteresis.

For a certain range of casting speeds there are therefore two solutions, i.e. two possible strip exit temperatures. These correspond to the physical conditions of the strip initially leaving the casting machine in the 'solid' or 'liquid' form. A localised decrease in the strip/roll heat transfer coefficient can cause the operating conditions to flip from the lower 'solid' curve to the higher 'liquid' curve – i.e. at a critical casting speed there is a catastrophic increase in strip exit temperature and above this speed heat lines can form in the exiting material.

The model indicates that one method of reducing the possibility of heat line formation is to increase the strip/roll heat transfer by, for example, using a roll material with a higher thermal diffusivity. Another method of increasing the heat transfer would be the use of an inert gas to help prevent a thick oxide layer from forming and so increase the strip/roll heat transfer coefficient. The model has also shown that the critical speed at which heat line formation first occurs can be increased by improving the feed-tip design so that a more uniform melt temperature distribution is obtained throughout the tip orifice and thus the formation of hot spots is reduced.

Acknowledgements

The authors would like to express their gratitude to Professor Sir Peter Hirsch FRS and the Oxford University Computing Service for the provision of office and computing facilities respectively, and to Alcan International Ltd and Hunter Engineering Ltd, Riverside, CA, for help with the experimental aspects of the work. One of the authors (MJB) would also like to thank the Science and Engineering Research Council for financial support.

References

1. H. Bessemer: British Patent No. 11, 317 (1846).
2. E.A. Bloch and G. Thym: Continuous casting of aluminium strip – an economical comparison. *Metals and Materials*, 6 (1972) 90–94.
3. B. Carnahan, H.A. Luther and O. Wilkes: *Applied Numerical Methods,* pp. 308–329.
4. T.W. Clyne: Heat flow, solidification, and energy aspects of d.c. and strip casting of aluminium alloys. *Metals Technology* Vol. II (1984) 350–357.
5. T.W. Clyne: Numerical modelling of heat transfer during solidification of metals. In: *Proc. 2nd Intern. Conf. on Numerical Methods in Thermal Problems,* Venice, Italy (1981) pp. 240–256.
6. R. Elliott: *Eutectic Solidification Processing.* Butterworths & Co, London, UK (1983) pp. 247–251.
7. I. Jin and J.G. Sutherland: Thermal analysis of solidification of aluminium alloys during continuous casting. In: *Solidification and Casting of Metals,* Metals Society, London, UK (1979) pp. 256–259.
8. D.M. Lewis: The production of non-ferrous metal slab and bar by continuous casting and rolling methods. *Metallurgical Reviews,* Vol. 6 No. 22 (1961) 143–192.
9. L.R. Morris: Private communication. Alcan International Ltd. (1980).
10. E. Scheil: *Z. Metallkunde* 42 (1942) 70.
11. S. Slevolden: The horizontal sheet casting of aluminium. *Metals and Materials* 6 (1972) 94–96.
12. G.D. Smith: *Numerical Solution of Partial Differential Equations: Finite Difference Methods,* pp. 227–236.

Applied Scientific Research 44: 175–195 (1987)

Modelling of liquid–liquid metal mixing

S. ROGERS, L. KATGERMAN, P.G. ENRIGHT & N.A. DARBY
Alcan International Limited, Southam Road, Banbury, Oxon, England

Abstract. Mathematical and physical modelling have been used interactively to predict the behaviour of two liquid metals mixing in the nosetip of a Roll-casting machine. Attention has been given to the transient development of density driven backflows due to incomplete mixing and forward transport of fluid. The results are compared with temperature and composition data obtained from metallurgical trials.

Mixing reactions which may involve a simultaneous phase change (partial solidification) are also discussed in relation to the local mixing conditions.

Introduction

When two liquid metals of different composition and temperature are mixed together a number of different things can happen, dependent upon the chemistry of the system (thermodynamics) and the degree of local turbulent mixing (kinetics). In the simplest case mixing results in nothing more than a homogeneous liquid alloy and the mixing process becomes one of simply alloying in the fully liquid state.

On the other hand, solidification and/or chemical reactions may take place on mixing. Such reactions are controlled by coupled heat and solute diffusion fields which in turn depend on the local hydrodynamic structure of the flow (turbulent diffusion). If mixing (flow) is carefully controlled, turbulent diffusion processes can be used to obtain very fine dispersion of precipitates homogeously distributed throughout the bulk of a subsequently cast product.

With this in mind it is the objective of the current study to describe the mixing and flow behaviour of liquid metals in the nosetip (feed system) of a continuous casting machine and to assess the viability of the $k-\epsilon$ turbulence model as an engineering design tool. The work involves both mathematical and physical simulations of the casting process together with quantitative metallurgical validation.

Metallurgical background

From a consideration of local phase equilibria during the mixing of two different alloys, three different idealised situations can be envisaged, viz
1. IDEAL MIXING
2. IDEAL QUENCHING
3. REACTION MIXING.

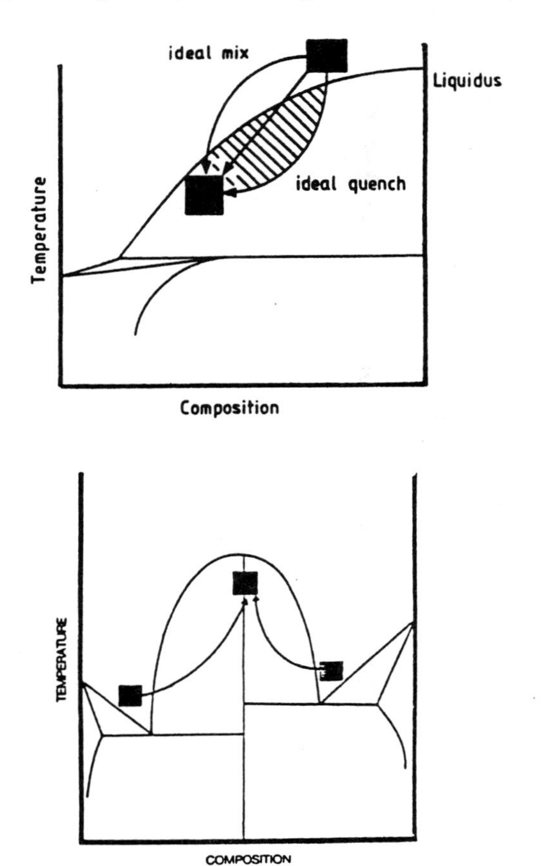

Fig. 1. Schematic of ideal mixing, ideal quenching and reaction mixing processes. (a) A volume element of feedstock rapidly undercools below its local liquidus temperature and follows a solidification path defined by local mixing conditions. Process control relies on understanding turbulent transport in a solidifying system. (b) Two low melting point fluids mix to form a high temperature intermediate compound. The process relies on turbulent transport and thermal control in the reactor vessel.

These processes are best described by reference to Fig. 1. IDEAL MIXING refers to the case where turbulent diffusion dominates the process and the system homogeneously mixes (with reference to solute) in the liquid state prior to any local phase change. IDEAL QUENCHING refers to the case where the system is dominated by heat conduction and a situation arises whereby a volume element of the injected fluid instantaneously finds itself undercooled with respect to its local liquidus, and begins to freeze rapidly. The third case, REACTION MIXING refers to the case where two alloys can exist at essentially the same temperature in the liquid state but on mixing react immediately to form a high temperature intermediate compound. Such a reaction is governed by the ability of the system to (a) lose internally

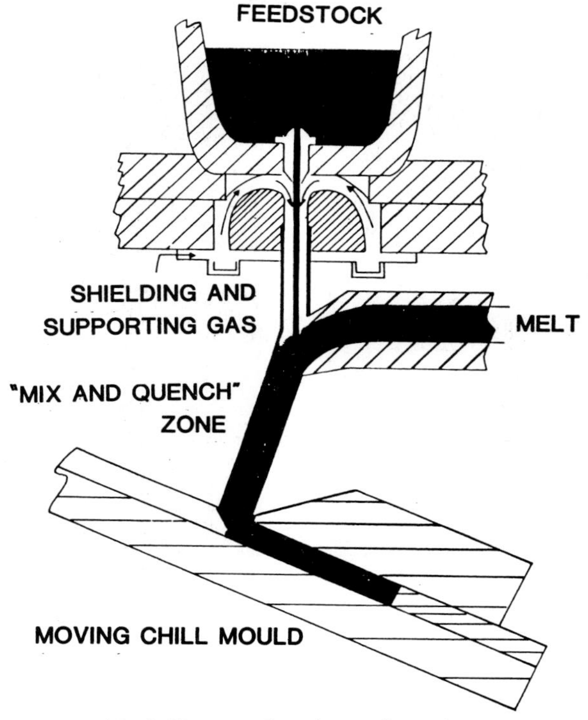

Fig. 2. Sheet metal casting configuration.

generated latent heat and (b) by the tendency of the system to form solid diffusion boundary layers at reacting surfaces.

Since the properties of metal alloys are largely controlled by their solidification microstructure this type of processing can lead to a wide range of potentially new materials with novel combinations of properties.

An example of this is given in Fig. 2. In this case, two different aluminium based alloys are mixed together using the jet mixing apparatus shown in the figure. Because of the elevated liquidus of the feedstock this material is quenched and frozen rapidly by the lower temperature melt. This results in the formation of a fine dispersion of intermetallic particles distributed throughout the mixing volume. Figure 3 illustrates the effect of increasing jet Reynolds number on this dispersion. It is clear that at certain critical values of N_{Re} the entire microstructure is refined as a result of the correct balance between solidification reactions and hydrodynamics. For this mixing system the microstructure is optimised at $N_{Re} \sim 8000$ (based on jet diameter), and Fig. 4 indicates the improvements in mechanical properties that result when this is achieved. The relationships between solidification and hydrodynamic reactions in this type of system are more fully described elsewhere [1,2].

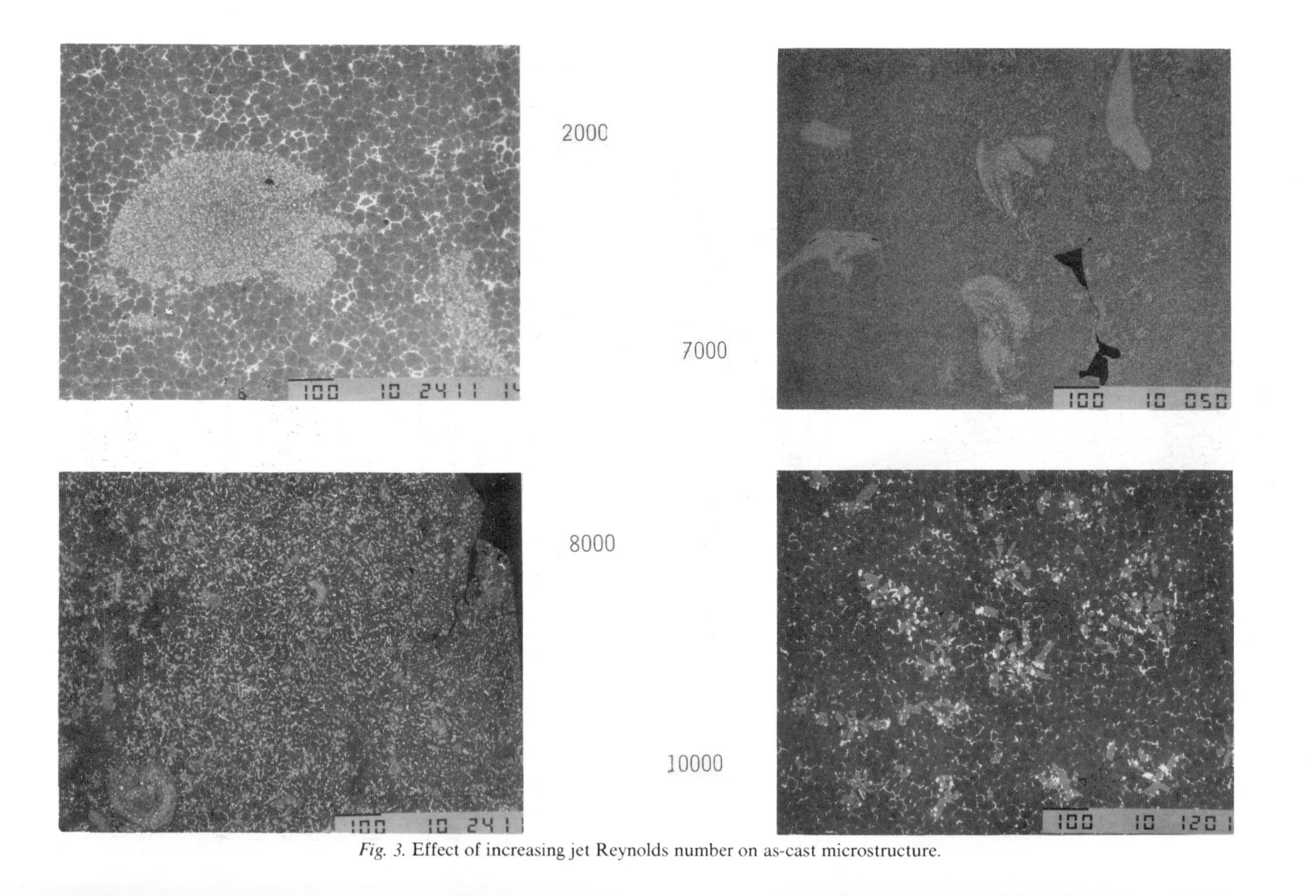

Fig. 3. Effect of increasing jet Reynolds number on as-cast microstructure.

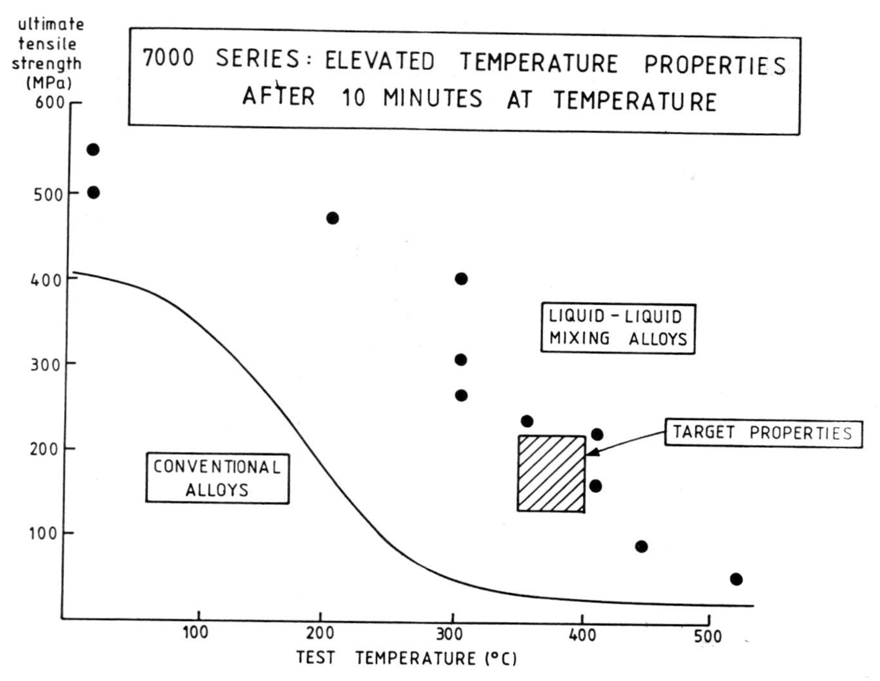

Fig. 4. Elevated temperature properties of liquid–liquid mixing alloys.

Other metallurgical process advantages can also be gained from this technique. In conventional casting operations it is often necessary to hold large volumes of metal at elevated temperatures before and during casting in order to prevent the premature nucleation and growth of intermetallic phases on cold surfaces. In some systems this metal transfer requirement is in direct conflict with the desire to cast material very close to its liquidus temperature in order to optimise the casting process itself. IDEAL MIXING relaxes these constraints by developing the final alloy composition and temperature requirement within the casting machine itself.

Experiments have been carried out using a Twin Roll Casting Machine, in which rapid solidification rates can be achieved.

Figure 5 is a schematic diagram of the casting machine. In the conventional mode of operation, liquid metal is supplied from a tundish via a refractory nosetip to the roll bite of the casting machine. The rolls extract heat very rapidly and provide some deformation as the material solidifies as a thin strip. In the liquid-liquid mixing mode a secondary incoming fluid is injected as a high velocity (turbulent) jet having a temperature and density different to that of the mainstream flow. The nosetip of the caster is necessarily tilted upwards at an angle of 15° to aid initial priming of the caster and ensure that metal freezes uniformly across the full width of the strip at the start of casting.

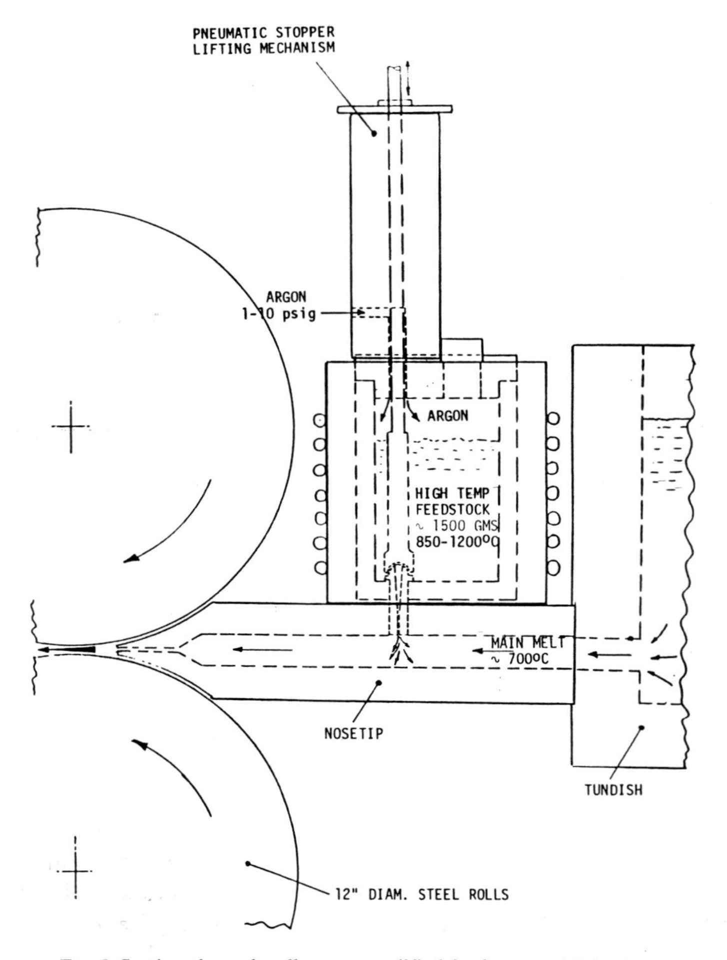

Fig. 5. Section through roll caster modified for hot metal injection.

Linear defects (heat lines) also occur during casting of the metal because the temperature at the roll bite is either too high or non-uniform across the strip width. Conversely, too low a metal temperature may cause premature local solidification resulting in either microstructural defects or strip breakdown. The casting 'window' for aluminium-manganese alloys is shown schematically in Fig. 6.

This alloy and other refractory metal systems are of interest because of the sensitivity of their mechanical properties to local solidification rate. Roll casting and liquid-liquid mixing techniques are expected to improve properties significantly.

To provide quantitative metallurgical data, local transient temperature distributions during casting were recorded using an in-line array of up to 17

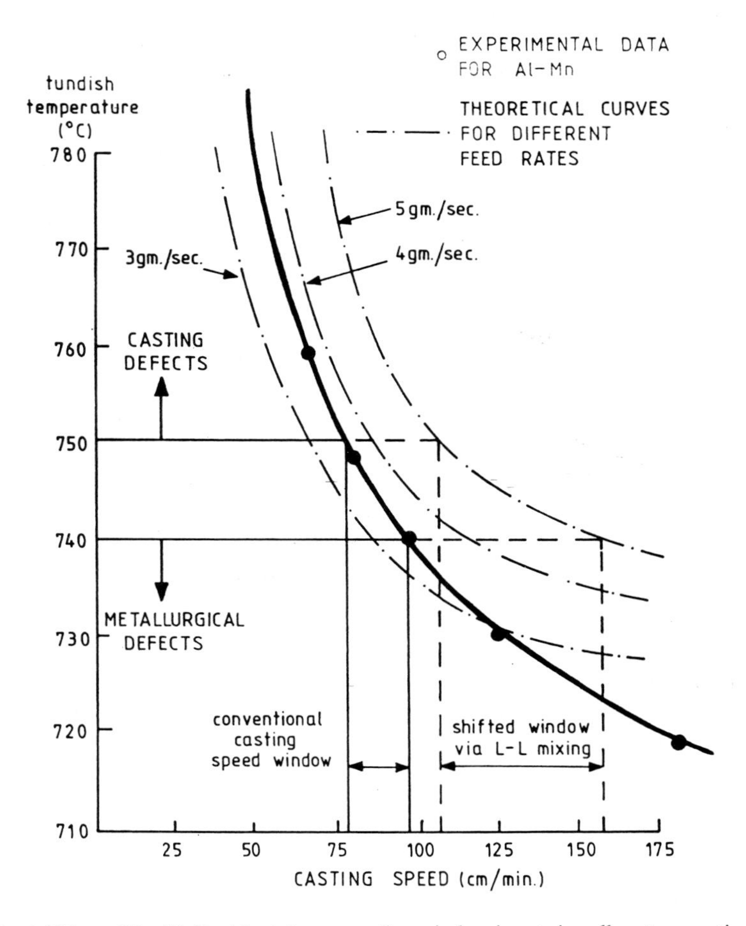

Fig. 6. Effect of liquid–liquid mixing on casting window in a twin roll caster nosetip.

thermocouples positioned in the nosetip (Fig. 7). Compositional information was obtained from chemical analysis of the cast product.

In order to develop an understanding of the liquid mixing behaviour in the nosetip of the caster, a dimensionally similar perspex model has been constructed. This model has been used interactively with data obtained from real metal trials and a mathematical model of turbulent flow, in order to design special nosetips optimised for liquid-liquid mixing experiments.

The physical model

Essential features of the model (Fig. 8) are the 15° inclination to the horizontal, and the outlet manifold containing an array of balancing nozzles.

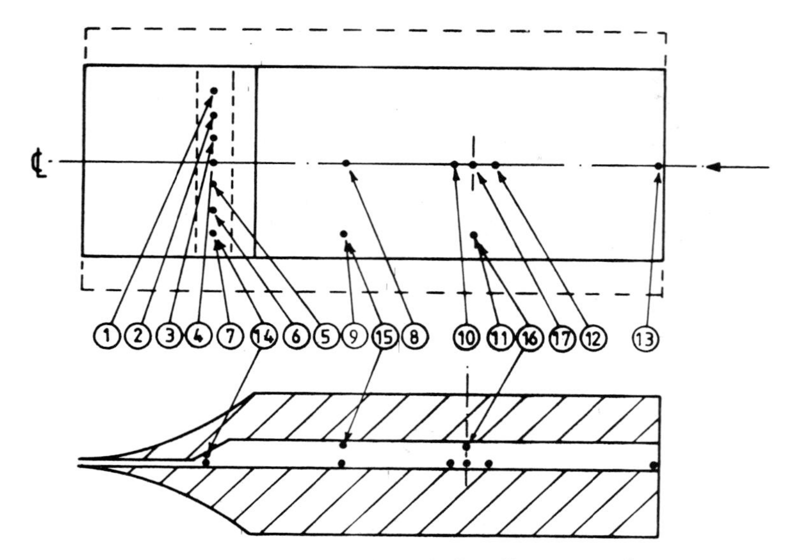

Fig. 7. Thermocouple positions in the roll caster nosetip.

These allow the flow profile at the outlet to be controlled in such a way to simulate uniform removal of solid metal across the width of the tip. Failure to do this results in forced internal recirculation of fluid and a perturbation to the mixing conditions.

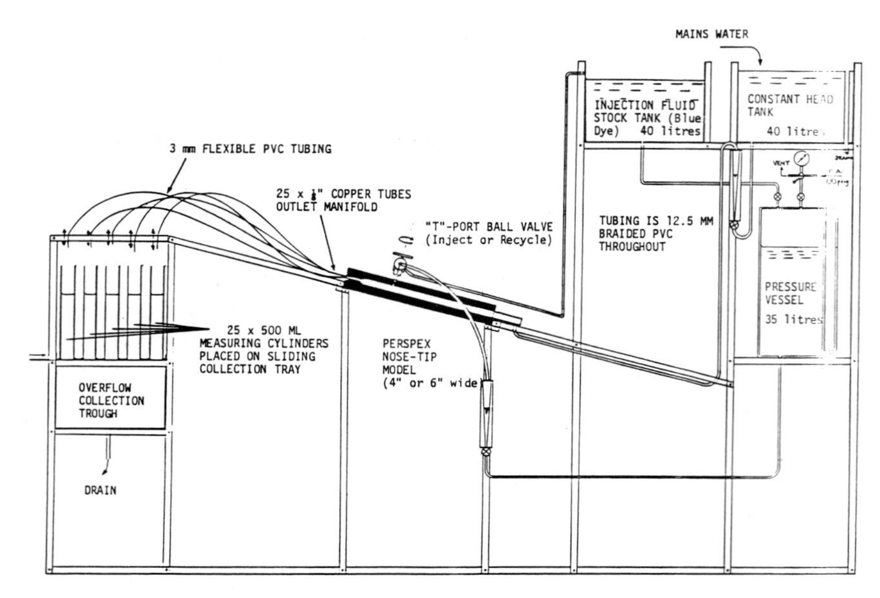

Fig. 8. Experimental apparatus for roll caster nosetip modelling.

To adequately simulate molten flow behaviour with hydraulic-based analogues it is necessary to select the most appropriate physical properties to the problem and convert them into useful dimensionless groups. For aqueous solutions physical property data is well documented and available – the same cannot be said for aluminium and its alloys – particularly molten alloy viscosities and surface tensions. In this study, where physical property data were unavailable, the values for pure aluminium were used. For dilute alloy systems, the errors incurred by this means are kept to a minimum.

Because mixing takes place in an enclosed volume the main criterion of similarity is the Reynolds number. Relative density similarity between main stream and jet was also taken into account.

Water was used for the main flow and in order to simulate the production of Al-Mn alloys in the roll caster, water-brine solutions (NaCl) were used as jet fluid. General flow visualisation was achieved using water soluble food dye added either to the jet fluid before injection, or added by hypodermic syringe to the main stream flow upstream of the injection point. To visualise large coherent structures within the turbulent regions of the flow, varnish grade aluminium flake in water suspension was injected via a syringe. Flow patterns were recorded either on high resolution video or by using a Nikon 35 mm rapid drive (5 frames/second) camera with either continuous illumination for transient development or flash for steady state flow recording.

The essential features of the nosetip design are shown in Fig. 5. The position of the jet inlet is an input variable in the model, and the jet fluid has a velocity two orders of magnitude greater than the main flow. Jet size (diameter), angle of inclination to the main flow and internal geometry of the nosetip are all design variables with the objectives of:

(a) Removing backflow and long residence time zones in the mixing chamber (nosetip) for all cases;
(b) optimising undercooling of jetted material in certain cases;
(c) ensuring uniform and even feeding of the outlet slot from the tip.

The mathematical formulation

In this analysis, only fluid mixing is considered and no solidification or phase change reactions are included. The solidification of a turbulent jet in a co-flowing system with axi-symmetry is described elsewhere [1,2].

The flow is treated as single phase, ith alloying constituents prescribed as inert contaminants in the fluid. Density changes with temperature and concentration are included dynamically. The fluids are assumed to be fully miscible and no jet fragmentation is allowed for. The Reynolds number of the jet is in the order of 10^4 so flow is consequently highly turbulent. This involves splitting the flow into two components; a mean component and a fluctuating component.

$$U_i = \overline{U}_i + u_i \quad i = 1, 3. \tag{1}$$

Here the mean component, $\overline{U}$, is essentially the flow averaged over a time scale such that the fluctuating component, u, is no longer present. It is the calculation of $\overline{U}$ that turbulence models are generally aimed at and it is this approach that is adopted here. Mathematically the equations governing the flow of a (incompressible, for simplicity) Newtonian fluid are the continuity and the Navier Stokes equations:

$$\frac{\partial U_i}{\partial x_i} = 0 \tag{2}$$

$$\frac{\partial U_i}{\partial t} + u_j \frac{\partial U_i}{\partial x_j} = -\frac{1}{\rho}\frac{\partial p}{\partial x_i} + F_i + \nu \nabla^2 U_i \quad i = 1, 3. \tag{3}$$

Here ρ is the density, p the pressure, ν the kinematic viscosity of the fluid and $\boldsymbol{F}$ is any body force (e.g. gravity) that is acting on the fluid. By substituting (1) into (2) and (3), averaging each equation over the turbulent time scale and subtracting the mean equation from the instantaneous equations ((2) and (3)), the following can be obtained, for the mean velocity $\overline{U}$:

$$\frac{\partial \overline{U}_i}{\partial x_i} = 0 \tag{4}$$

$$\frac{\partial \overline{U}_i}{\partial t} + \overline{U}_j \frac{\partial \overline{U}_i}{\partial x_j} = -\frac{1}{\rho}\frac{\partial p}{\partial x_i} + F_i - \frac{\partial}{\partial x_j}\left(\overline{u_i u_j}\right) + \nu \nabla^2 \overline{U}_i \quad i = 1, 3. \tag{5}$$

The third term on the right represents the influence of the fluctuating component on the mean (the bars represent time averaged quantities). The turbulence model used to represent these terms is the well known $k - \epsilon$ model [3] and is summarised as follows:

$$-\overline{u_i u_j} = \nu_t \left(\frac{\partial \overline{U}_i}{\partial x_j} + \frac{\partial \overline{U}_j}{\partial x_i}\right) - \tfrac{2}{3}\delta_{ij} k \tag{6}$$

$$\nu_t = C_\mu \frac{k^2}{\epsilon} \tag{7}$$

$$k = \frac{\overline{u_i u_i}}{2} \tag{8}$$

$$\epsilon = \nu_t \overline{\frac{\partial u_m}{\partial x_k}\frac{\partial u_m}{\partial x_k}} \tag{9}$$

$$\frac{\partial k}{\partial t} + \overline{U}_j \frac{\partial k}{\partial x_j} = P_r - \epsilon + \frac{\partial}{\partial x_j}\left(\frac{\nu_t}{\sigma_k}\frac{\partial k}{\partial x_j}\right) \tag{10}$$

$$\frac{\partial \epsilon}{\partial t} + \overline{U}_j \frac{\partial \epsilon}{\partial x_j} = C_{\epsilon_1}\frac{P_r \epsilon}{k} - C_{\epsilon_2}\frac{\epsilon^2}{k} + \frac{\partial}{\partial x_j}\left(\frac{\nu_t}{\sigma_\epsilon}\frac{\partial \epsilon}{\partial x_j}\right) \tag{11}$$

$$P_r = -\overline{u_i u_j}\frac{\partial \overline{U}_i}{\partial x_j} \tag{12}$$

and the general transport equation for energy and mass (Q for temperature T or concentration C)

$$\frac{\partial Q}{\partial t} + \overline{U}_j \frac{\partial Q}{\partial x_j} = \frac{\partial}{\partial x_j}\left[\left(\frac{\nu_t}{\sigma_t} + \frac{\nu}{\sigma}\right)\frac{\partial Q}{\partial x_j}\right]$$

where σ is the Prandtl (temperature) or Schmidt number (concentration).

In eq. (6) the concept of an isotropic turbulent viscosity, ν_t, is introduced and is defined in terms of k and ϵ in eq. (7). k, the turbulent kinetic energy, and ϵ, the dissipation rate are defined in (8) and (9) respectively. Transport equations for k and ϵ are defined by (10) and (11) and P_r, the production of turbulent kinetic energy, is defined by (12). The semi-empirical constants, C_μ, σ_k, σ_ϵ, $C_{\epsilon_1}V$ and C_{ϵ_2} have been obtained by comparing results of calculations using this model with experiments [3].

It should be stated that there are a multiplicity of methods [4,5,6] for looking at turbulent flow and each claims its own advantages. The $k-\epsilon$ model used here is widely used in engineering problems [7] however it is being replaced at research level. Although the mean component of turbulence is actually solved for, the variables k and ϵ do give an indication of the size of the fluctuating component ($k^{1/2}$), the turbulent viscosity (k^2/ϵ) and the length scale of the turbulence ($k^{1/2}/\epsilon$).

Because the roll caster feed tip is inclined at an angle of 15° to the horizontal (in order to achieve initial filling) it is important that buoyancy (gravity) effects due to density differences are taken into account. The gravity term included as a source in the Navier Stokes equation is the Boussinesq term $(\rho - \rho_0)g$ where ρ_0 is a reference density and ρ is the following function of temperature (T) and composition (c).

$$\rho_1 = \rho_1(1 - {}_\alpha(T - T_0))(1 - c) + \rho_2 c.$$

Here ρ_1, α and T_0 are the density, thermal expansion coefficient and melting point of pure aluminium and ρ_2 is the density of the alloying constituent.

Boundary conditions

(1) At the entrance to the rolls, where in practice the liquid metal will solidify and hence exit at constant speed, it will be assumed that there is fixed flux of mass exiting the system. The value of this flux being determined by the velocity of the rolls multiplied by the density of the (assumed uniform) product. All other variables (e.g. temperature) will have zero gradient.

(2) At the jet position, it will be assumed that there is a given mass inflow with prescribed temperature, velocity, concentration, turbulent kinetic energy and its dissipation rate. The turbulent kinetic energy at this inlet is assumed to be 0.1% of the inlet velocity squared and its dissipation rate is given by the

following:

$$\epsilon = 0.1643 \frac{k^{1.5}}{L}.$$

Here L is the turbulent length scale and is taken to be 7.5% of the jet diameter This data is based on known data for free jets.

(3) At the main inlet to nosetip, it is not known, a priori, what the velocity conditions are as this boundary adjusts itself to match the flow prescribed at the other boundaries. Also, as can be seen later, backflow through this boundary can occur and hence the boundary condition at this boundary will be one of uniform external pressure. Also, at positions of inflow on this boundary, it will be assumed that material flows in with prescribed temperature, velocity, concentration, turbulent kinetic energy and its dissipation rate.

(4) At all other boundaries (walls) the velocity is zero and the gradient of the other variables is zero.

For turbulent flows, this involves the use of wall functions (see [3] for details) which relate the value of the variables at the near wall node to the condition on the boundary. This is necessary because it is impractical to computationally model to viscous sublayer that occurs near a wall.

The governing equations have been solved using the PHOENICS code.

Results and discussion

The modelling approach is exemplified by consideration of two metallurgical cases. The first is the production of an aluminium 2.5 wt% manganese alloy by the mixing of Al-10 Mn with pure aluminium. The process is represented by reference to the relevant phase diagram in Fig. 9(a). It can be seen that this system is essentially an IDEAL MIXING case. The second example compares the difference in local behaviour between Al-Mn and Al-Zr alloys. The Al-Zr system is shown in Fig. 9(b) and because of the different shape of the phase

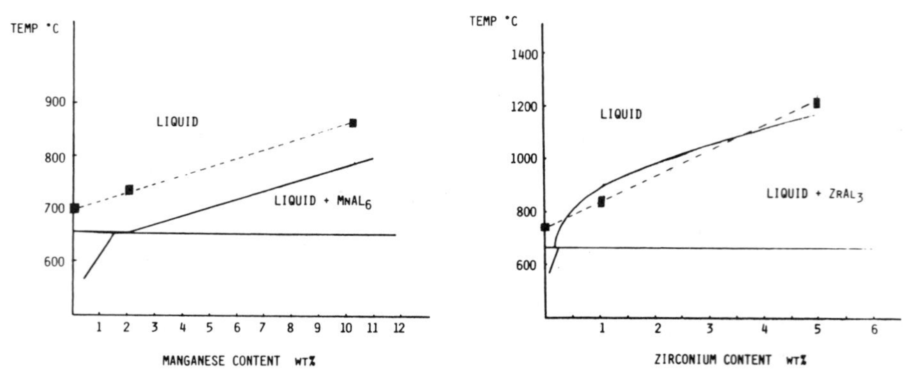

Fig. 9. Solidification track through (a) Al-Mn (b) Al-Zr phase diagram.

diagram compared to Al-Mn this alloys represents a case approaching IDEAL QUENCHING.

CASE 1. Injection of Al-10% Mn into high purity aluminium to form an Al-2.5% Mn alloy

Despite a controlled injection rate of feedstock the average composition of the product is only 1.4% Mn suggesting severe Mn rich liquid build-up or backflow in the nosetip. This was confirmed by simulation in the mathematical model as shown in Fig. 10. In order to gain a more quantitative understanding of the phenomenon the case was simulated in the physical model using the analogue conditions. The sequence of photographs shown in Fig. 11 clearly shows the transient development of recirculation zones on the upstream side of the injection point which grow a steady state position after some $3\frac{1}{2}$ secs of injected flow. This is followed by a tongue of dense fluid dropping out of the back of the zone and backflowing to the tundish with a measured interface velocity of 6.5 mm/s.

A transient mathematical simulation of this case show in Fig. 12 a similar effect and the development of back flow near the jet is shown in Fig. 12.

Considering again the simulation of the metal trials, Fig. 13 compares mathematical and metallurgical data for transient metal temperature distributions over the first 50 seconds of the run.

The greatest deviations occur in the near jet field and it is probable that these are caused by non-optimised empirical constants in the $k-\epsilon$ model and the absence of any fragmentation and jet break-up effects in the current calculations. It has been shown that such events can significantly influence the temperature fields through local latent heat and turbulent transport phenomena [1,2].

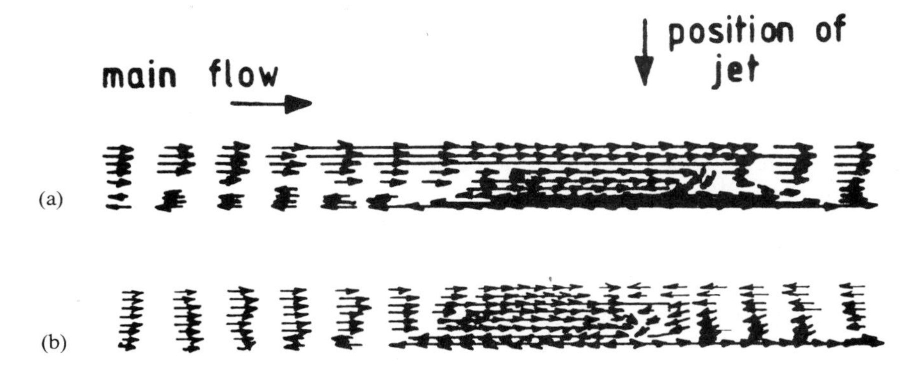

Fig. 10. Effect of gravity on back flow: (a) angled at 15° to horizontal; (b) angled at 0° to horizontal.

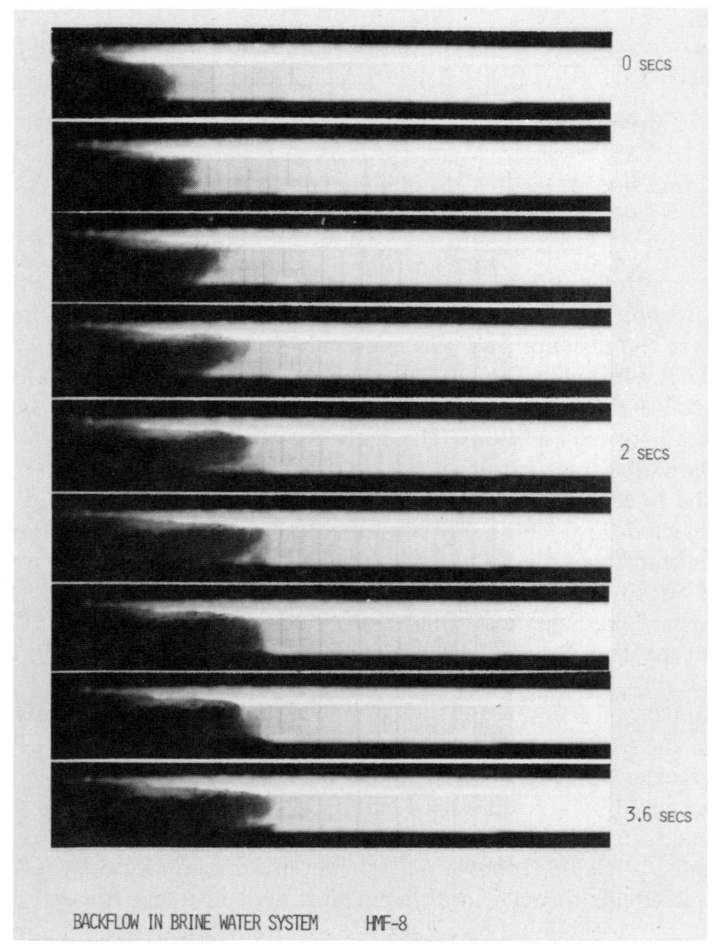

Fig. 11. Development of back flow in the physical model.

Observations from the water model reveal large recirculation zones on the downstream side of the injection point caused by the strong entrainment effects of the high velocity jet, and the presence of the narrow flow restricting channel at the outlet of the tip, but these are not expected to influence (metal) temperature profiles in the near jet field directly because of the high degree of turbulent diffusion in this region and the high thermal conductivity of liquid metal.

The PHOENICS vector patterns obtained for the metal case shown in Fig. 14, support the water model predictions and the concentration and temperature calculations show complete mixing within a short distance downstream of the jet inlet.

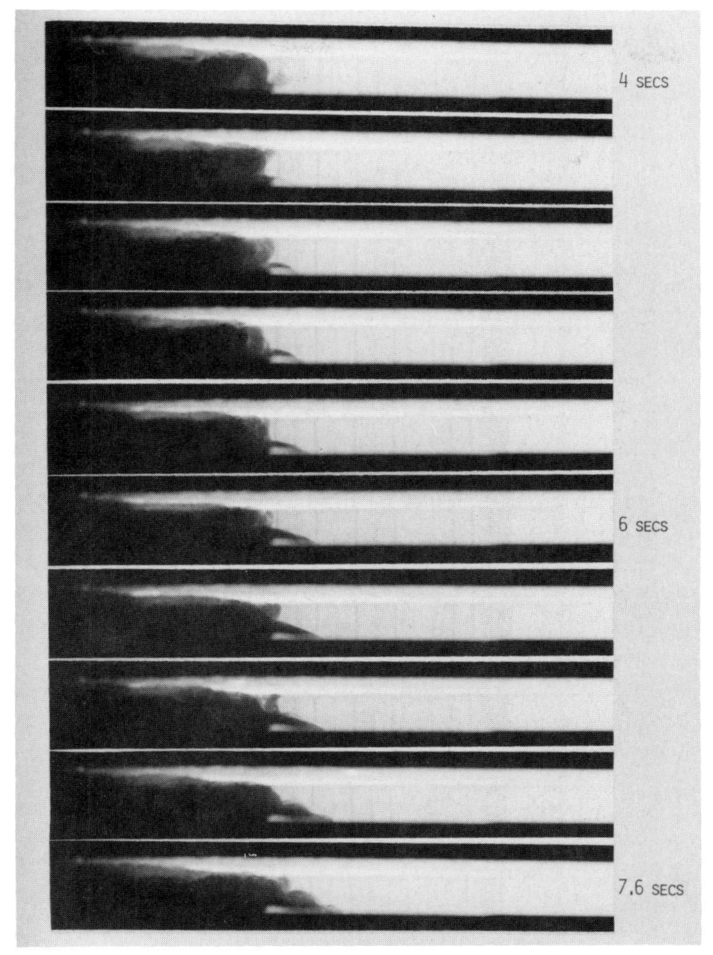

Fig. 11(b)

At the nosetip outlet both metal data (Table 1) and mathematical predictions are in close agreement and show essentially a uniform temperature profile across the width of the roll caster.

By iterating between qualitative water model data and quantitative PHOENICS predictions for metal systems an improved geometry was established in which both recirculation and backflow are eliminated. The design of this tip is given in Fig. 15. Essentially, the main stream fluid has been increased in velocity by constriction of the flow channel. To reduce pressure losses in the system a venturi type restriction is used and the hold-up volume in the near jet region is reduced to a minimum using an inclined base plate. This increases local turbulent kinetic energy but also gives forward thrust to

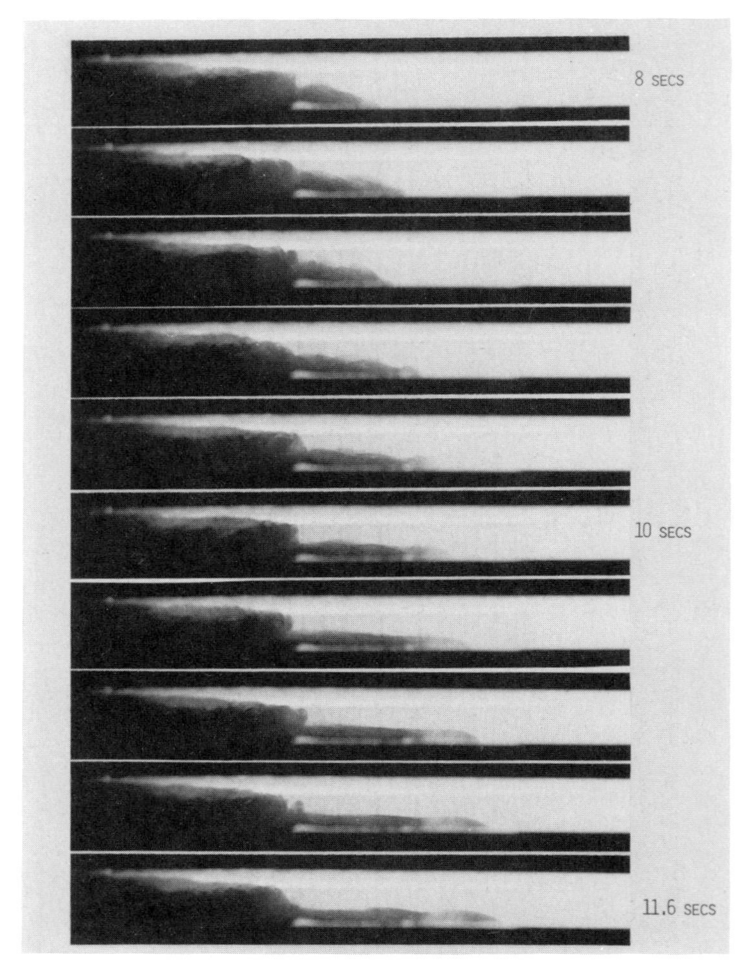

Fig. 11(c)

the jet and overcomes tendency to back-flow. A top surface tapered plate and 'filled in' bottom surface have been produced on the downstream side of the jet to increase flow rate, minimise recirculation and reduce hold-up volume. Also indicated in Fig. 15 are the water model flow patterns obtained. Experimental trials with the modification nosetip on the twin roll caster showed a uniform casting temperature and concentration. The composition of the final product was 2.5% Mn indicating that no back flow had occurred and compositional differences between top and bottom of the cast slab were less than 0.2% Mn which is an improvement over the conventional casting practice.

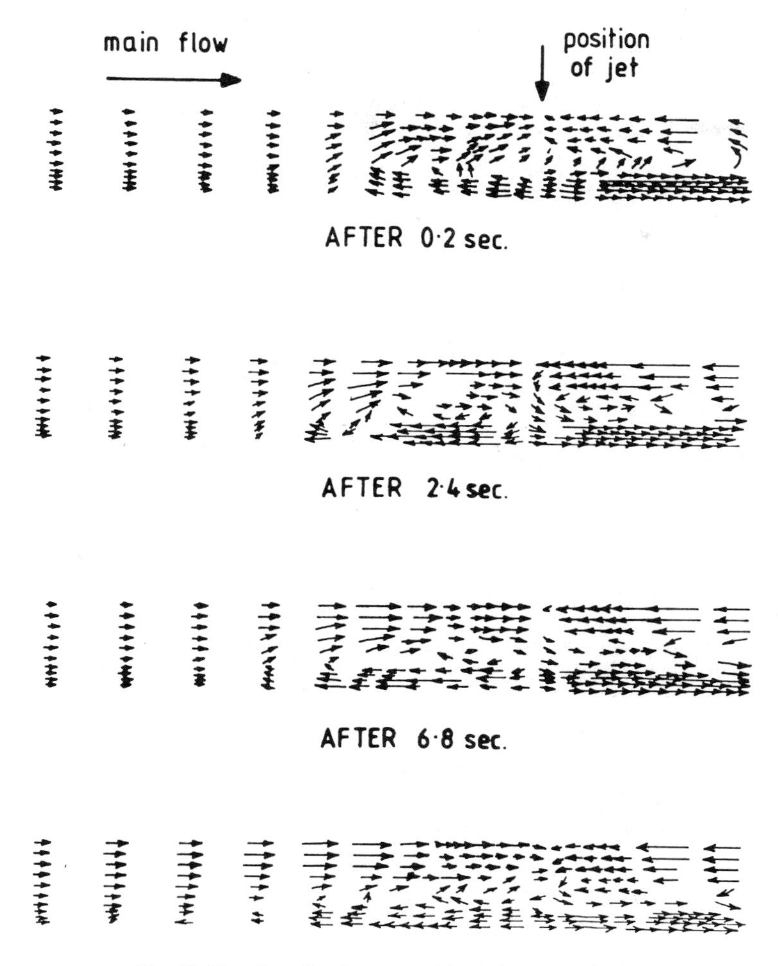

Fig. 12. Transient development of back flow near the jet.

CASE 2. Comparison of undercooling (solidification potential) in Al-Mn and Al-Zr injection systems

The velocity vectors are qualitatively the same for both cases showing the same recirculation patterns, (see Fig. 14). The temperature and concentration contours are similar and with non-dimensionalising they would be identical. This is a result of the dominating influence of turbulent diffusion at the velocity considered although energy is dissipated faster than the concentration due to the high thermal conductivity.

Consideration of local undercoolings however, reveals the metallurgical difference between the processing of these two systems. Undercooling is defined here as the amount (in degrees centigrade) any particular point in the

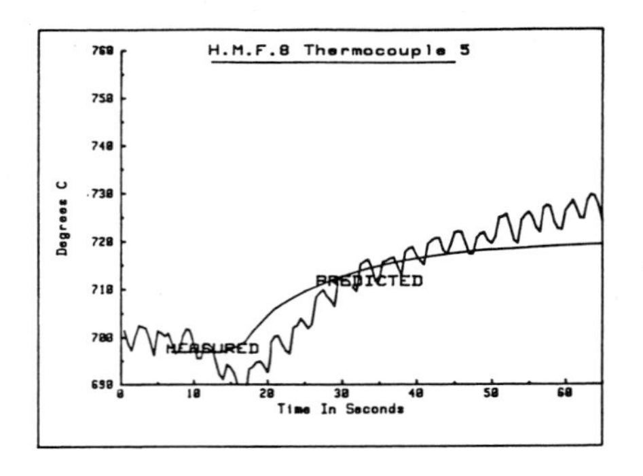

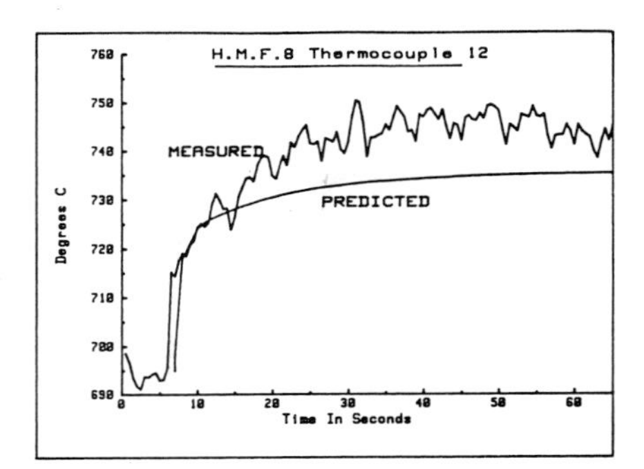

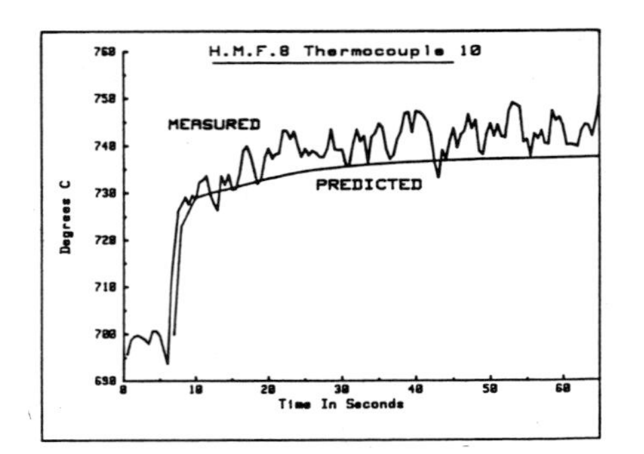

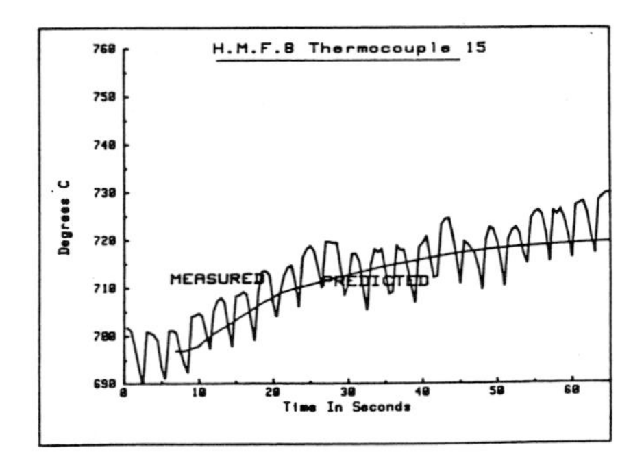

Fig. 13. Comparison between measured and predicted temperatures.

Table 1. HMF-3 roll caster nosetip temperature data.

Thermocouple	Experimental temperatures/°C					Maths model predictions
	At start (pre-injection)	Peak	Low	Range	Average	
1	–	721	720	1	720	721
2	698	721	718	3	720	720
3	–	724	720	4	722	721
4	703	734	728	6	731	721
5	702	738	722	11	731	734
6	707	728	722	6	724	722
7	709	721	715	6	715	708
8	–	726	715	11	715	715
9	708	728	724	4	726	720
10	705	722	717	5	720	715

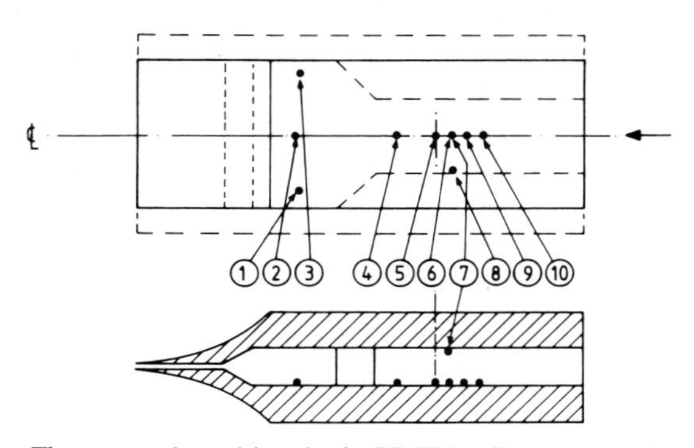

Thermocouple positions in the HMF-3 roll caster nose tip.

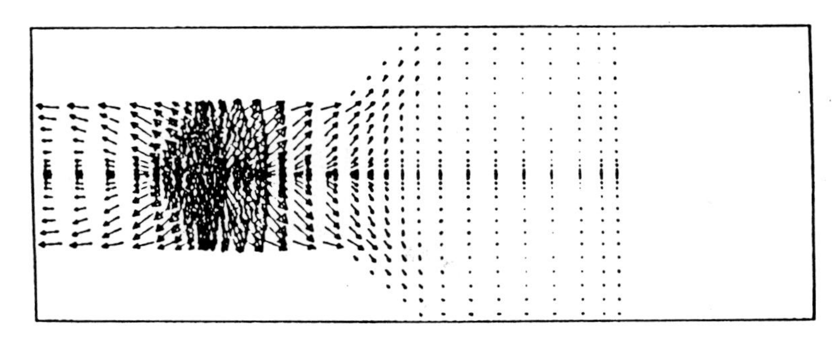

PLAN VIEW – BOTTOM PLANE

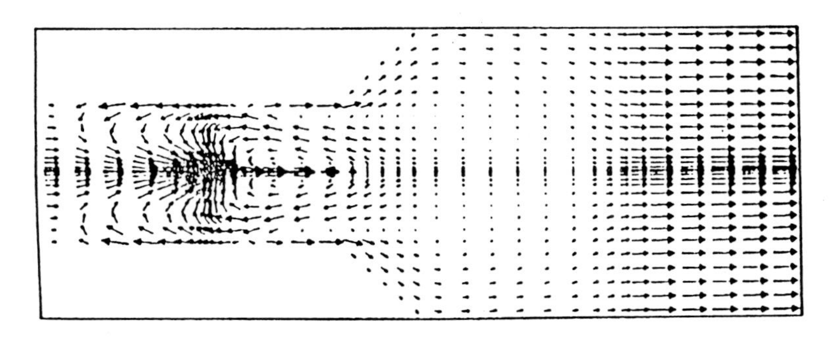

PLAN VIEW – MIDDLE PLANE

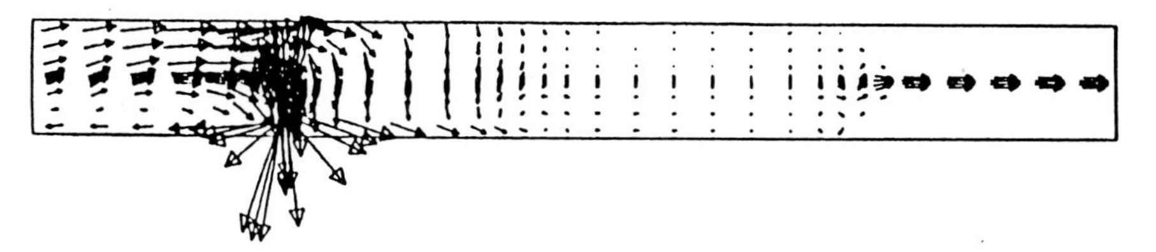

SIDE VIEW – NEAR CENTRAL PLANE

Fig. 14. Velocity vectors in the nosetip.

mixing zone lines below the local liquidus temperature. The results for the bottom lane of the nosetip are shown in Fig. 16. Whereas the Al-Zr system exhibits a considerable degree of undercooling, (and hence possesses a local solidification potential), the Al-Mn system shows no undercooling at all and will therefore simply mix with no solidification reaction.

Solidification of the Al-Zr system has been modelled separately as a 2-D, two phase, steady state problem [1,2], but this model has not yet been incorporated into the full 3-D simulations.

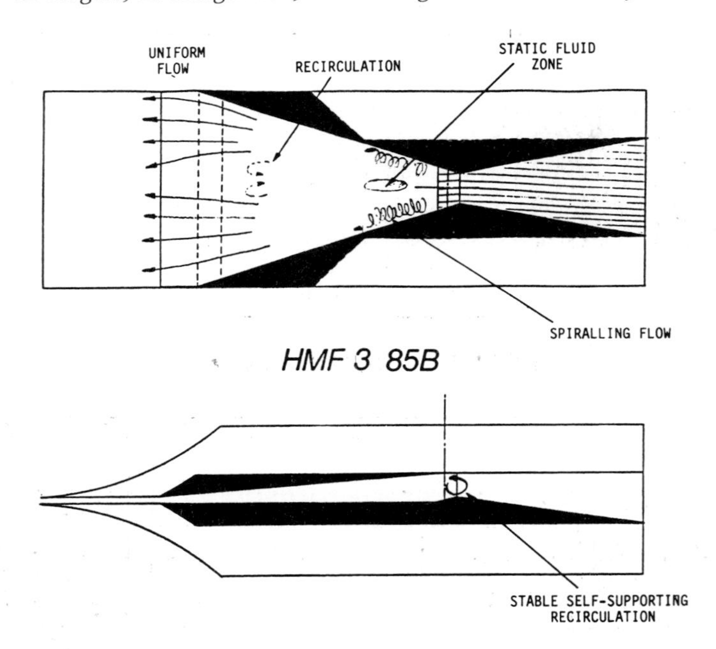

Fig. 15. Schematic of modified nosetip.

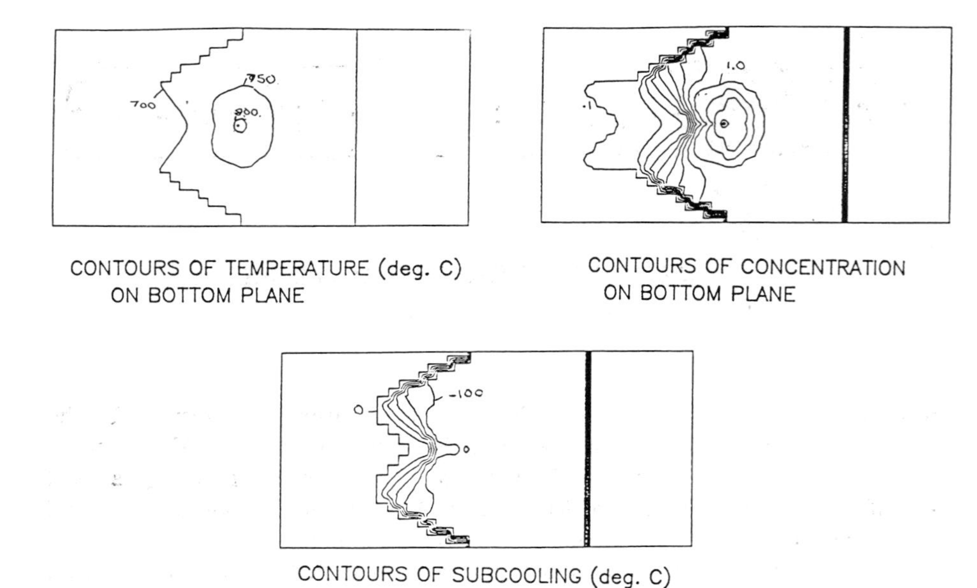

Fig. 16. Contour maps along bottom plane of roll caster nosetip.

Conclusions

Interactive use of a physical and mathematical model of a twin roll caster have lead to the design of a modified liquid metal distributor (nosetip) suitable for in-situ mixing of two different liquid metals during casting. Subsequent metallurgical trials have shown that the new design removes any internal backflow of unmixed fluid and results in uniform temperature and composition distribution at the outlet.

Simulation of the local mixing behaviour of Al-Mn and Al-Zr alloys has shown that although the flow patterns for the two fluid systems are qualitatively the same, a high degree of undercooling is obtained in the Al-Zr case. This means that a high local solidification potential exists within the mixing zone and should be modelled explicitly.

Acknowledgements

This work was carried out as part of the advanced materials programme at Alcan International Limited: Banbury. The authors would like to acknowledge Mr C.E. Jeffs for fabrication of essential equipment, Mr I.R. Hughes for the quantitative metal data and analysis.

Collaboration with Dr. J.C. Ludwig (CHAM Limited) in the first stages of the programme was greatly appreciated.

References

1. P.G. Enright, J.C. Ludwig, S. Rogers and L. Katgerman: Mixing and Solidification of a turbulent liquid metal jet. In: N.C. Markatos, D.G. Tatchel, M. Cross and N. Rhodes (eds) *Numerical Simulation of Fluid Flow and Heat/Mass Transfer Processes.* Springer Verlag, Berlin, (1986), pp. 397–407.
2. P.G. Enright, L. Katgerman, J.C. Ludwig and S. Rogers: Mixing and Solidification of a turbulent liquid jet in a co-flowing stream. *Int. J. for Numerical Methods in Engineering* (in press).
3. B.E. Launder and D.B. Spalding: The numerical computation of turbulent flows. *Comp-Meth Appl. Mech. Eng.* 3 (1974) 269–289.
4. J.L. Lumley: prediction methods for turbulent flows VKI. Lecture Series 76 (1975).
5. A. Leonard: Review of vortex dynamics for flow simulation. *J. Comp. Physics* 37 (1980) 289.
6. P. Bradshaw: Heat and Mass Transfer, Vol. 12. *Topics in Applied Physics: Turbulence*. Springer Verlag, Heidelberg.
7. W. Rodi: Turbulence models and their applications in hydraulics. *IAHR* (1984).

Applied Scientific Research 44: 197–207 (1987)

Mathematical modelling of bubble driven flows in metallurgical processes

S.T. JOHANSEN [1], F. BOYSAN [2] & W.H. AYERS [2]
[1] *Norwegian Institute of Technology, Department of Metallurgy, Trondheim, Norway;*
[2] *University of Sheffield, Department of Chemical Engineering and Fuel Technology, Sheffield S1 3JD, UK*

Abstract. The complex fluid dynamics of two-phase bubbly flows in metallurgical reactors is modelled numerically by using a $k-e$ turbulence model for the liquid phase, with a driving force determined by considering the motion of the bubbles. The latter are affected by the buoyancy forces and the drag caused by their relative motion with the mean and turbulent motions of the liquid, the turbulent component being obtained by random sampling to give an ensemble of bubble trajectories. The two-way coupling between the two phases is resolved by an iterative procedure which converges on a stable overall solution. The results are compared with measurements carried out on an air–water model and show good overall agreement.

Introduction

In recent years, gas-stirring in metallurgical reactions has become increasingly popular due to its effectiveness and low cost. The gas can be introduced into a bath of molten metal either through porous plugs, or using nozzles or lances. For flow rates below 300 kg/m^2-s and bottom injection, the resulting flow field is classified as a bubbling jet [1]. In this flow regime, large unstable bubbles break up into smaller ones shortly after detachment leading to a plume-like two-phase flow region with high voidage. The bubbles which rise due to gravity exchange momentum with their surroundings, which in turn induce a large toroidal vortex in the melt due to the confinement of the flow.

The metallurgical events which then take place in the reactor are largely governed by the velocity and turbulence fields. This is exemplified by the deposition and re-entrainment of inclusions, as well as refractory wear, which are controlled by the fluid dynamics, while homogenization and chemical reaction rates depend on the structure of turbulence. In circumstances where the reactor is operated on a continuous basis, the performance is governed by the degree of stirring, agitation and residence time. Short-circuiting is therefore to be avoided when dissolved species are to be removed from the melt by inert gas injection. When bubbles contain reactive gases, it is important to avoid high local void fractions otherwise chemical equilibrium is difficult to attain and gas is wasted.

In spite of the important role that the fluid dynamics plays in gas-stirred reactors, the analysis of the two-phase flow field from first principles has so far been only partially successful. Previous studies on this subject have mainly been directed towards obtaining the field distributions of velocity components and turbulence quantities in the melt without attempting to model the two-phase region. Szekely et al. [2] have assumed that the bubble column was cylindrical in shape and imposed measured values of velocities as boundary conditions on its periphery. Although the authors could obtain reasonable qualitative agreement between the calculations and the measured data, unsatisfactory quantitative agreement was reported. The importance of the void fraction distribution in the two-phase region in the prediction of the melt velocity distributions has been emphasized in recent studies by Guthrie and co-workers [3], who developed an elaborate phenomenological model of the bubble column, and by He Qinglin et al. [4] who used the measured void fraction and inter-phase force distributions in the governing equations of the conservation of mass and momentum in the liquid phase.

Although semi-empirical models of the two-phase region can lead to reliable predictions of the flow field in the melt for a given configuration and operating conditions, such models conspicuously lack generality and hence cannot be used to analyse a new configuration or operating point with any confidence. The key to the development of a general model of gas-stirring therefore lies in a detailed representation of the motion of the gas bubbles and their interaction with the liquid phase.

The present study addresses itself to the development of such a mathematical model, which differs from earlier contributions in attempting to model the motion of the bubbles, as well as that of the liquid, from first principles. The sets of partial differential equations so obtained are solved numerically using a minicomputer, and the results of these calculations are compared with both axisymmetric and asymmetric bubble jet configurations which have been studied experimentally.

Liquid phase equations

The equations which describe the motion of the liquid phase are the time averaged balances of mass and momentum which will be given here in compact tensor notation for brevity.

Continuity

$$\frac{\partial}{\partial x_i}\alpha_l \rho_l u_i = 0. \tag{1}$$

Momentum conservation

$$\frac{\partial}{\partial x_i}\alpha_l \rho_l u_i u_j = -\alpha_l \frac{\partial p}{\partial x_j} + \frac{\partial}{\partial x_i}\alpha_l \mu_l \left(\frac{\partial u_i}{\partial x_j} + \frac{\partial u_j}{\partial x_i}\right) - \frac{\partial}{\partial x_i}\alpha_l \rho_l \langle u_i' u_j' \rangle + F_j. \tag{2}$$

Where, u_j and u'_j are the mean and fluctuating parts of the velocity component in the direction x_j, p is the pressure, ρ_l is the density, μ_l is the viscosity and α_l is the liquid volume fraction. The j-component of momentum exchange between the gas and the liquid is contained in the term F_j. These equations are obtained from the Navier-Stokes equations by velocity decomposition and time averaging (denoted by angled brackets) assuming that the fluctuations in volume fraction can be neglected. They cannot be solved (numberically or otherwise) unless the pair correlations of the fluctuations are related to known or calculable quantities by a turbulence model, and the distribution of the inter-phase force F_j is provided in some way.

Many models of turbulence of varying complexity have been proposed in the past which range from the simple mixing length model to more sophisticated second order closures and large eddy simulations. The widely used k–ϵ model [5] provides a compromise between these two extremes which uses as its starting point a Boussinesq type of relationship between the Reynolds stresses and the rate of mean strain:

$$\rho_l \langle u'_i u'_j \rangle = \tfrac{2}{3} \rho_l k \delta_{ij} - \mu_t \left(\frac{\partial u_i}{\partial x_1} + \frac{\partial u_j}{\partial x_i} \right). \tag{3}$$

Where k is the kinetic energy of the fluctuating motion, μ_t is a turbulent viscosity which is a property of the particular flow situation rather than of the fluid and δ is the kronecker delta. The spatial distribution of μ_t is related to the kinetic energy of turbulence k and its rate of dissipation ϵ through

$$\mu_t = \rho_l C_\mu \frac{k^2}{\epsilon}. \tag{4}$$

Where C_μ is a constant. The fields of k and ϵ are provided by the solution of the following transport equations:

$$\frac{\partial}{\partial x_i} \alpha_l \rho_l u_i k = \frac{\partial}{\partial x_i} \alpha_l \mu_t \frac{\partial k}{\partial x_i} + \alpha_l \rho_l \left(-\langle u'_i u'_j \rangle \frac{\partial u_i}{\partial x_i} - \epsilon \right) \tag{5}$$

and

$$\frac{\partial}{\partial x_i} \alpha_l \rho_l u_i \epsilon = \frac{\partial}{\partial x_i} \alpha_l \frac{\mu_t}{\sigma_\epsilon} \frac{\partial \epsilon}{\partial x_i} + \alpha_l \rho_l \left(-C_1 \langle u'_i u'_j \rangle \frac{\partial u_i}{\partial x_j} - C_2 \frac{\epsilon^2}{k} \right). \tag{6}$$

Where the constants C_μ, C_1 and C_2 and σ_ϵ are given the values

$$C_\mu = 0.09, \quad C_1 = 1.44, \quad C = 1.92 \quad \text{and } \sigma_\epsilon = 1.3.$$

The above model is probably only a crude approximation to bubbly flow because the turbulence is scaled on the gradients of the mean flow rather than on the bubble size, and is employed here in the absence of anything more suitable.

The motion of the gas bubbles

Two different approaches are currently available for the analysis of the behaviour of a dispersed phase in turbulent flows. These are termed the continuum (Eulerian) and discrete (Lagrangian) methods. In the continuum approach the problem is formulated in terms of mass and momentum conservation equations for each phase in an Eulerian reference frame. In the discrete method on the other hand, the trajectories of individual bubbles are tracked in time by solving ordinary differential equations and the momentum interchange between the phase is accounted for by recording what is gained or lost by the bubbles as they pass through the liquid and using this information in the equations of the continuous phase.

Due to the overwhelming advantages offered by the Lagrangian method in terms of simplicity of formulation, ability to accommodate complicated exchange processes and computational effort, it has been adopted in the present study.

The rate of change of velocity of a discrete bubble with respect to time can be expressed as [6].

$$\frac{\mathrm{d}V_i}{\mathrm{d}t} = -\frac{3}{4}\frac{\mu_l}{\rho_g d_g^2} C_D R_e (V_i - U_i) + \frac{\rho_l}{\rho_g}\frac{DU_i}{Dt} - \frac{1}{2}\frac{\rho_l}{\rho_g}\left(\frac{\mathrm{d}V_i}{\mathrm{d}t} - \frac{DU_i}{Dt}\right) + \left(1 - \frac{\rho_l}{\rho_g}\right) g_i . \tag{7}$$

Where V_i and U_i are the instantaneous components of bubble and liquid velocities respectively in the i-direction, t is the time, ρ_g is the gas density, d_g is the bubble diameter,

$$R_e = \rho_l d_g \,|V - U|/\mu_l$$

is the relative Reynolds number and C_D is the drag coefficient which is empirically determined. The above equation is supplemented by the following kinematic relationship which defines the bubble trajectory

$$\mathrm{d}x_i/\mathrm{d}t = V_i . \tag{8}$$

The instantaneous liquid velocity is evaluated by decomposing it into a mean and a fluctuating part. The spatial distributions of the mean velocity and the r.m.s. fluctuations are obtained from the liquid phase equations. The values of the fluctuating velocities u' associated with the particular eddy that the bubble is traversing, are sampled by assuming that the fluctuations are isotropic and that these possess a Gaussian probability distribution:

$$u_i' = \phi\sqrt{\tfrac{2}{3}\,k} \tag{9}$$

where ϕ is a normally distributed random variable [7]. A bubble is assumed to interact with an eddy for a time equal either to the eddy lifetime

$$\tau_e = L_e/\sqrt{\tfrac{2}{3}k} \tag{10}$$

or the bubble transit time

$$\tau_t = -\tau_R \ln(1 - Le/(\tau_R |V - U|)) \quad (11)$$

whichever is the smaller [8].

Here, L_e is the eddy dissipation length scale given by

$$L_e = C_\mu^{3/4} k^{3/2}/\epsilon \quad (12)$$

and τ_R is the bubble relaxation time expressed as

$$\tau_R = \frac{4}{3} \frac{\rho_g d_g^2}{\mu_l} \frac{1}{C_D R_e}. \quad (13)$$

The above ordinary differential equations can be solved in time by simple numerical integration, starting with the given initial values. At the end of each interaction time, a new value of u_i' is sampled using Eq. (9), while u_i is updated at every time step of the integration process.

It is of course impractical to track stochastically every individual bubble released into the system, but only a statistically representative sample is needed. The residence times of bubbles in a network of control volumes superimposed on the reactor can be readily worked out from the trajectory calculations by summing the residence times of all the bubbles which pass through a given control volume ΔV and dividing by the sample size. The void fraction distribution is then obtained from

$$\alpha_g = \frac{Q}{N\Delta V} \sum_{m=1}^{n} t_{R,m}. \quad (14)$$

Where Q is the volumetric flow rate of the gas, n is the number of bubbles which pass through the control volume in question, t_R is the residence time and N is the sample size. The momentum interaction terms F_i in the liquid momentum balances are deduced from similar arguments based on the stipulation that the drag force experienced by the bubbles acts in equal magnitude but opposite direction on the liquid.

$$F_i = \frac{Q}{N\Delta V} \sum_{m=1}^{n} \int_0^{t_{R,m}} \frac{3}{4} \frac{\mu_l}{\rho_g d_g^2} C_D R_e (V_i - U_i)\, dt. \quad (15)$$

The above relations can be easily generalized to take into account a spectrum of bubble sizes or gas injection from several locations.

Boundary conditions and solution procedure

The elliptic nature of the governing equations of the liquid phase requires the specification of the conditions at all the boundaries. In the regions near the solid walls the turbulence model outlined in the preceding sections does not apply. To preclude fine grid calculation the outer solutions are matched by the empirically based log-law of the wall [9]. The values of the turbulence kinetic energy and its dissipation rate at near wall points are also deduced from wall functions. At the free surface, which is assumed to remain flat, all the stresses vanish and zero normal-gradient type conditions apply to all the dependent variables except the component of velocity normal to the surface which is zero.

The gas bubbles enter the solution domain from locations dictated by the particular configuration and at a rate determined by the gas flow rate. These are assumed to be removed from the system once they reach the free surface.

The equations to be solved are made up of 3 equations for the conservation of liquid phase momentum, the continuity equation which is transformed into an equation for pressure correction [10] and the transport equations for the turbulence kinetic energy and its dissipation rate. Coupled with these partial differential equations are the equations of motion of the bubbles which are solved to obtain the void fraction and the momentum interchange between the phases.

The finite difference grid which is superimposed onto the solution domain consists of a set of orthogonal lines in the z, r and θ directions of a cylindrical polar system of co-ordinates best suited to the geometry of the problem in hand. The usual staggered arrangement of the variables is employed where the velocity components are calculated and stored mid-way between the pressure nodes which lie at the intersections of the grid lines. The finite difference analogues of the partial differential equations are obtained by integrating these over a typical control volume which encompasses the point where the value of a particular dependent variable is to be calculated. While central differencing is used for the diffusion terms, a quadratic upstream difference scheme is employed for the convective fluxes in order to be able to minimize false diffusion problems [11].

The set of algebraic simultaneous equations thus obtained are solved iteratively by using a tri-diagonal matrix algorithm. The dependent variables are solved sequentially within each iteration and at regular intervals of the outer iteration loop, a large number of bubble trajectories are tracked stochastically to obtain the field distributions of the void fraction and the momentum exchange terms F_i. This process is continued until the equations for both phases converge to a solution.

Applications

In this section, the mathematical model is applied to the situations studied experimentally by Johansen et al. [12,13] in an air-water system. The physical ladle is shown in Fig. 1 which consists of a slightly conical perspex reactor of free surface height 1.327 m and top and bottom diameters of 1.1 m and 0.93 m respectively. Air is supplied through a porous plug which was located on the base plate, and the flow rate is varied between 1.33×10^{-4} Nm^3/s to 7.5×10^{-4} Nm^3/s.

Fig. 1. The air-water model of the metallurgical ladle.

In the case where the porous plug is located centrally, the problem becomes axially symmetric and all derivatives with respect to the azimuthal direction can be neglected. The calculations in this case were carried out using a 16×16 non-uniform grid in both the axial and radial directions. The conical vessel was approximated by a cylindrical vessel of equal volume. The interphase force field and the void fraction distributions were updated at every 10 iterations of the liquid-phase solution procedure by ensemble averaging 100 stochastic bubble trajectories.

The qualitative feature of the flow field corresponding to a gas flow rate of 6.1×10^{-4} Nm^3/s are shown in the vector plot of Fig. 2. Here the vectors indicate both direction and magnitude. It can be seen that the flow has the appearance of an impinging submerged jet which, because of its entrainment appetite, gives rise to the characteristic toroidal vortex. The eye of the vortex is towards the top of the chamber and close to the vessel walls. The axial velocities are highest in the neighbourhood of the symmetry axis, while the radial velocities attain a comparable magnitude only in the vicinity of the free surface. The measured velocity vectors in the ladle are displayed in Fig. 3 which agree both qualitatively and quantitatively with the predictions.

The shape of a typical bubble column as calculated by the model is presented in Fig. 4. It can be seen that the model predicts the expected conical shape of the bubble column. The average cone angle for the range of bubble diameters between 6 mm and 12 mm was found to be around 15° which agrees fairly well with the experimental values quoted in the literature [3,4].

In addition to the above axi-symmetric calculation a simple 3-D example was also attempted. In this case, the geometry of the ladle remained unaltered

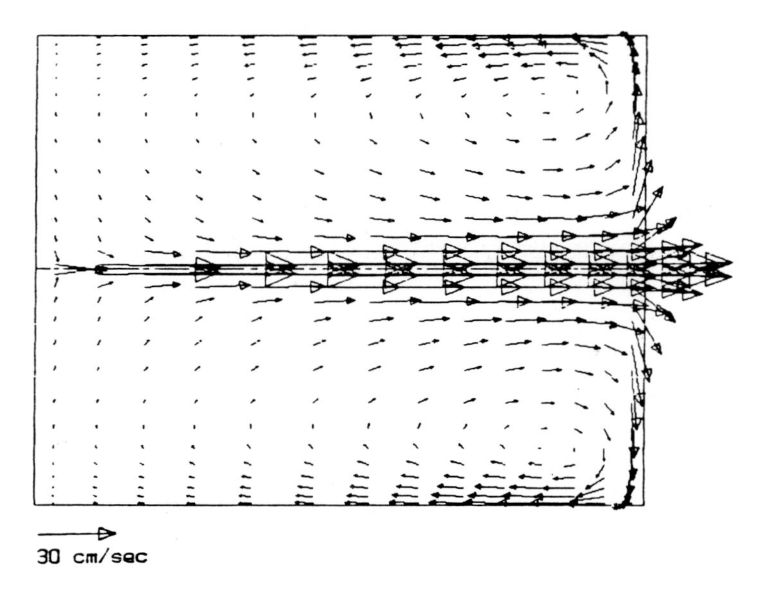

Fig. 2. Calculated velocity vectors in the ladle showing both direction and magnitude.

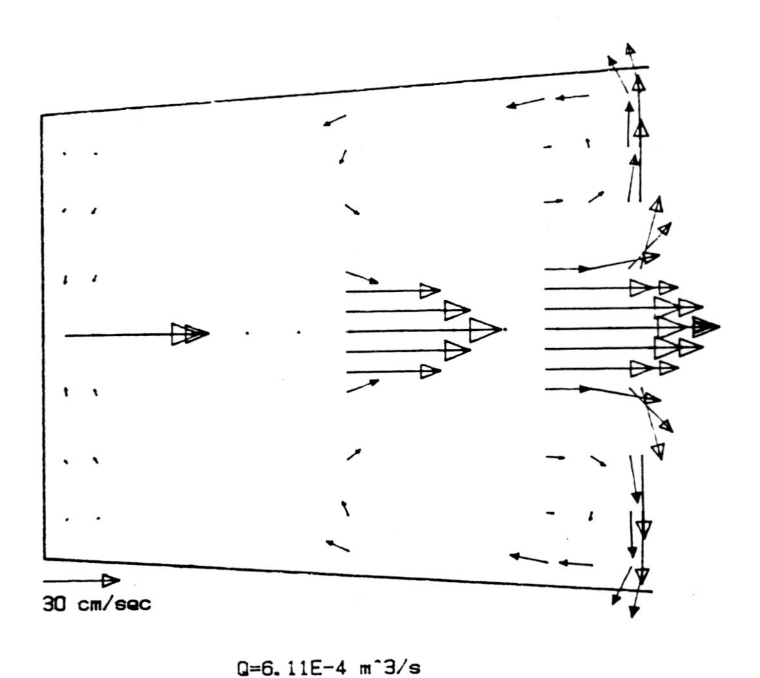

Fig. 3. Measured velocity vectors in the ladle corresponding to a gas flow rate of 6.11×10^{-4} m^3/s.

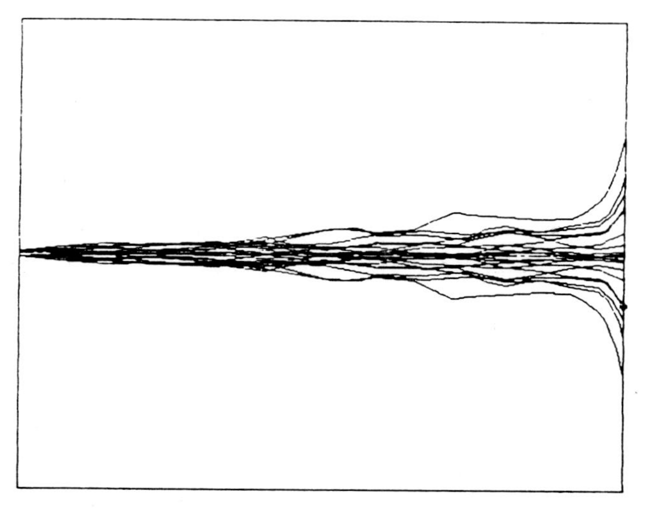

Fig. 4. Typical stochastic trajectories of 6-mm-diameter bubbles.

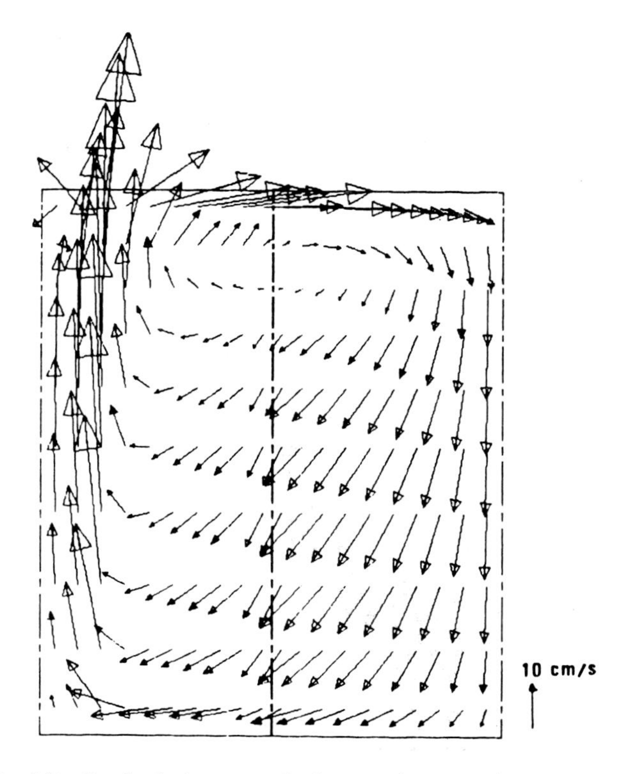

Fig. 5. Predicted velocity vectors in the case of asymmetric gas injection.

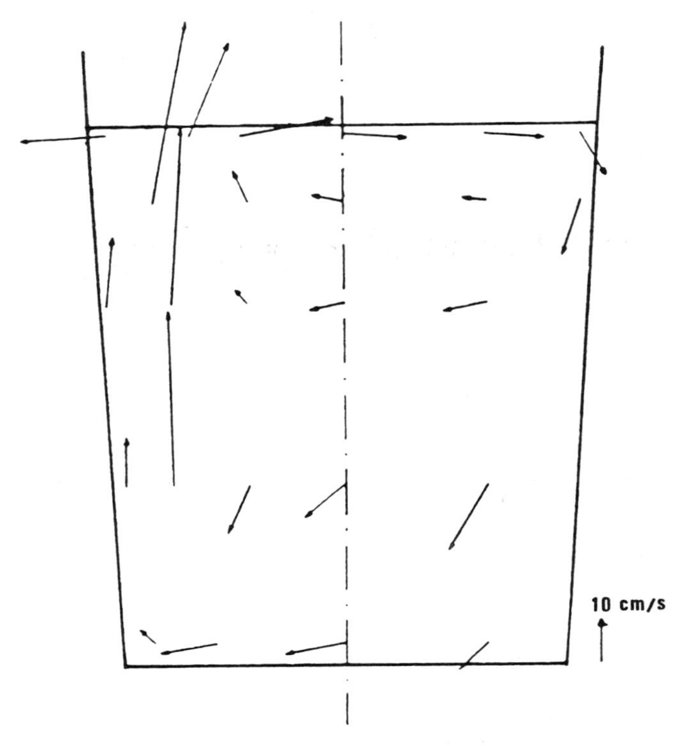

Fig. 6. Measured velocity vectors in the case of asymmetric gas injection for a gas flow rate of 5.5×10^{-4} m^3/s.

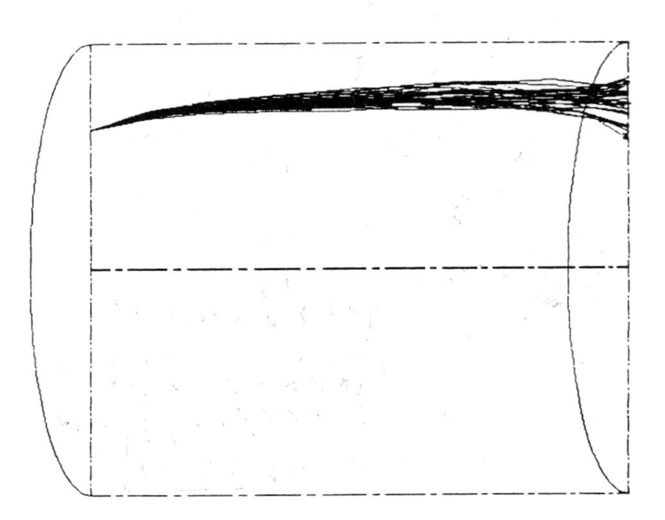

Fig. 7. Calculated trajectories of 6 mm diameter bubbles.

except that the porous plug is now moved from the axis to a position $\frac{2}{3}R$ from it. This renders the problem three dimensional. The calculations were performed on a $12 \times 12 \times 12$ grid in the z, r and Θ directions of the cylindrical polar system of coordinates. Because of the symmetry around the plane which passes through the centerline and the plug location only a $180°$ sector was considered in the asymmetric case. The gas flow rate was 5.5×10^{-4} Nm^3/s.

Figures 5 and 6 show the predicted and measured velocity vectors on the plane passing through the porous plug. It can be noted that in this case there is a single recirculation region the eye of which lies in the upper half of the vessel close to the free surface. The magnitude of the maximum axial velocity in this case was calculated to be 0.5 m/s which compared reasonably well with the measured value of 0.4 m/s.

The predicted shape of the bubble column is displayed in Fig. 7 which shows that it leans towards the adjacent wall and is in full agreement with experimentally observed behaviour displayed in Fig. 1.

Conclusions

A new, more fundamental approach to the mathematical modelling of gas-stirred metallurgical reactors has been described in which the complexities of the two phase bubbling jet flow have been fully addressed. By completing the mathematical picture in this way, a good degree of agreement with experimental results has been obtained, while the fundamental basis of the technique allows us a degree of confidence in applying the results to cases lying outside the compass of the experimental validations.

Work is currently underway to improve the model still further and to provide additional experimental verification under different conditions of operation.

References

1. M.J. McNallan and T.B. King: *Metallurgical Transactions B*, Vol. 13B (1982).
2. J. Szekely, H.J. Wang and K.M. Kiser: *Metallurgical Transactions B*, Vol. 7B (1976).
3. Y. Sahai and S.I.L. Guthrie: *Metallurgical Transactions B*, Vol. 13B (1982).
4. He Qinglin, Pen Yichvan and Hsiao Tse-Chiang: Shenyang Symposium of Injection Metallurgy, September (1984).
5. B.E. Launder and D.B. Spalding: *Mathematical Models of Turbulence*. Academic Press, London (1982).
6. M.R. Maxey and J.J. Riley: *Phys. Fluids* 26 (1983) 883–889.
7. A.D. Gosman and E. Ioannides: *AIAA*, paper No. 81-323 (1981).
8. J.S. Shuen, L.D. Chen and G.M. Faeth: *AICHE Journal* 29 (1983) 167–170.
9. B.E. Launder and D.B. Spalding: *Comp. Meths. Appl. Mech. Eng.* 3 (1974).
10. S.V. Patankar: *Numerical Heat Transfer and Fluid Flow*. Hemisphere, New York (1982).
11. P.G. Huang, B.E. Launder and M.A. Leschziner: UMIST Report No. TFD/83/1 (1983).
12. S.T. Johansen, D. Robertson, K. Woje and T.A. Engh: unpublished work.
13. S.T. Johansen and T.A. Engh: *Scand. J. Metallurgy* 14 (1985) 214–223.

Applied Scientific Research 44: 209–224 (1987)

A simplified model of bubble-driven flow in an axisymmetric container

W. KUNDA & G. POOTS
Centre for Industrial and Applied Mathematics, University of Hull, Hull HU6 7RX, UK

Abstract. A simplified mathematical model simulating a gas bubble agitation system is here examined for the case when the orifice is located at the centre of the base of a cylindrical vessel. The two phase flow which is confined to a cone region is approximated by the drift flux model. The governing equations for the recirculating liquid flow are quai-linearized and the flow domain is transformed by a simple transformation into a cylindrical region. Using standard finite difference techniques numerical solutions are obtained for Reynolds numbers in the range $0\text{–}10^4$; $\mathrm{Re} = \rho U_0 R_0/\mu_{\mathrm{eff}}$, where R_0 is the radius of the vessel, U_0 the velocity of injection of the gas, ρ the density of the liquid phase & μ_{eff} the constant effective turbulent viscosity. For large Re it is shown that the primary recirculating flow is confined to a narrow region adjacent to the two phase/liquid interface.

1. Introduction

Numerous theoretical and experimental investigations of bubble flow systems have been carried out so as to obtain information on multiphase heat and mass and momentum transfer. Particular attention has been given to the important industrial process of stirring of molten steel in a ladle by injection with an inert gas and to the nature of recirculating liquid flow that results, see Szekely et al. [1] and Deb Roy and Majumdar [2]. In these investigations the gas bubble agitation system considered was for the case of an orifice located at the centre of the base of a cylindrical vessel. In the numerical solutions for the liquid recirculating flow conditions at the two phase/liquid interface were taken from experiment. To avoid this formulation Deb Roy et al. [3] developed a mathematical model in which the void fraction was predicted from the gas flow rate. Two idealized cases were examined: the case when gas and liquid move with the same speed and the case when the gas moves at a higher speed than the liquid so allowing for slip. In Grevet et al. [4] a more detailed experimental and theoretical study of gas bubble driven circulation systems was given. Here the two phase region was confined to a cone region whose dimensions are specified by experimental data; see also the theoretical model of Aldham et al. [5], and the recent experimental results on bubble driven laminar flow given by Durst et al. [6].

It is the purpose of this paper to present theoretical results on the liquid recirculating flow for a range of gas flow rates. It is assumed that the two phase flow is confined to a cone region and is approximated by the drift flux model as already employed in the theoretical investigations [3] and [4]. The governing equations for the recirculating liquid flow are quasi-linearized and the flow domain is transformed by a simple transformation into a cylindrical region. Standard central difference techniques are employed to solve the transformed governing equations for increasing Reynolds number; if ρ_0 is the density of the liquid, μ_{eff} the constant effective turbulent viscosity, R_0 the radius of the vessel and U_0 a representative gas velocity near the orifice the Reynolds number is defined as $\mathrm{Re} = \rho_0 U_0 R_0/\mu_{eff}$. The computer algorithm developed here is capable of producing results within the Reynolds number range 0–10^4. This is in contrast to the earlier numerical studies of recirculating flow in a square cavity as reviewed by Olson [7] in which results for various available methods are summarized for Reynolds number of 1 and 400.

2. Mathematical model

A cylindrical polar coordinate system (r, z) is chosen with origin at the centre of the base of the cylindrical vessel of radius R_0 and height H; let (u_r, u_z) be the velocity components in the (r, z) directions. The equations to be solved for the liquid flow are:

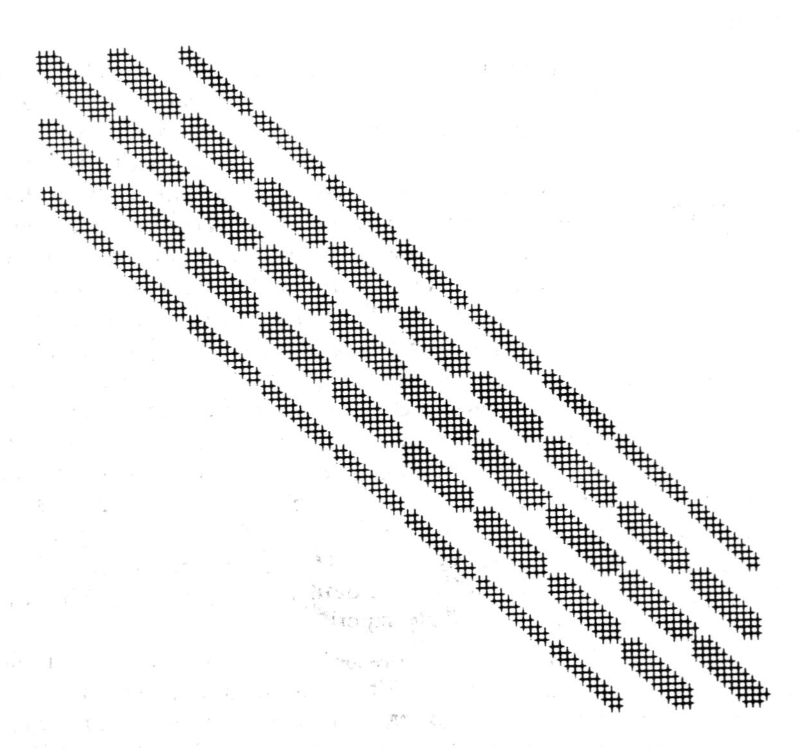

Fig. 1. Structure of the matrix W appearing in equation (25).

Continuity:

$$(1/r)\ \partial(ru_r)/\partial r + \partial u_z/\partial z = 0; \tag{1}$$

Momentum:

z-direction

$$u_r\ \partial u_z/\partial r + u_z\ \partial u_z/\partial z$$
$$= -(1/\rho)\ \partial p/\partial z + (\mu_{\text{eff}}/\rho)$$
$$\times\left[2\partial^2 u_z/\partial z^2 + (1/r)(\partial/\partial r)\{r(\partial u_z/\partial r + \partial u_r/\partial z)\}\right]; \tag{2}$$

r-direction

$$u_r\ \partial u_r/\partial r + u_z\ \partial u_r/\partial z$$
$$= -(1/\rho)\ \partial p/\partial r + (\mu_{\text{eff}}/\rho)$$
$$\times\left[(2/r)\ \partial/\partial r(r\ \partial u_r/\partial r) + \partial^2 u_r/\partial z^2 + \partial^2 u_z/\partial r\ \partial z - 2u_r/r^2\right]; \tag{3}$$

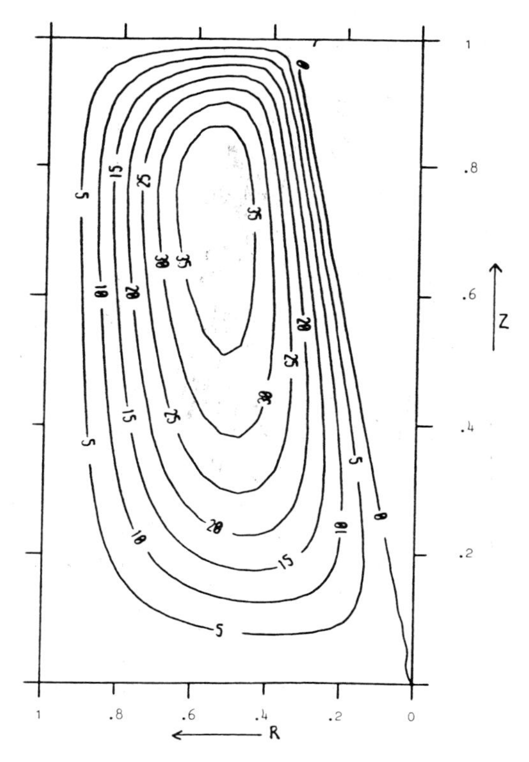

Fig. 2. Streamlines for $\text{Re} = 0$, constant velocity along the cone generator. Streamline value $\times 10^{-3}$.

Boundary conditions:

$$u_r = u_z = 0 \quad \text{on } z = 0, \quad 0 < r \leqslant R_0 \quad \text{and on } r = R_0, \quad 0 \leqslant z \leqslant H; \tag{4}$$

$$u_z = \partial u_r/\partial z = 0 \quad \text{on } z = H, \quad 0 \leqslant r \leqslant R_0; \tag{5}$$

$$u_z = U_0(\text{const}) \quad \text{at } r = 0, \quad z = z_0 \text{ (to be specified)}. \tag{6}$$

It remains now to specify the boundary conditions at the two phase/liquid interface. The two phase flow occurs within the (right circular) cone region

$$0 \leqslant r \leqslant z \tan\lambda, \quad z_0 \leqslant z \leqslant H, \tag{7}$$

where 2λ is the angle of the cone whose vertex is at $r = z = 0$.

It is assumed that the two phase flow at any height $z \geqslant z_0$ has uniform vertical velocity $u_z = U_z(z)$. Following [4] a momentum balance over a volume element in the axial z-direction yields

$$\mathrm{d}(\rho U_z^2 A)/\mathrm{d}z = \mathrm{d}(\rho \bar{\alpha} g A z)/3\mathrm{d}z \tag{8}$$

where $A(z) = \pi(z \tan\lambda)^2$ is the cross-sectional area at height z. The void fraction $\bar{\alpha}$ is computed using the drift flux model. For no slip between the gas

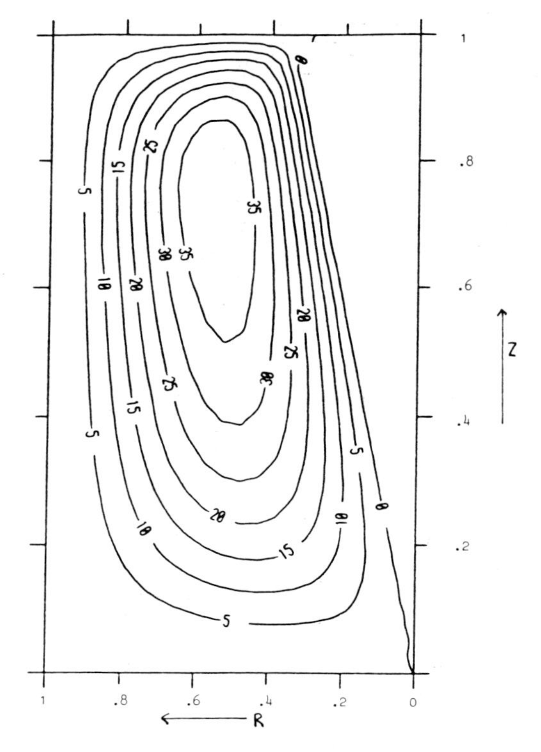

Fig. 3. Streamlines for Re = 1, constant velocity along the cone generator. Streamline value $\times 10^{-3}$.

and liquid phase this yields

$$\bar{\alpha} = Q_g / A U_z, \tag{9}$$

where Q_g is the volumetric flow rate of gas taken at $z = z_0$. The solution of (8) subject to the condition $U_z = U_0$ at $z = z_0$, such that $Q_g = \pi z_0^2 \tan^2\lambda U_0$, is given by

$$U_z/U_0 = \left\{ \left(3\pi g/4U_0^2\right) z_0^2 \left(z^2 - z_0^2\right) + z_0^3 \right\}^{1/3} / z. \tag{10}$$

Thus at any height z in the cone region

$$u_z = U_z(z), \quad u_r = 0. \tag{11}$$

For λ small these lead to the velocity conditions along and normal to a generator of the cone given by

$$U_t = U_z \cos\lambda \simeq U_z(z), \quad U_n = 0. \tag{12}$$

Consequently the model for the recirculating liquid flow is now reduced to a 'belt-driven' model. This simplification greatly reduces the complexity of the

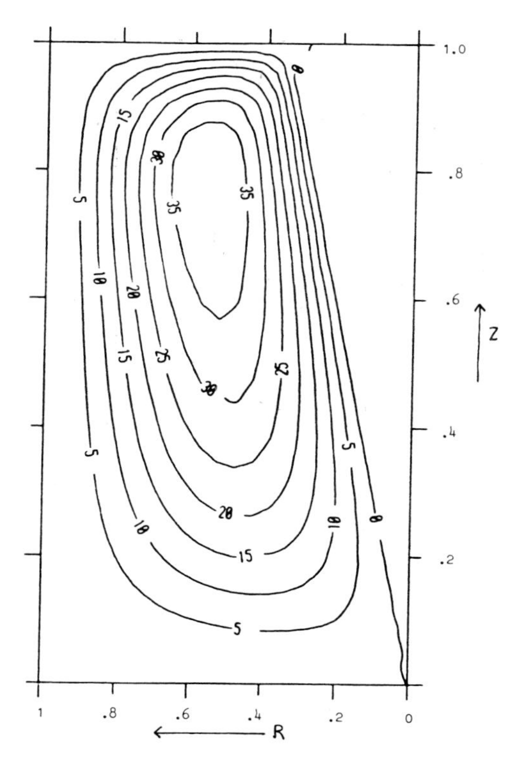

Fig. 4. Streamlines for Re = 10, constant velocity along the cone generator. Streamline value $\times 10^{-3}$.

computation of the liquid flow. Moreover, see Section 3, on using a simple spatial transformation the problem further reduces to a form related to the 'belt-driven' square cavity recirculating liquid flow problem, as reviewed by Olson [7]. Here the belt is a conical surface moving with variable speed along the cone generator; the case of constant speed is also computed for comparison.

Note that the constant effective turbulent viscosity μ_{eff} may be deduced using the ad-hoc viscosity hypothesis of Pun and Spalding [8], see ref. [3].

3. Calculation of the liquid flow

A numerical algorithm is now developed for solving the above governing equations (1)–(7) and (12) for the liquid phase.

Introduce the dimensionless variables

$$R = r/R_0, \quad Z = z/H, \quad \Psi = \psi/R_0^2 U_0, \tag{11}$$

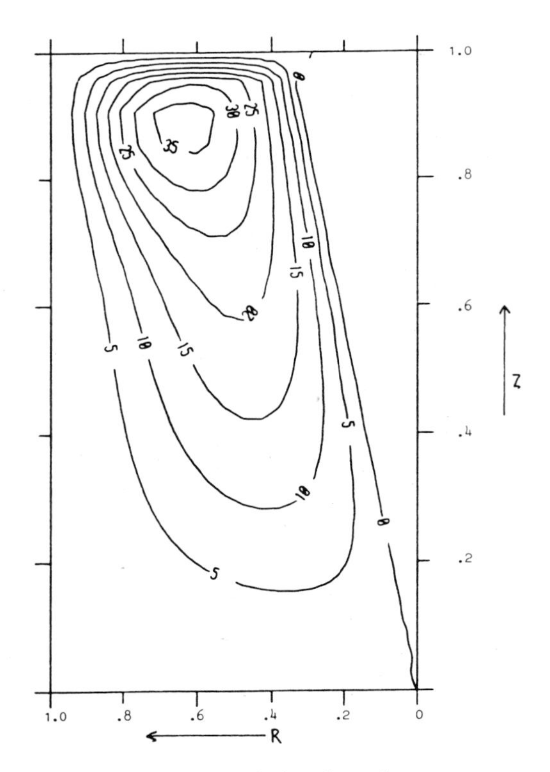

Fig. 5. Streamlines for Re = 100, constant velocity along the cone generator. Streamline value $\times 10^{-3}$.

and the parameter $\gamma = R_0/H$. In the usual fashion the stream function ψ is defined by

$$ru_r = -\partial\psi/\partial z, \quad ru_z = \partial\psi/\partial r. \tag{12}$$

Introduce the co-ordinate transformation

$$\xi = (1 - R)/(1 - Z \tan \lambda/\gamma), \quad \eta = Z. \tag{13}$$

The liquid region between the cone and the vessel walls is then transformed into a square domain $\xi \in [0, 1]$ and $\eta \in [0, 1]$.

The governing equations are quasi-linearized on writing

$$\Psi = \overline{\Psi} + \Psi^*, \tag{14}$$

where $|\Psi^*|/|\overline{\Psi}| \ll 1$. The linearized transformed equations, for $\xi \in [0, 1]$ and $\eta \in [0, 1]$, now become

$$-N\Psi + \mathrm{Re}\left[D_1\overline{\Psi}L\Psi + L\overline{\Psi}D_1\Psi + D_2\overline{\Psi}M\Psi + M\overline{\Psi}D_2\Psi\right]$$
$$= \mathrm{Re}\left[D_1\overline{\Psi}M\Psi + D_2\overline{\Psi}L\Psi\right]. \tag{15}$$

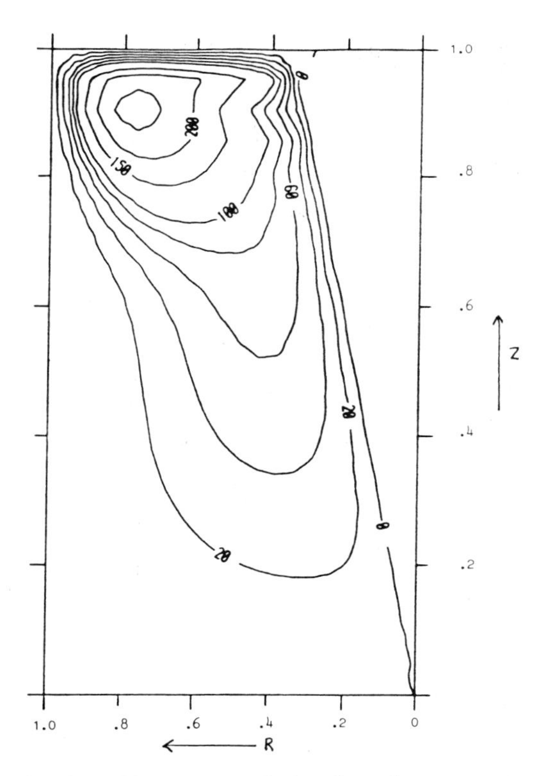

Fig. 6. Streamlines for Re = 500, constant velocity along the cone generator. Streamline value $\times 10^{-4}$.

The differential operators L, M, N, D_1 and D_2 are defined in the (R, Z) plane and transformed into the (ξ, η) plane in the Appendix.

The boundary conditions for (15) are

$$\Psi = \partial\Psi/\partial\eta = 0 \quad \text{on } 0 \leqslant \xi \leqslant 1, \quad \eta = 0; \tag{16}$$

$$\Psi = \partial\Psi/\partial\xi = 0 \quad \text{on } \xi = 0, \quad 0 \leqslant \eta \leqslant 1; \tag{17}$$

$$\Psi = \partial^2\Psi/\partial\eta^2 = 0 \quad \text{on } 0 \leqslant \xi \leqslant 1, \quad \eta = 1; \tag{18}$$

and

$$\psi = 0, \quad \partial\Psi/\partial\xi = \beta(\beta\eta - 1)\left[\left(3\pi\gamma^2 Hg/4U_0^2\right)\left(\eta^2 - k^2\right) + k^3\right]^{1/3}$$
$$\text{on } \xi = 1, \quad 0 \leqslant \eta \leqslant 1, \tag{19}$$

where

$$\beta = -\tan\lambda/\gamma \quad \text{and } k = z_0/H. \tag{20}$$

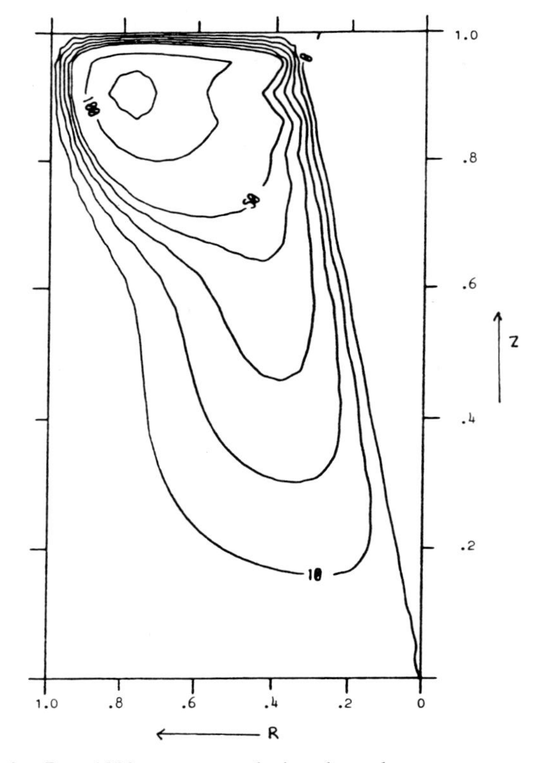

Fig. 7. Streamlines for Re = 1000, constant velocity along the cone generator. Streamline value $\times 10^{-4}$.

In the special case of the tangential velocity along a generator of the cone being taken as U_0 then (19) is replaced by

$$\psi = 0, \quad \partial\psi/\partial\xi = \beta\eta(\beta\eta - 1) \quad \text{on } \xi = 1, \quad 0 \leqslant \eta \leqslant 1. \tag{21}$$

Discretization in the ξ and η directions by means of standard central differences in the square domain $0 \leqslant \xi \leqslant 1$ and $0 \leqslant \eta \leqslant 1$ leads to mn nodes (m in the ξ-direction and n in the η-direction). Of the mn nodes $2(m+n)-4$ are boundary nodes at which $\Psi = 0$. For convenience define the mesh lengths h and k such that

$$h = 1/(m-1), \quad k = 1/(n-1) \tag{22}$$

and hence the axial distance z_0 appearing in (6), (19) and (20) is now specified as $H/(n-1)$; also define the values of Ψ at the nodes by

$$\Psi_{i,j} = \Psi(ih, jk), \tag{23}$$

$$\text{for } i = 0, 1, \ldots, (m-1) \text{ and } j = 0, 1, \ldots, (n-1). \tag{24}$$

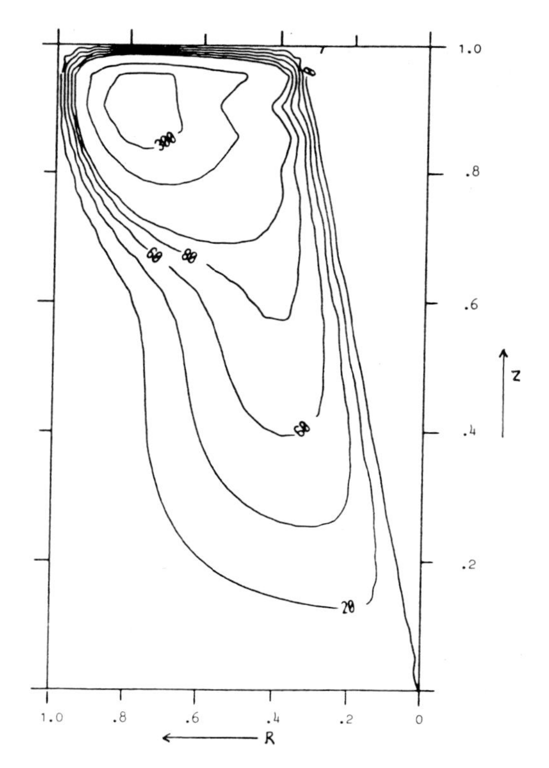

Fig. 8. Streamlines for Re = 5000 (a) constant velocity, (b) drift flux velocity, and (c) $U_t(z)$, velocity along the cone generator for model (b). Streamline value $\times 10^{-5}$.

The banded matrix equation for the $\Psi_{i,j}$ is of the form

$$WX = Y, \tag{25}$$

where X is a column vector the transpose of which is given by

$$X^T = (\Psi_{11}, \Psi_{21} \ldots \Psi_{m-1,1}, \Psi_{12}, \Psi_{22} \ldots \Psi_{m-2,2}, \Psi_{13}, \\ \Psi_{2,3} \ldots \Psi_{m-3}, \Psi_3 \ldots \Psi_{m-2,n-2}) \tag{26}$$

where W is a banded matrix with $2m-3$ bands on either side of the main diagonal and has the structure as displayed in Fig. 1.

Numerical solutions have been obtained for a range of values of the Reynolds number on a grid with $m = n = 22$. The matrix equation (25) for the 400 unknowns was solved by a routine specially written for the particular matrix W. The method of triangular decomposition was used; the largest array declared in the program was 400 by 83.

For Re $\leqslant$ 5000 the belt driven model with constant velocity along the generator, given by boundary condition (21), is employed. For Re $\geqslant$ 5000 the solutions using both condition (19) and (21) are given for comparison.

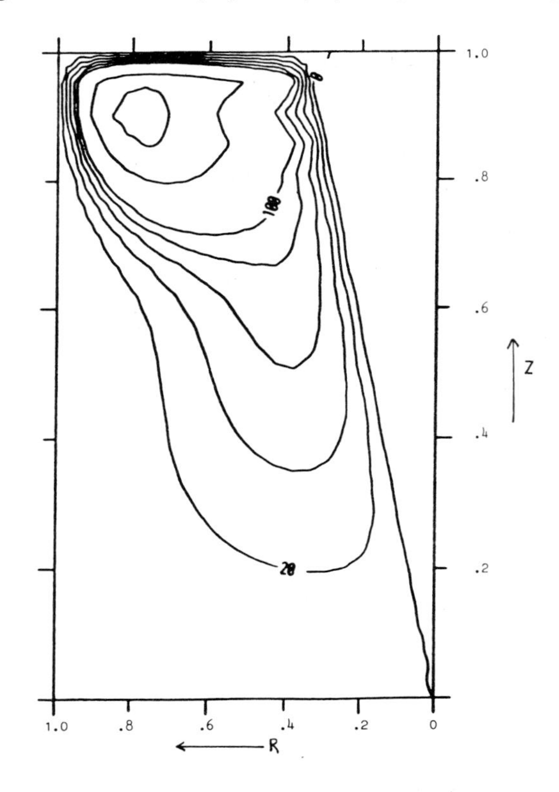

Fig. 8b.

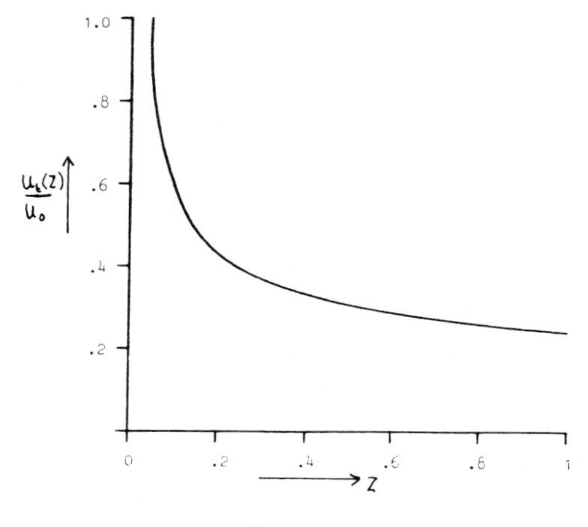

Fig. 8c.

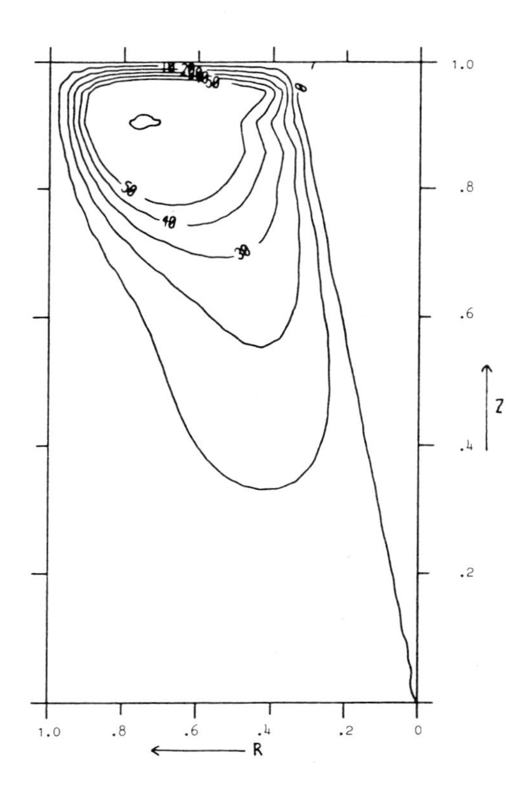

Fig. 9. Streamlines for Re = 10,000 (a) constant velocity, (b) drift flux velocity, and (c) $U_t(z)$, velocity along cone generator for model (b). Streamline value $\times 10^{-5}$.

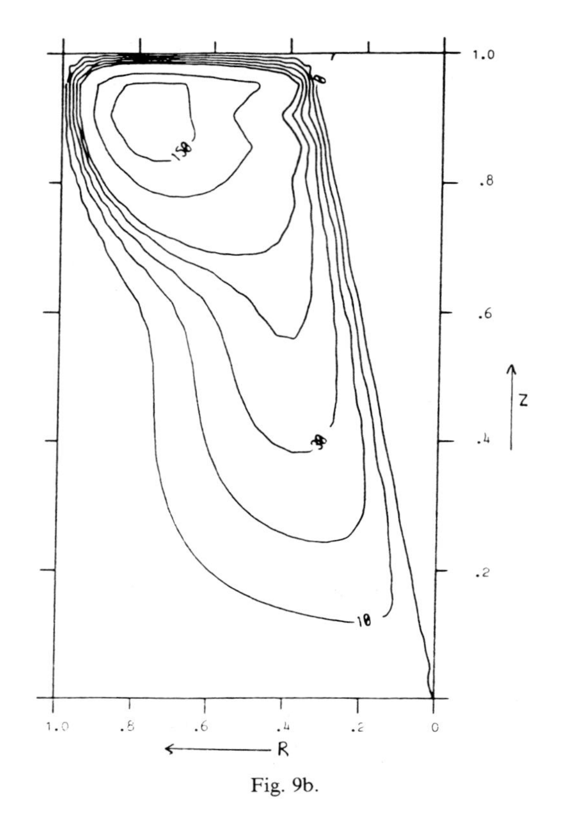

Fig. 9b.

Fig. 9c.

The program is first employed to obtain a solution for the linear Stokes problem when Re = 0. The stream function so obtained is then used as $\overline{\Psi}$ in the calculation for Re = 10. At the end of each solution of (25) $\overline{\Psi}$ is updated. Convergence of Ψ to five significant figures was achieved in 3 interactions. In this fashion the Reynolds number was increased to Re = 500 taking Δ Re = 10, 3 iterations per case being required. From Re = 500 – 1000 with Δ Re = 50 each case converged in 5 or 6 iterations; from Re = 1000 – 4000 with Δ Re = 200 each case converged in 4 iterations; finally from Re = 4000 – 10,000 with Δ Re = 500 convergence was achieved in 2 iterations and clearly at these Reynolds numbers the boundary layer structure of the recirculating flow was established.

4. Discussion of numerical solutions

The purpose of this investigation was to develop a rapid computational procedure to examine the broad features of the recirculating liquid motion in a gas bubble agitation system. Of practical interest is the change in the circulation pattern as the Reynolds number increases. Unfortunately the number of figures necessary to convey the full extent of the theoretical information available is excessive and so only a selection of figures can be displayed.

Prior to discussion of these results it is important to emphasize some of the implications of the various approximations made. The use of the drift flux model yields a useful simplification for computing the two phase flow field in the cone region. In particular it is clear that the streamlines, see for example Fig. 2, in the neighbourhood of the cone surface must pass into the cone region (on the right) be accelerated vertically upwards to the top and pass out of the cone region. Clearly this is a result of our model and the over simplification of the boundary conditions at the cone surface. It will of course lead to obvious errors in the transport of material in the cone region and at the upper free surface of the vessel.

Figures 2–7 give the streamlines of Re = 0, 1, 10, 100, 500 and 1000, respectively; in these cases the liquid velocity along the cone generator is taken as constant. These figures are self explanatory with the centre of the primary vortex moving towards the top left hand part of the vessel as the Reynolds number increases; in the lower left hand part of the vessel the liquid motion becomes more and more stagnant as Re increases. For Re = 5000 Figs. 8(a) and 8(b) display the streamlines for (a) constant velocity along the cone generator and (b) variable velocity along the cone generator as predicted by the drift-flux model; in Fig. 8(c) a graph of U_z/U_0 against z as used in the computation of 8(b) is given. The streamlines in models (a) and (b) are almost identical except in the upper left hand part of the flow field where the circulation in (a) is more vigorous than that in (b) as is expected.

These results for Re = 10,000, see Figs. 9(a), (b) and (c), show that the final structure of the recirculating flow at large Reynolds number is now established. The stagnant region at the lower left hand portion of the vessel is now quite extensive and stirring is confined to be liquid/two phase interface and near the upper free surface of the vessel. Note that the main features of the flow are well predicted using the simple model (a).

5. Conclusions

A simplified model has been developed to examine the extent of recirculation in a gas bubble agitation system as a function of the Reynolds number. The computed flow fields will be of considerable use in further heat and mass transfer studies of this system.

Acknowledgement

One of the authors (W.K.) is grateful to the University of Zambia for the award of a Research Fellowship.

Appendix

The partial differential operators appearing in the linearized stream function equation (15) are defined as follows:

$$L = \gamma^3\, \partial^3/\partial R\, \partial Z^2 - 2(\gamma^3/R)\, \partial^2/\partial Z^2 - 3(\gamma/R)\, \partial^2/\partial R^2 + \gamma\, \partial^3/\partial R^3, \tag{A1}$$

$$M = (\gamma/R)\, \partial^2/\partial R\, \partial Z + 3(\gamma/R^2)\, \partial/\partial Z - \gamma\, \partial^3/\partial R^2\, \partial Z - \gamma^3\, \partial^3/\partial Z^3, \tag{A2}$$

$$N = (3/R^3)\, \partial/\partial R - (3/R^2)\, \partial^2/\partial R^2 + 2/R\, \partial^3/\partial R^3 + 2\gamma^2\, \partial^3/\partial R\, \partial Z^2 - \partial^4/\partial R^4 - 2\gamma^2\, \partial^4/\partial R^2\, \partial Z^2 - \gamma^4\, \partial^4/\partial Z^4, \tag{A3}$$

$$D_1 = \partial/\partial Z, \tag{A4}$$

and

$$D_2 = \partial/\partial R. \tag{A5}$$

In terms of the $(\xi,\ \eta)$ coordinates (13) the above differential operators transform to

$$L = L_1\, \partial/\partial\xi + L_2\, \partial^2/\partial\xi^2 + L_3\, \partial^2/\partial\xi\, \partial\eta + L_4\, \partial^2/\partial\eta^2 + L_5\, \partial^3/\partial\xi^3 + L_6\, \partial^3/\partial^2\xi\, \partial\eta + L_7\, \partial^3/\partial\xi\, \partial\eta^2; \tag{A6}$$

$$M = M_1\, \partial/\partial\xi + M_2\, \partial/\partial\eta + M_3\, \partial^2/\partial\xi^2 + M_4\, \partial^2/\partial\xi\, \partial\eta + M_5\, \partial^3/\partial\xi^3 + M_6\, \partial^3/\partial\xi^2\, \partial\eta + M_7\, \partial^3/\partial\xi\, \partial\eta^2 + M_8\, \partial^3/\partial\eta^3; \tag{A7}$$

$$N = N_1\, \partial/\partial\xi + N_2\, \partial^2/\partial\xi^2 + N_3\, \partial^2/\partial\xi\, \partial\eta + N_4\, \partial^3/\partial\xi^3 + N_5\, \partial^3/\partial\xi^2\, \partial\eta + N_6\, \partial^3/\partial\xi\, \partial\eta^2 + N_7\, \partial^4/\partial\xi^4 + N_8\, \partial^4/\partial\xi^3\, \partial\eta + N_9\, \partial^4/\partial\xi^2\, \partial\eta^2 + N_{10}\, \partial^4/\partial\xi\, \partial\eta^3 + N_{11}\, \partial^4/\partial\eta^4; \tag{A8}$$

$$D_1 = \partial/\partial\eta - (\beta\xi/(-1+\beta\eta))\, \partial/\partial\xi; \tag{A9}$$

and

$$D_2 = (1/(-1+\beta\eta))\, \partial/\partial\xi. \tag{A10}$$

For convenience introduce a new variable

$$s = 1/(-1+\beta\eta). \tag{A11}$$

The coefficients L_i, M_i, N_i appearing in (A6)–(A8) are evaluated as follows:

$$L_1 = 2\gamma^3\beta^3 s^3 - 4\gamma^2\beta^2 s\xi/(\xi - s), \tag{A12}$$

$$L_2 = 4\gamma^3\beta^2\xi s^3 - 2\gamma^2\beta^2\xi^2 s^2/(\xi - s) - 3\gamma s^3/(\xi - s), \tag{A13}$$

$$L_3 = 4\gamma^3\beta s^2\xi/(\xi - s) - 2\gamma^3\beta s^2, \tag{A14}$$

$$L_4 = -2s\gamma^3/(\xi - s), \tag{A15}$$

$$L_5 = \gamma^3\beta^2\xi^2 s^3 + \gamma s^3, \tag{A16}$$

$$L_6 = -2s^2\gamma^3\beta^2\xi, \tag{A17}$$

$$L_7 = \gamma^3 s; \tag{A18}$$

$$M_1 = 6\gamma^3\beta^3 s^3\xi - \gamma\beta s^3/(\xi - s) + 3\gamma s^3\beta\xi/(\xi - s)^2, \tag{A19}$$

$$M_2 = 3\gamma s/(\xi - s), \tag{A20}$$

$$M_3 = 2\gamma\beta s^3 + 6\gamma^3\beta^3 s^3 - \gamma\beta s^3\xi/(\xi - s), \tag{A21}$$

$$M_4 = -6\gamma^3\beta^2\xi s^2 + \gamma s^2/(\xi - s), \tag{A22}$$

$$M_5 = \gamma\beta\xi s^3 + \gamma^3\beta^3\xi^3 s^3, \tag{A23}$$

$$M_6 = -\gamma s^2 - 3\gamma^3\beta^2\xi^2 s^2, \tag{A24}$$

$$M_7 = 3\gamma^3\beta\xi s, \tag{A25}$$

$$M_8 = -\gamma^3; \tag{A26}$$

$$N_1 = 3s^4/(\xi - s)^3 + 4\gamma^2\beta^2 s^4/(\xi - s) - 24\gamma^4\beta^3\xi s^4, \tag{A27}$$

$$N_2 = -3s^4/(\xi - s)^2 + 8\gamma^2\beta^2\xi s^4/(\xi - s) - 12\gamma^2\beta^2 s^4 - 36\gamma^4\beta^4\xi^2 s^4, \tag{A28}$$

$$N_3 = 24\gamma^4\beta^3\xi s^3 - 4\gamma^2\beta s^3/(\xi - s), \tag{A29}$$

$$N_4 = 2s^3/(\xi - s) + 2\gamma^2\beta^2\xi^2 s^3/(\xi - s) - 12\gamma^2\beta^2\xi s^4 - 12\gamma^4\beta^4\xi^3 s^4, \tag{A30}$$

$$N_5 = 24\gamma^4\beta^3\xi^2 s^3 + 8\gamma^2\beta s^3 - 4\gamma^2\beta\xi s^3/(\xi - s), \tag{A31}$$

$$N_6 = 2\gamma^2 s^2/(\xi - s) - 12\gamma^4\beta^2\xi s^2, \tag{A32}$$

$$N_7 = -2s^4 - 3\gamma^2\beta^2\xi^2 s^4 - \gamma^4\beta^4\xi^4 s^4, \tag{A33}$$

$$N_8 = 4\gamma^2\beta\xi s^3 + 4\gamma^4\beta^3\xi^3 s^3, \tag{A34}$$

$$N_9 = -2\gamma^2 s^2 - 6\gamma^4\beta^2\xi^2 s^2, \tag{A35}$$

$$N_{10} = 4\gamma^4\beta\xi s, \tag{A36}$$

and finally

$$N_{11} = -\gamma^4 \tag{A37}$$

References

1. J. Szekely, N.H. El-Kaddah and J.H. Grevet: *Proc. Int. Conf. Injection Metallurgy*, Lulea, Sweden, 5, 1–5: 32 (1980).
2. T. Deb Roy and A.K. Majumdar: *Turbulent recirculating flows in a cylindrical reactor agitated by gases* (Unpubl. Rep.) Imperial College London (1977).

3. T. Deb Roy, A.K. Majumdar and D.B. Spalding: *Appl. Math. Modelling* 2 (1978) 146–150.
4. J.H. Grevet, J. Szekely and N.H. El-Kaddah: *Int. Journ. Heat Mass Transfer* 25 (1982) 487–497.
5. C. Aldham, M. Cross and N.C. Markatos: In: P.C.Hudson and M.J. O'Carroll (eds.) *Mathematical Modelling of Industrial Processes*, Emjoc Press (1982).
6. F. Durst, A.M.K.P. Taylor and J.H. Whitelaw: *Int. J. Multiphase Flow* 10 (1984) 557–569.
7. M.D. Olsen: Comparison Problem No. 1, Recirculating flow in a square cavity. *Structural Research Series*, Report 22, University of Columbia (1979).
8. W.M. Pun and D.B. Spalding: Rep. HTS 76/2 Heat Transfer Section, Imperial College, London (1977).

Applied Scientific Research 44: 225–239 (1987)

Stirring phenomena in centrifugal casting of pipes

G. MARTINEZ, M. GARNIER & F. DURAND
GIS MADYLAM, INPG, CNRS, BP 95, F.38402 - Saint Martin D'Heres Cedex, France

Abstract. In the centrifugal horizontal casting of steel pipes, the normally used rotational speed results in an acceleration of 100 times gravity within the liquid metal layer. Due to such centrifugal forces liquid metal seems to have a body rotation. However metallurgical analysis of pipes show spatial variations in solidification structure which can only originate from recirculating flows. The present study is concerned with the analysis of such stirring motions. Both theoretical and experimental approaches are presented.

Résumé. Dans le procédé d'élaboration par coulée centrifuge horizontale de tubes d'acier, la vitesse de rotation utilisée conduit à une accélération de 100 g dans le métal liquide. A cause des forces centrifuges qui en résultent, le métal semble animé d'une rotation en bloc. Cependant, des analyses métallurgiques effectuées sur des tubes ainsi élaborés montrent des variations spatiales de la structure de solidification dont l'origine est la présence d'écoulements de recirculation. La présente étude concerne l'analyse de ces écoulements. Les résultats de ces deux approches, théorique et expérimentale, sont présentés.

1. Introduction

Centrifugal horizontal casting means rotating the cylindrical mold by means of a motor, roller tracks and carrying rollers while filling it with liquid metal (Fig. 1). The commonly used rotational velocity of the mold leads to a radial acceleration, within the liquid metal layer, whose intensity is 100 times gravity.

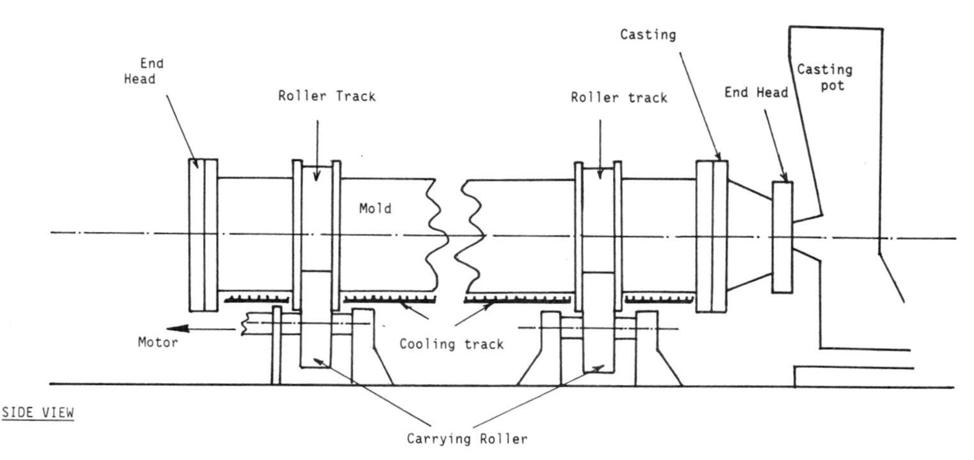

Fig. 1. Centrifugal casting machine.

The effect of such high centrifugal forces is first to regularly fill the mold with metal in order to give the elaborated pipes a uniform wall thickness; secondly to improve the cleanliness of the pipes: particles of various densities are submitted to different forces and consequently higher elements, such as non metallics or slags gather inside the pipe [1,2]. Recent centrifugal castings use either gaseous or solid-liquid protections which limit the zone of impurities to a few millimeters easily removed by machining. Solidification in chilled molds under high pressure would lead to a good control of a desired crystallization and of the grain size. But stirring motions exists, which modify heat and mass transfer and consequently solidification process.

Among the various causes of recirculating flows are:

- the presence of gravity: during one revolution a fluid particle experiences an oscillating force varying from 99 g at the top of the mold and 101 g at the bottom.
- some defects in the equilibrium of the mold with respect to rotation.
- some deformations of the mold due to the thermal gradient through the thickness of the cooled wall.

Hydrodynamical aspects of centrifugal casting have been investigated theoretically [7] [3] and experimentally [5] [6]. Experimental works are only concerned with observations of the free surface deformation. No correlation have been made between the wavy shape of liquid free surface and recirculating flow pattern, temperature field and metallurgical structure of elaborated pipes. Only one interesting work [4] analyses the influence of centrifugal casting conditions, for example vibrations of the mold, upon the structure and the final properties of metals.

The aim of this paper is to study experimentally the hydrodynamical features of a centrifugal liquid layer with respect to external sollicitations. An analytical approach tends to explain some experimental results.

2. Experimental study

Experiments have been done with water to simulate liquid steel in a transparent plexiglas rotating mold which is 1.30 m long and whose internal diameter is 0.225 m. Rotational velocity may vary from 0 to 1000 revolutions per minute.

In order to analyse the free surface shape stroboscopic light is used. To determine the size of recirculating eddies a very simple tracor is used: saw dust. Saw dust is floating on water free surface and gather along lines where fluid velocity is directed towards the external radial direction. It is thus possible to reach the typical size of stirring motions (Fig. 2).

Equilibrium free surface shapes do not depend upon the way of filling the mold with water. It is to be noticed that only a limited number of equilibrium

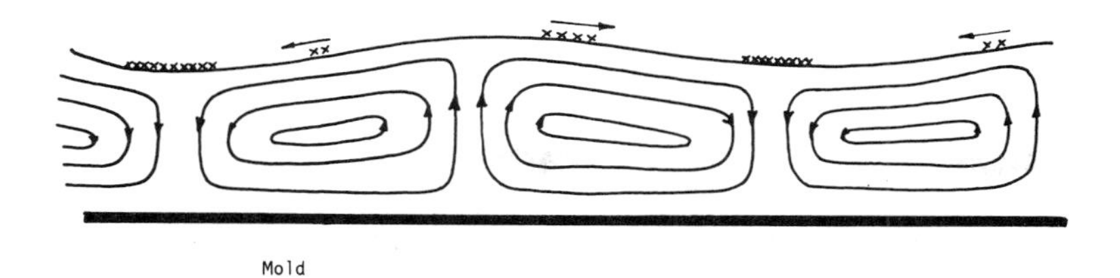

Fig. 2.

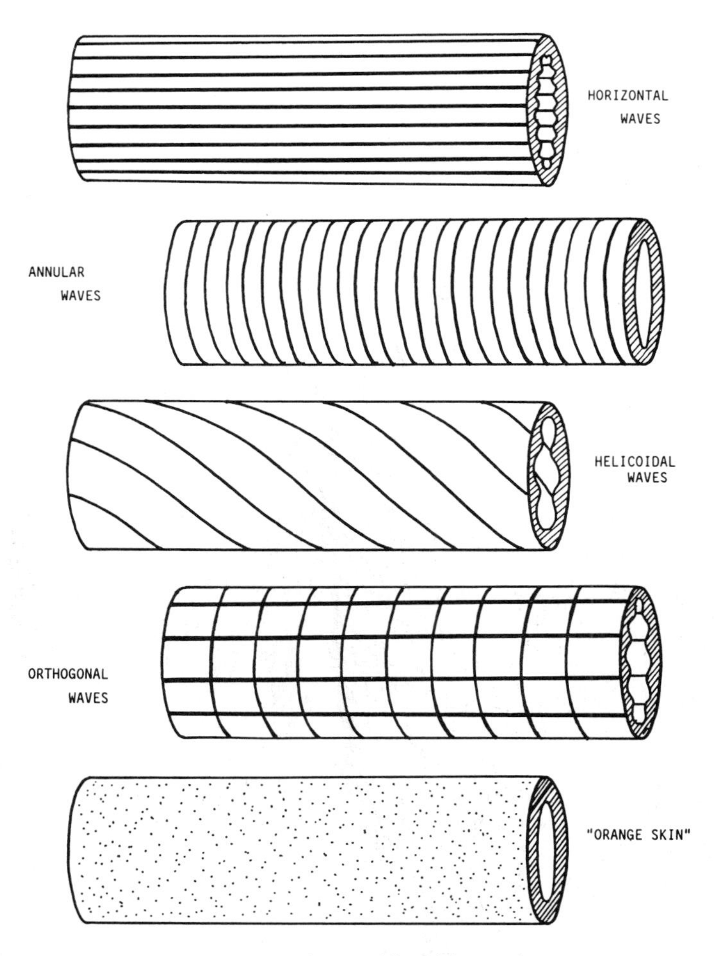

Fig. 3. Various shapes of liquid free surface.

Table 1. Free surface shapes

Ω rps / Ve	10,8	11,4	11,9	12,4	13,0	13,5	14	14,5
2	C	C	C	A	O_s	O_s	A	A
4	C	C	H	H	A	O_s	O_s	O_s
6	C	C	H	H	C	O	O_s	O_s
8	C	C	C	H	H	H	O	O_s
10	C	C	C	H	H	H	H	O

C = Purely CYLINDRICAL free surface; A = AXIAL Horizontal waves; H = HELICOIDAL waves; O = ORTHOGONAL waves; O_s = 'ORANGE SKIN'.

shapes can exist, i.e. (Fig. 3)

- Purely cylindrical waves
- Horizontal waves
- Annular waves
- Helicoïdal waves
- Orthogonal waves

Except those wavy shapes, a particular configuration of the free surface may occur which corresponds to very small holes every-where on the surface which then looks like an 'orange skin'.

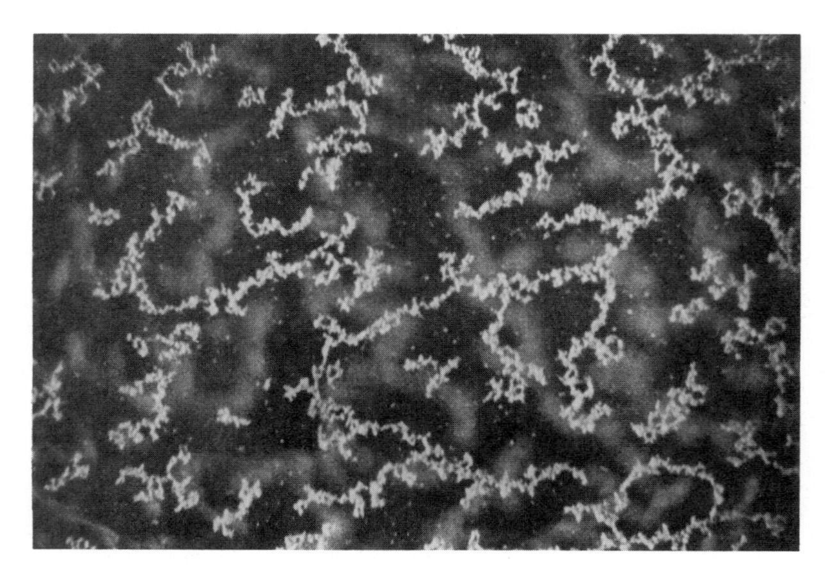

Picture 1.

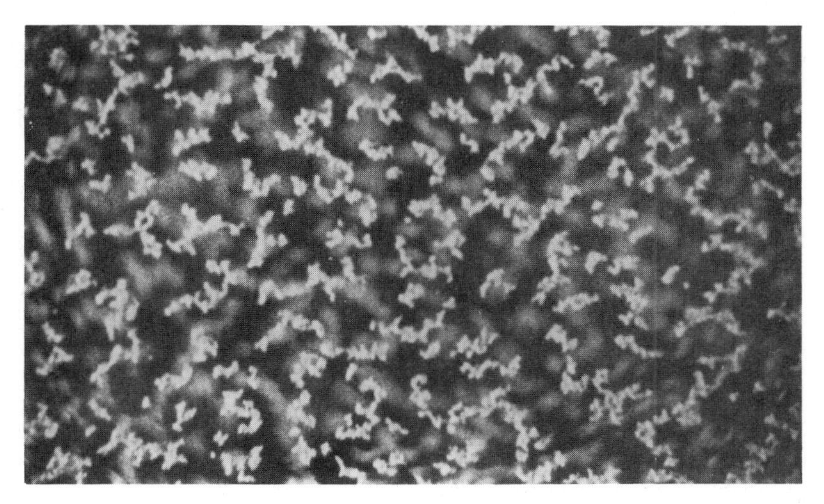

Picture 2.

These phenomena are quite reproducible: for a given volume of water, whatever the way of reaching a given rotational velocity may be (from superior or inferior values), the equilibrium shape is always the same and is very stable with time. In a frame of coordinate rotating with the same velocity as the mold (simulated by stroboscopic light) the wavy free surface is rotating very slowly (some cm/s).

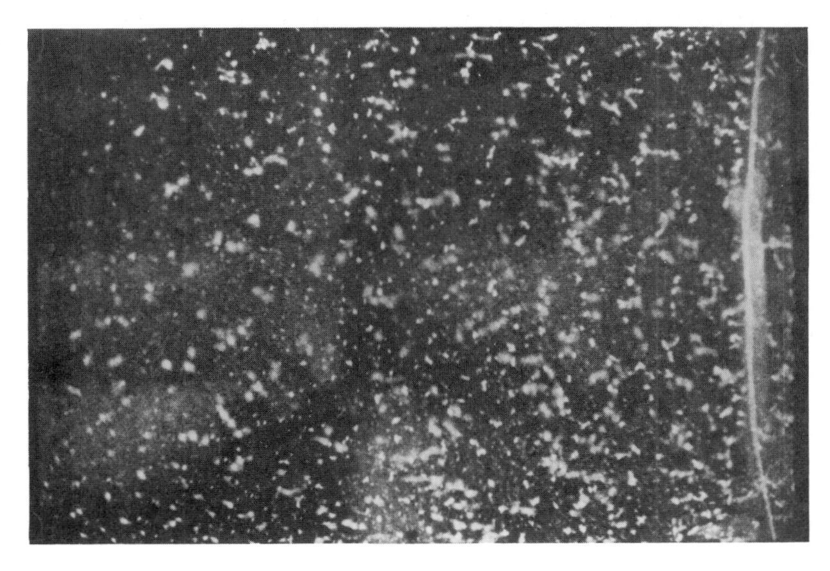

Picture 3.

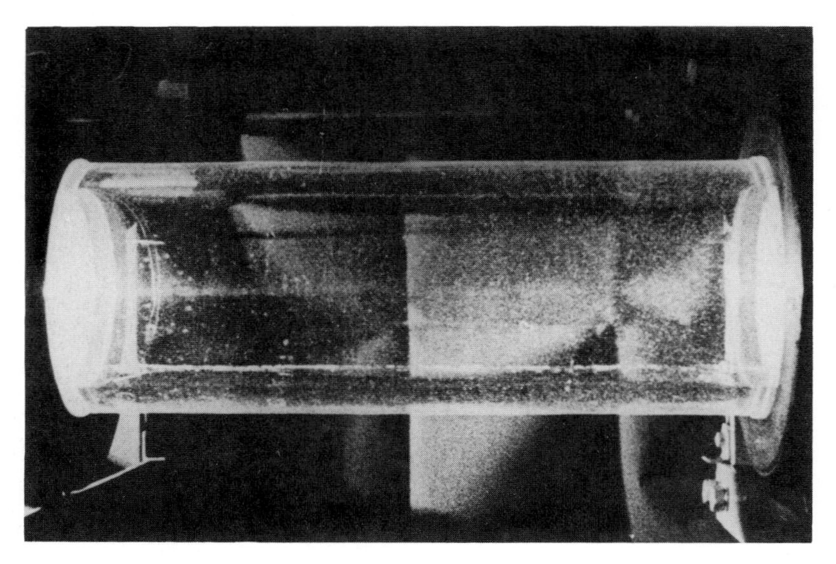

Picture 4.

Table 1 gives the relationship between the free surface shapes and both volume of water and rotational velocity Ω. The maximum value for Ω: 14,5 rps corresponds to an acceleration of 95 g along the internal wall of the mold; 1 liter of water corresponds to a liquid layer thickness of 1.5 mm.

What may be surprising in the behaviour of the free surface is that the higher the rotational velocity is (and therefore the higher the centrifugal force) the more disturbed the free surface: cylindrical waves only occur for low

Table 2. Structures of wood particles

V_1 \ Ω rps	10,8	11,4	11,9	12,4	13,0	13.5	14.0
2	R	R / T	R / T	T	D	D	D
4	R	R / T	T	S / T	S	D	D
6	R	R	S	S	s / D	D	D
8	R / T	R / T	R / T	s / T	S / D	D	D
10	S	S	S	S / D	D	D	D

R: RAMIFICATIONS; T: 'Twigs'; S: 'Stars'; D: Dispersion.

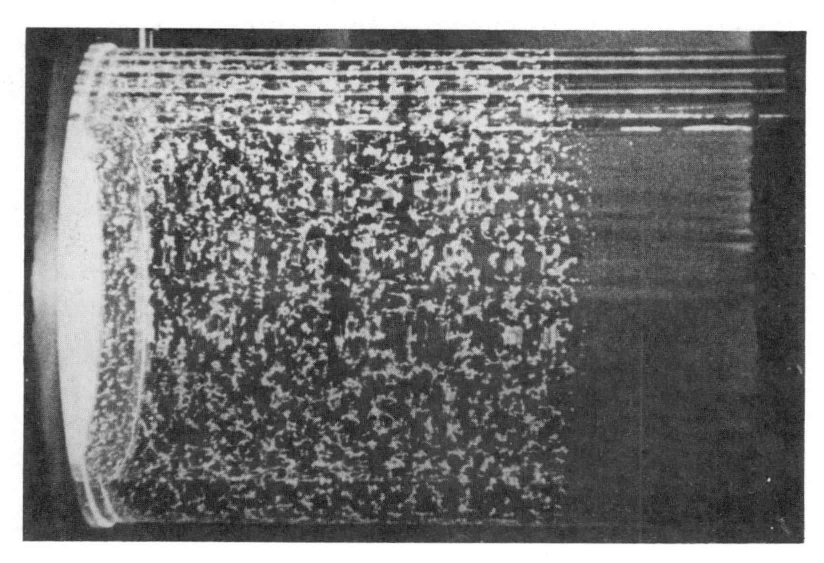

Picture 5.

values of Ω. This can be explained through experimental observations: for weak values of Ω the mold is perfectly rotating; on the contrary when Ω increases, vibrations appear: the waves are the dynamical response of the fluid layer to the mechanical sollicitations.

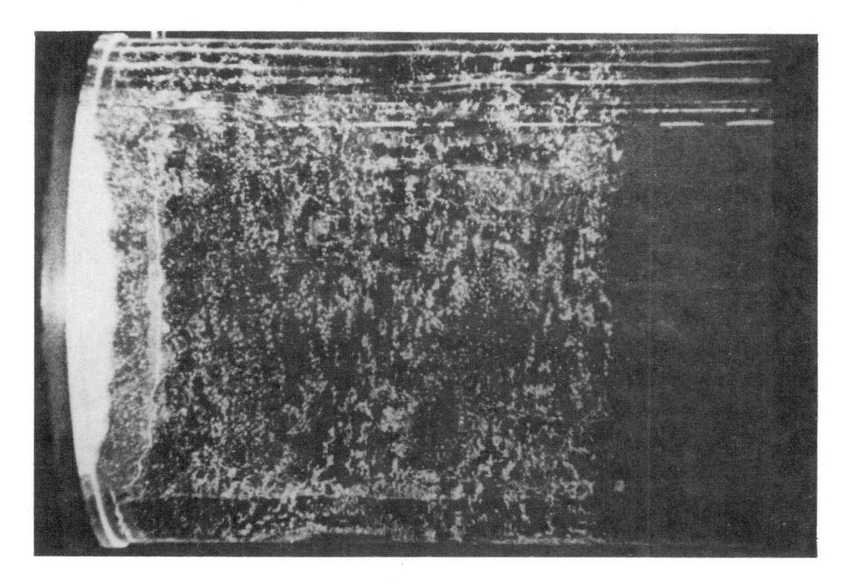

Picture 6.

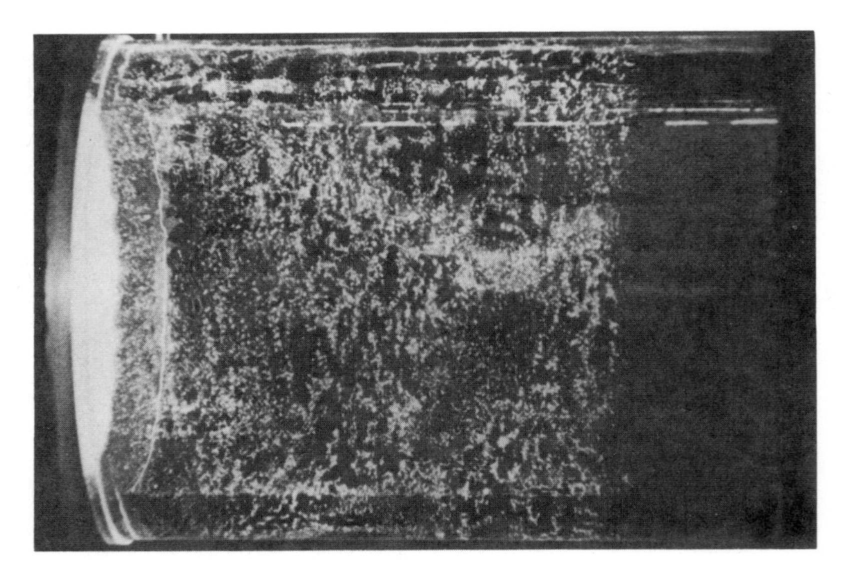

Picture 7.

Interesting informations concerning the recirculating flow pattern can be deduced from stroboscopic observation of saw dust distribution on the free surface. Saw dust particles gather into only four possible configurations which

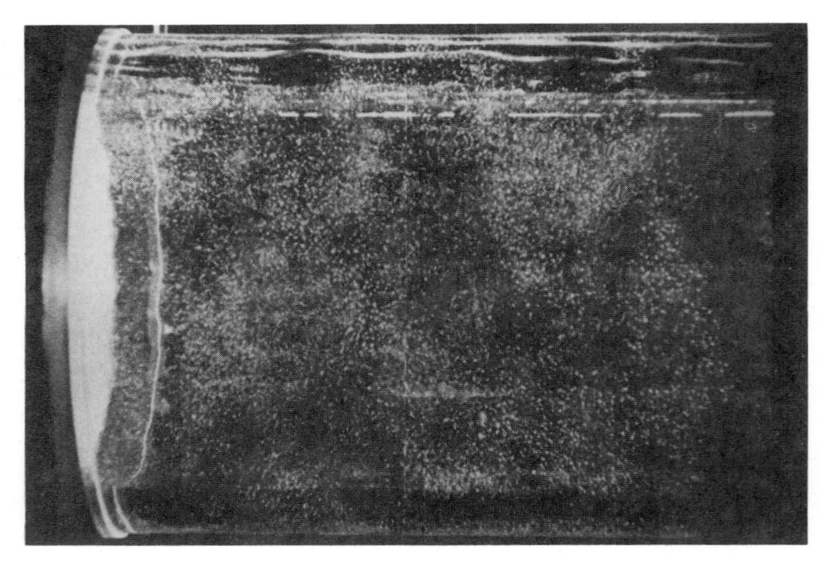

Picture 8.

Table 3

V_L \ Ω rps	10,8	11,4	11,9	12,4	13,0	13.5	14.0
2	0,3	0,7	0,5	2,4	H.	?	?
4	0,3	0,8	3	3,5	5,5	?	?
6	0,7	0.85	H.	1,75	1,6	?	?
8	1,1	1,4	1,2	14	H.	H.	H.
10	2,8	3,5	5,6	H.	V.H.	V.H.	V.H.

H :High velocity > 20 cm/s
V.H. :Very high velocity ≫ 20 cm/s
} Impossible to measure with our stroboscopic technique.

? :High Velocity
Large amplitude waves
} no possible measurement.

look like 'ramifications' (Picture 1) 'Twigs' (Picture 2), 'Stars' (Picture 3) and 'Dispersion' (Picture 4). Typical size of wood particles structures varies from some centimeters for ramifications, one centimeter for Twigs to some millimeters for Stars and less than one millimeter for Dispersion corresponding to separate grains of wood.

Table 2 shows the connection between wood particle structure and liquid volume, and rotational velocity of the mould. Low values of Ω and V lead to large scale recirculating flows (ramifications). When both rotational velocity and liquid volume increase the scale of recirculating flows decreases.

Pictures (5) (6) (7) and (8) successively show the modification of saw dust structure for $4l$ water when rotational velocity increases from 10,8 rps to 13 rps: second line of Table 2. The velocity of wood particles is different from the velocity of the mould and the difference may be important. This difference is given in Table 3 with respect to V and Ω. Cylindrical waves corresponding to small values of V and Ω lead to large structure recirculating flows and low velocities. Relative velocity of saw dust may be high: 14 cm/s, for example, with $\Omega = 12{,}4$ rps and $V = 8l$. When Ω and V increases the relative velocity becomes high (> 20 cm/s) but very difficult to measure because of the large amplitude wavy shape of the free surface.

3. Analytical approach

Recirculating flows are caused by perturbations in axisymmetry due to:

a)– gravity
b)– vibration induced in the rotating mold because of non circularity of rolling tracks or carrying rollers

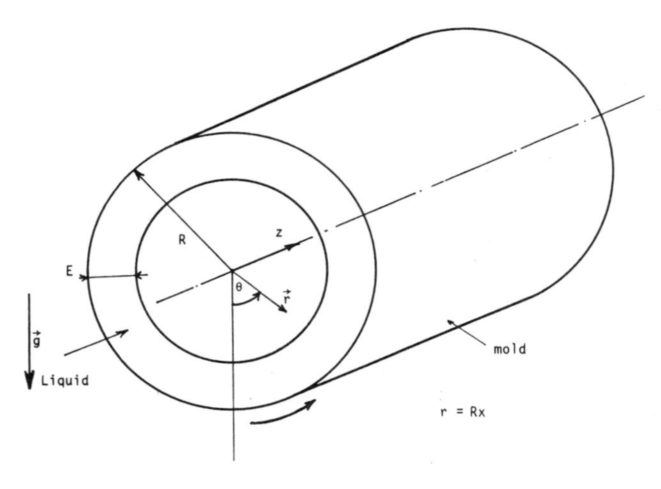

Fig. 4.

c)– axial deformation of the mold due to thermal effects when pouring liquid metal.

Geometry of the problem is defined on Fig. 4.

Methods for analysis of the resulting velocity field are the same for the three cases. Gravity acceleration (a) and fictitious acceleration (b) (c) are introduced in inviscid Navier Stokes equations. All these accelerations are very small compared to centrifugal acceleration $\Omega^2 R$. Thus perturbation methods are used. They consist in expanding the solution in power series of the small parameters $A/\Omega^2 R$, where A is the typical value of acceleration disturbance.

Vibrations are simulated by a time depending perturbation of gravity: $g(1+\xi \cos \omega t)$, where ξ is a non zero real number. The following fictitious acceleration gives an account of the axial deformation of the mold:

$$\vec{A} \left| \begin{array}{l} -\eta_0 \sin\dfrac{2\pi}{\lambda}(z-l)\cos(\theta-\Omega t) \\ +\eta_0 \sin\dfrac{2\pi}{\lambda}(z-l)\sin(\theta-\Omega t) \\ 0 \end{array} \right.$$

where (Fig. 5) η_0 denotes the amplitude of the mold deformation, l the distance between the carrying roller and the end of the mold, λ the distance between the carrying rollers. Expansion parameters are:

$$\frac{g}{\Omega^2 R};\frac{\xi g}{\Omega^2 R};\frac{\eta_0}{\Omega^2 R}.$$

Velocity field: $\vec{V}=(u_r, u_\theta, u_z)$, pressure distribution p and liquid thickness

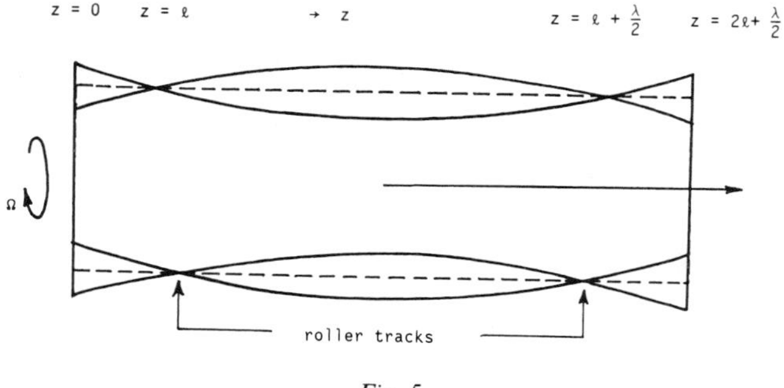

Fig. 5.

e are defined as follows:

$$
\begin{array}{llll}
u_r = 0 & + \;\Omega R u & + \;\Omega R u_\lambda \\
u_\theta = \Omega R x & + \;\Omega R v & + \;\Omega R v_\lambda \\
u_z = 0 & + \;0 & + \;\Omega R w_\lambda \\
p = \rho \dfrac{\Omega^2 R^2}{2}\left[x^2 - \left(1 - \dfrac{e}{R}\right)^2\right] & + \;\rho \dfrac{\Omega^2 R^2}{2} q & + \;\rho \dfrac{\Omega^2 R^2}{2} q_\lambda \\
e = E & + \;Ef & + \;Ef_\lambda
\end{array}
$$

each variable is the sum of three different terms:

- The first is the basic state ($g = \eta_0 = 0$)
- The second introduces the effect of gravity or vibrations
- The third corresponds to axial deformation.

ρ is the fluid density, u, v, u_λ, v_λ, w_λ, q, q_λ, f, f_λ denote respectively the non dimensional perturbations of velocity, pressure and liquid thickness. Each of them is written as a product of three terms depending on only one among the three following variables: radius x, time τ and angle θ.

4. Analytical results

Gravity alone

The expression of the stream function ψ relating to flow disturbance is given by:

$$\psi = -\frac{\Omega R^2}{2}\left(x^2 + \frac{g}{\Omega^2 R}\left(1 - \frac{E}{R}\right)^2\left(x - \frac{1}{x}\right)\cos\theta\right).$$

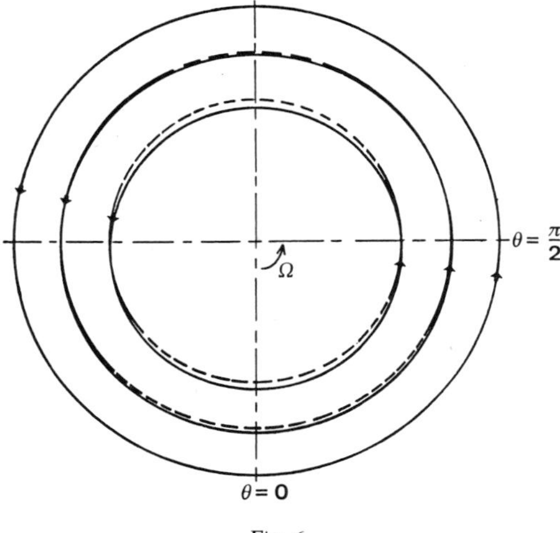

Fig. 6.

In the laboratory referential streamlines are close to concentric circles and the free surface is lowered from its initial position (Fig. 6). This is due to mass conservation: when fluid particles are going upward, gravity force tends to brake them; since flowrate is the same in any cross section, the thickness of liquid has to increase. The opposite phenomenon occurs when fluid particles go downward.

In the mold referential (Fig. 7) the basic state is concealed and the free surface shape becomes time dependent. It is the same with the other variables:

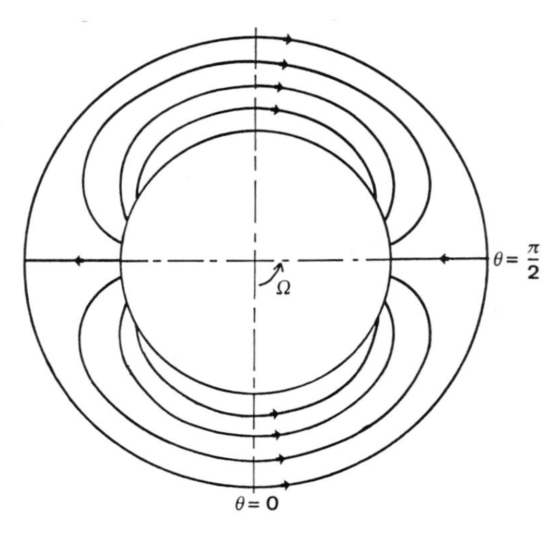

Fig. 7.

pressure, velocity... So streamlines are different from trajectories and because the free surface is moving, it is possible for streamlines to intersect it.

Liquid velocity near the wall is depending on its radius: for $R = 0{,}10$ m maximum velocity is about 4 cm/s and for $R = 1$ m, to day the largest radius for centrifugal casting molds, velocity reaches 31 cm/s. These values are in good agreement with measurements.

Vibrations

The streamfunction corresponding to the velocity perturbation is given by:

$$\psi = -\frac{2\epsilon\xi\left(1+\frac{\omega}{\Omega}\right)\left(1-\frac{E}{R}\right)^2\left(x-\frac{1}{x}\right)}{2+4\frac{\omega}{\Omega}+\left(\frac{\omega}{\Omega}\right)^2\left(1+\left(1-\frac{E}{R}\right)^2\right)}\cos\left(\frac{\omega}{\Omega}\tau+\theta\right).$$

The free surface is the same as for gravity alone, but it is moving round in the mold with a speed which is different from the mold velocity. The denominator involved in the expression of ψ can never be zero: so no resonance is possible.

Since ψ is time dependent, trajectories and streamlines are distinct, as in the case studied above. It is possible to draw them by computation (Fig. 8): they are distorted gravity trajectories; distortion rate is depending on the ratios between gravity perturbation, and amplitude and period.

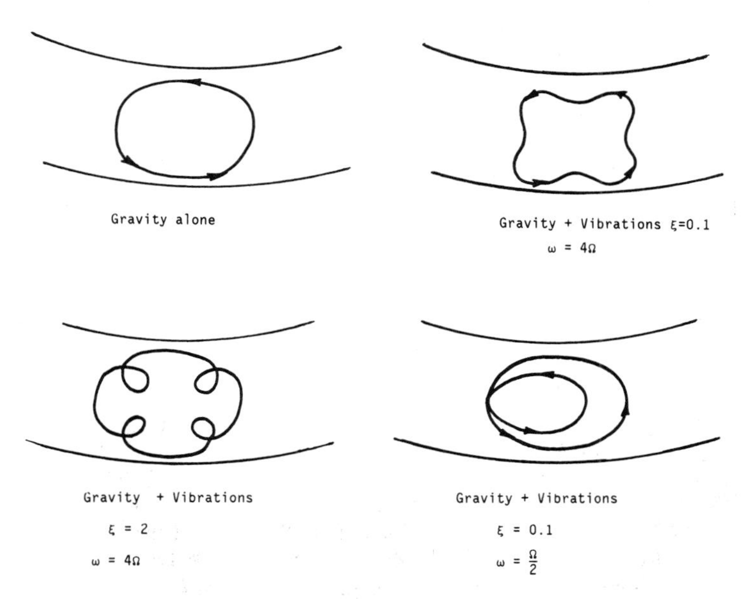

Fig. 8.

Deformation of the mold

Rather long calculations lead to the following expression for the velocity field:

$$u = \alpha(x)\,\sin\left[\frac{2\pi R}{\lambda}\left(\xi - \frac{e}{R}\right)\right]\,\sin(\theta + \tau)$$

$$v = \beta(x)\,\sin\left[\frac{2\pi R}{\lambda}\left(\xi - \frac{e}{R}\right)\right]\,\cos(\theta + \tau)$$

$$w = \gamma(x)\,\cos\left[\frac{2\pi R}{\lambda}\left(\xi - \frac{e}{R}\right)\right]\,\sin(\theta + \tau)$$

where:

$$\xi = \frac{z}{R}, \quad x = \frac{r}{R}, \quad \tau = \Omega t$$

and

$$\alpha(x) = \frac{x^2}{4}\frac{\eta_0}{\Omega^2 R}\left(\frac{2\pi R}{\lambda}\right)^2 - (\mathrm{Ln}\ x)\left[A\left(\frac{2\pi R}{\lambda}\right)^2\right] + \frac{A}{2x^2} + B$$

$$\beta(x) = \frac{x^2}{4}\frac{\eta_0}{\Omega^2 R}\left(\frac{2\pi R}{\lambda}\right)^2 - (\mathrm{Ln}\ x)\left[A\left(\frac{2\pi R}{\lambda}\right)^2\right] - \frac{A}{2x^2} + B$$

$$\gamma(x) = x\frac{\pi R}{\lambda}\frac{\eta_0}{\Omega^2 R} + \frac{A}{x}\frac{2\pi R}{\lambda}$$

with:

$$\frac{\eta_0}{\Omega^2 R} \ll 1, \quad A = \frac{\eta_0}{\Omega^2 R}\frac{H_1}{H_2}, \quad B = -\frac{\eta_0}{2\Omega^2 R}\frac{H_1}{H_2} - \frac{\eta_0}{\Omega^2 R}\left(\frac{\pi R}{\lambda}\right)^2$$

$$\frac{H_1}{H_2} = \frac{-2 + \left(\frac{\pi R}{\lambda}\right)^2 \frac{E}{R}\left(\frac{E}{R} - 2\right)}{\frac{7}{2\left(1 - \frac{E}{R}\right)^2} + \frac{1}{2} + \left(\frac{2\pi R}{\lambda}\right)^2 \mathrm{Ln}\left(1 - \frac{E}{R}\right)}.$$

In this case, it is difficult to give a 3D representation of streamlines and trajectories are not easily obtained by computation.

Some interesting result is to be noticed: resonance is possible. The critical wavelength λ_c corresponding to resonance is given by:

$$\lambda = \lambda_c = \frac{2\pi R\sqrt{2\ \mathrm{Ln}(1/(1 - E/R))}}{\sqrt{7(1 - E/R)^{-2} + 1}}.$$

λ_c is only depending on values of R and E/R, so in the study of a centrifugal casting device the rolling tracks position is very important. As concerned with our experimental device ($R = 0{,}11$ m and $E = 0{,}04$ m) the critical wavelength is $\lambda_c = 0.15$ m. It is clear that any wavelength λ defined by $\lambda = \lambda_c/n$ where n is a non zero integer may be critical for the system.

From experimental observations, typical wavelengths of the free surface deformation is between 5 and 10 cm for any equilibrium shape (orthogonal waves, helicoïdal waves, annular waves). Even if it is difficult to get with very good precision the value of wavelength, and if the roller tracks are not lines but have a finite width, wavelengths near $\lambda_c/2$ occur in the system.

5. Conclusion

Casting conditions have a large influence upon the structure of thick pipes prepared in chill molds by the centrifugal casting process. Vibrations are known to modify cristallization in liquid metal during solidification.

The present study gives informations, deduced from experimental simulation with water and theoretical analysis, concerning flow patterns in the rotating liquid. Stroboscopic observations show that, with respect to volume liquid and rotational speed of the mold, liquid metal free surface can only have a limited number of shapes. The observed phenomena are quite reproducible and have no dependence upon initial conditions. A very simple tracor gives an estimation of recirculating cells. Theoretical approach of the dynamical response of liquid to external unsteady sollicitations (gravity, vibrations, deformation of the mold) leads to typical flow patterns. Trajectories of fluid particles which are very important for metallurgical interpretation of hydrodynamical effects are computed.

This work is only a first approach of stirring phenomena in centrifugal casting and cannot yet bring precise responses to the questions of metallurgists. It is necessary to go on these complementary experimental and theoretical ways, to analyse resulting heat and mass transfer in the liquid during solidification.

Acknowledgement

This work was supported by: Centre de Recherches de Pont à Mousson Maidieres, France.

References

1. A. Royer: Horizontal centrifugal casting: a technique for the manufacture of large diameter, heavy wall pipes. *Bul. Cercle d'Etudes des Métaux*. 18 (1981) 1–12.
2. A. Royer, B. Dumas and M. Gantois: Spun steel pipes for the offshore industry. *J. of Energy Resources Technology* 105 (1983) 97–102.
3. K.J. Ruschak and L.E. Scriven: Rimming flow of liquid in a rotating horizontal cylinder. *J. of Fluid Mechanics* 76 (1976) 113–125.
4. L. Northcott and V. Dickin: The influence of centrifugal casting (horizontal axis) upon the structure and properties of metals. *J. Institute of Metals* 70 (1944) 301–323.
5. M.J. Karweit and S. Corrsin: Observation of cellular patterns in a partly filled, horizontal, rotating cylinder. *Physics of Fluids* 18 (1975) 111–112.
6. O.M. Phillips: Centrifugal waves. *J. of Fluid Mechanics* 21 (1959) 340–352.
7. T.J. Pedley: The stability of rotating flows with a cylindrical free surface. *Journal of Fluid Mechanics* 30 (1967) 127–147.

Applied Scientific Research 44: 241–259 (1987)

The importance of secondary flow in the rotary electromagnetic stirring of steel during continuous casting

P.A. DAVIDSON [1] & F. BOYSAN [2]

[1] *Department of Engineering, Cambridge University, Cambridge, UK;*
[2] *Department of Chemical Engineering and Fuel Technology, Sheffield University, Sheffield, UK*

Abstract. This paper considers some aspects of the flow generated in a circular strand by a rotary electromagnetic stirrer. A review is given of one-dimensional models of stirring in which the axial variation in the stirring force is ignored. In these models the magnetic body force is balanced by shear, all the inertial forces being zero (except for the centripetal acceleration).

In practice, the magnetic torque occurs only over a relatively short length of the strand. The effect of this axial dependence in driving force is an axial variation in swirl, which in turn drives a secondary poloidal flow. Dimensional analysis shows that the poloidal motion is as strong as the primary swirl flow.

The principle force balance in the forced region is now between the magnetic body force and inertial. The secondary flow sweeps the angular momentum out of the forced region, so that the forced vortex penetrates some distance from the magnetic stirrer. The length of the recirculating eddy is controlled by wall shear. This acts, predominantly in the unforced region, to diffuse and dissipate the angular momentum and energy created by the body force.

Notation

$\boldsymbol{B}$	magnetic field strength
$\boldsymbol{E}$	electric field
$\hat{e}$	unit vector
F_θ	force
$f(z/R)$	dimensionless variation of force with depth
$\boldsymbol{J}$	current density
k	turbulence kinetic energy
k'	wall roughness
L	axial length scale in unforced region
p	pressure
R	radius
Re	Reynolds number
r	radial coordinate
$\boldsymbol{T}$	torque
t	time
$\boldsymbol{u}$	velocity
$\overline{V}$	characteristic velocity $B\omega R\sqrt{(\sigma/\rho\omega)}$
V_*	shear velocity
$\boldsymbol{v}$	fluctuating velocity
z	axial coordinate
Γ	angular momentum $u_\theta r$
δ	boundary layer thickness
ϵ	viscous dissipation rate
θ	angular coordinate
ν	viscosity
ν_t	eddy viscosity
ρ	density
σ	conductivity
τ_{ij}	shear stress
Ω	angular velocity
$\boldsymbol{\omega}$	vorticity
ω	frequency

Subscripts

R	wall
r	radial
θ	azimuthal
z	axial
0	core
p	poloidal

1. Introduction

Continuous casting has become an increasingly common means of producing steel ingots. The process is shown schematically in Fig. 1. There is a metallurgical requirement to stir the melt as it solidifies, and this has led to the use of electromagnetic stirring [3]. Stirrers have been placed around the mould, below the mould and near the point of final solidification. Typically these stirrers resemble the stator of an induction motor, producing a travelling or rotating magnetic field. The field induces motion in the melt with peripheral velocities of the order of 20 cm/s. We shall be concerned with rotary stirrers, whose primary purpose is to induce swirl in the melt.

The cost of implementation of magnetic stirring is considerable; yet the optimum configuration for stirring is often assessed on an empirical basis [4]. The question of how many stirrers are required, and where they should be placed, frequently arise in the literature [3,4]. In order to give general answers to these, it is necessary to understand not only the metallurgical processes at work, but also the nature of the velocity field induced by stirring. In particular, the following hydrodynamic questions arise.

(i) How does the magnitude of the induced swirl scale on magnetic field strength, mould size and melt properties?
(ii) How far beyond the stirrer does the induced vortex extend?
(iii) Do secondary flows develop and are they important?

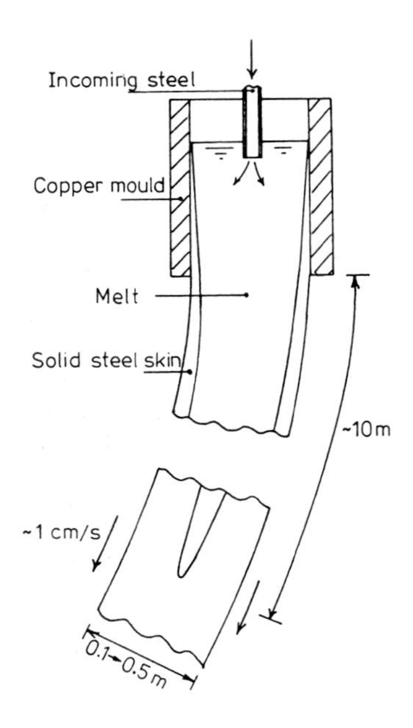

Fig. 1. Diagrammatic representation of the continuous casting process.

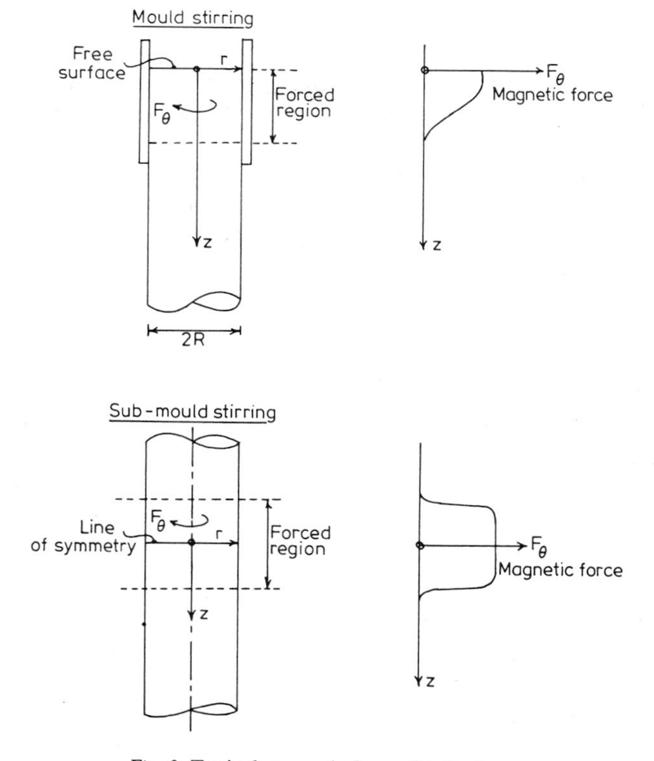

Fig. 2. Typical magnetic force distributions.

In order to simplify the problem, attention is restricted to flow in a circular strand, and entrance effects of the melt in the mould are ignored. Typical magnetic force distributions are shown in Fig. 2. Often rotation rates are sufficiently low that the surface of the melt remains flat. In this situation the surface may be treated as a plane of symmetry and the problems of mould and sub-mould stirring become hydrodynamically identical.

We shall consider two models of stirring. Firstly, a one-dimensional model is reviewed. In this analysis the axial variation in the magnetic force, and hence swirl, is ignored. In such a situation the secondary flows are, by definition, zero. This flow has been analysed several times [1, 5, 7, 9] and is a popular model of stirring. We shall show, however, that it is misleading in the context of continuous casting. A more realistic two-dimensional axisymmetric model is then considered, in which all three velocity components are non-zero.

The coordinate system used is shown in Fig. 2, and notation is given at the start of this paper. The radius R, shown in Fig. 2, refers to the outer radius of the melt. It is assumed to be constant, the taper resulting from the increasing shell thickness being ignored.

II. The magnetic body force

The magnetic field within the melt is governed by the advection diffusion equation,

$$\frac{\partial \boldsymbol{B}}{\partial t} = \nabla \times (\boldsymbol{u} \times \boldsymbol{B}) + \frac{1}{\mu\sigma}\nabla^2 \boldsymbol{B}.$$

The relative size of the advection to diffusion terms is given by the magnetic Reynolds number,

$$\mathrm{Re}_m = uR\mu\sigma.$$

We shall assume that Re_m is small. This is generally true in both laboratory and industrial situations, and allows us to ignore advection of the magnetic field. In this approximation the melt is treated as a solid conductor, and the magnetic field determined by the standard eddy-current equation,

$$\frac{\partial \boldsymbol{B}}{\partial t} = \frac{1}{\mu\sigma}\nabla^2 \boldsymbol{B}.$$

It is worth noting that, in this approximation, there is no charge distribution within the melt, and that charges will not be deposited on the free surface of the melt by eddy currents, since there is no component of current normal to the surface [2].

The ratio of the time derivative term to the diffusion term in the equation for $\boldsymbol{B}$ is given by the skin depth parameter,

$$\Delta = R^2\omega\sigma\mu = 2(R/\delta)^2$$

where

$$\delta = (2/\mu\sigma\omega)^{1/2}.$$

Our second assumption is that Δ is also small, although larger than Re_m.

$$\mathrm{Re}_m \ll \Delta \ll 1.$$

This requires the skin depth to be large relative to R, and is referred to as a low frequency approximation. It is shown in [9] that, in this context, a low frequency analysis is a good approximation for values of Δ as high as unity. This covers most 50 Hz laboratory experiments and some industrial applications.

Since we have assumed that $\mathrm{Re}_m \ll \Delta$, a condition almost invariably met in practice, we may deduce,

$$u \ll \omega R$$

This implies that the field advection term in the advection-diffusion equation for $\boldsymbol{B}$ is much smaller than the time derivative term. This expression is sometimes taken as an alternative condition for field advection to be negligible.

The assumption $\mathrm{Re}_m \ll \Delta \ll 1$ greatly simplifies the calculation of the electromagnetic body force since the applied magnetic field remains unperturbed by either advection or diffusion.

The magnetic body force induced in an infinitely long metal column by a uniform, transverse, rotating magnetic field is well known [9]. For low frequency, the mean azimuthal force has the simple form,

$$F_\theta = \tfrac{1}{2}\sigma\omega B^2 r.$$

There is also an oscillatory component of force, of frequency 2ω. However, this is irrotational and consequently drives no fluid motion [5].

In this expression the magnetic field $\boldsymbol{B}$ is assumed to be uniform and independent of axial position. To allow for the fact that the field strength, and hence stirring force, will vary axially, it is convenient to introduce a dimensionless depth function $f(z/R)$ and take as an idealised force distribution,

$$F_\theta(r, z) = \left[\tfrac{1}{2}\sigma\omega B^2 r\right] f(z/R). \tag{1}$$

In this equation B now has the meaning of a characteristic field strength which is a constant. The function $f(z/R)$ depends on the geometry of the applied magnetic field and must be determined for each stirrer by detailed calculation.

When the applied magnetic field strength varies axially, the oscillatory component of the force need not be irrotational. However, it may still be neglected since the induced oscillatory component of velocity, u', will be significantly smaller than the mean component, u. This may be seen by comparing inertial and electromagnetic terms in the Navier-Stokes equations.

$$\rho\omega u' \sim \rho u^2/R \sim F_\theta$$

from which,

$$\frac{u'}{u} \sim \frac{u}{\omega R} \ll 1$$

and,

$$u \sim \sqrt{F_\theta R/\rho}\,.$$

Clearly, it is consistent with the initial assumptions to ignore the oscillatory components of force and velocity.

Equation (1) does not necessarily give the true distribution of F_θ when the magnetic field varies axially, because the z-independent solution is no longer locally valid for a particular value of z. However, it is shown below, using dimensional arguments, that equation (1) will be correct to first order in r. It is useful, therefore, to adopt this expression as an idealised 'model equation' in order to investigate the hydrodynamic consequences of forcing over a relatively short length of the liquid metal column.

The essence of the low frequency approximation is to ignore the magnetic field generated by eddy currents and consider only an imposed irrotational field. The first order current density is then obtained by substituting this field into Faradays equation, and the first order force distribution follows [2].

$$\nabla \times \boldsymbol{E} = -\frac{\partial \boldsymbol{B}}{\partial t} \quad \text{and } \boldsymbol{J} = \sigma \boldsymbol{E}$$

$$\Rightarrow J \sim \sigma \omega B R.$$

Also,

$$\boldsymbol{F} = \boldsymbol{J} \times \boldsymbol{B}$$

$$\Rightarrow F_\theta \sim \sigma \omega B^2 R.$$

We may therefore write F_θ in the form,

$$F_\theta = \left[\tfrac{1}{2}\sigma\omega B^2 R\right] F(r/R,\ z/R).$$

The function $F(r/R,\ z/R)$ may be expanded as a Taylor series about the axis $r = 0$.

$$F_\theta = \left[\tfrac{1}{2}\sigma\omega B^2 R\right]\left[f_0\left(\frac{z}{R}\right) + \frac{r}{R} f_1\left(\frac{z}{R}\right) + \dots\right].$$

Many fields have symmetry about the axis, in which case the first term in the expansion must be zero. It follows that, to first order, the body force may be approximated by,

$$F_\theta = \left[\tfrac{1}{2}\sigma\omega B^2 r\right] f_1(z/R).$$

This is the same as equation (1), where $f_1 = f$. It is convenient to introduce a characteristic velocity, defined as,

$$\overline{V} = B\sqrt{\frac{\sigma}{\rho\omega}}\ \omega R. \tag{2}$$

The idealised force distribution then becomes,

$$F_\theta = \left[\tfrac{1}{2}\rho \overline{V}^2 r/R^2\right] f(z/R). \tag{3}$$

Since this force distribution is axisymmetric, and only circular strands are considered, the resulting flow is itself axisymmetric.

III. A review of one-dimensional models of stirring

We now consider the flow which results when the axial variation in the body force is neglected. The body force is,

$$F_\theta = \tfrac{1}{2}\rho \overline{V}^2 r/R^2.$$

This force drives an axisymmetric, steady, one-dimensional, azimuthal flow, the radial velocity being zero for reasons of continuity.

$$\boldsymbol{u} = u_\theta(r)\hat{\boldsymbol{e}}_\theta.$$

The radial and azimûthal components of the time averaged Navier-Stokes equations for axisymmetric swirl flow are,

$$-\rho\frac{u_\theta^2}{r} = -\frac{\mathrm{d}p}{\mathrm{d}r} + \frac{1}{r}\frac{\mathrm{d}}{\mathrm{d}r}\left[r\left(-\rho\overline{v^2}_r\right)\right] - \frac{1}{r}\left(-\rho\overline{v^2}_\theta\right)$$

$$F_\theta = -\frac{1}{r^2}\frac{\mathrm{d}}{\mathrm{d}r}\left(r^2\tau_{r\theta}\right)$$

where

$$\tau_{r\theta} = -\rho\overline{v_r v_\theta} + \rho\nu r\frac{\mathrm{d}}{\mathrm{d}r}\left(\frac{u_\theta}{r}\right)$$

($\boldsymbol{v} \equiv$ fluctuating velocity, $\boldsymbol{u} \equiv$ time mean velocity).

The first of these shows that the radial gradient of the total pressure $p + \rho\overline{v^2}_r$ balances the centripetal acceleration. The second may be integrated to give the shear stress distribution. Substituting for F_θ and integrating we obtain,

$$\tau_{r\theta}/\rho = \overline{-v_r v_\theta} + \nu r\frac{\mathrm{d}}{\mathrm{d}r}\left(\frac{u_\theta}{r}\right) = -\tfrac{1}{8}\overline{V}^2\left(\frac{r}{R}\right)^2. \tag{4}$$

The flow considered here is similar to that of axial flow in a pipe. In both cases we wish to determine the flow resulting from a known imposed shear stress. In the case of laminar flow, equations (4) may be integrated to give the well known result [5],

$$u_\theta = \frac{1}{16}\frac{\overline{V}^2}{\nu}r\left[1-(r/R)^2\right].$$

In this case the velocity scales on $\overline{V}^2R/\nu$. For turbulent flow, however, we expect u_θ to scale on the shear velocity, and hence $\overline{V}$. Solution of equation (4) in the case of turbulent flow requires some estimate of the Reynolds stress term. Two equation closure models of turbulence have been applied to this problem [9]. This involves solving several subsidiary equations which require extensive computation. It is suggested in [1], however, that this shear flow is largely controlled by events near the wall, as in axial pipe flow. Consequently, the computed flow is insensitive to the turbulence model used, provided that it correctly models the wall region. At high Reynolds number, curvature effects are negligible near the wall and a simple mixing length model may be applied. This results in an explicit equation for the core angular velocity [1].

$$\Omega_0 = \frac{\overline{V}}{R}\left\{0.88\ \ln\frac{\overline{V}R}{\nu} + 1.0\right\}. \tag{5}$$

This equation is consistent with the results of computations using higher order turbulence models, and with experimental data. It is derived assuming a smooth wall. If the wall is rough, then equation (5) must be replaced by,

$$\Omega_0 = \frac{\overline{V}}{R}\left\{0.88\ \ln\frac{R}{k'}\right\}. \tag{6}$$

For a Reynolds number of 10^5, a radius of 100 mm, and a roughness k' of 1 mm, which is typical of a continuous casting application, equation (6) predicts only a half of the velocity given by equation (5). Thus roughness is an important parameter in this flow.

The essential feature of this type of flow is that the inertial forces are zero, except for the centripetal acceleration. It follows that the only forces available to balance the magnetic torque are shear stresses. In any real stirrer, however, the magnetic forcing occurs within a relatively short length of the cylinder (see Fig. 2). This results in differential rotation between forced and unforced regions, which, in turn, produces secondary flow. The inertial forces are then non-zero and are available for balancing the magnetic body force. This leads to a quite different type of flow, and a different scaling law for the velocity. We shall see that the magnitude of the swirl is largely determined by inertia and not shear.

IV. A two-dimensional axisymmetric model of stirring – some qualitative features of the flow

In this section we shall examine the laminar equations of motion for axisymmetric flow. In practice, however, all real flows of interest are turbulent. In order to interpret the discussion in terms of a turbulent flow, the viscosity ν must be considered as a mean 'eddy viscosity'. Such a procedure is reasonable in this case as many features of the flow are controlled by the inertia of the mean flow, and are not sensitive to the details of the shear.

To determine why an axial variation in the azimuthal body force gives rise to poloidal motion, it is convenient to split the velocity field into poloidal and azimuthal parts, and examine the interaction between them. (Poloidal motion is that in the $r-z$ plane.)

$$\boldsymbol{u} = \boldsymbol{u}_p + \boldsymbol{u}_\theta$$

where

$$\boldsymbol{u}_p = u_r\hat{e}_r + u_z\hat{e}_z.$$

The vorticity may be similarly divided

$$\boldsymbol{\omega} = \boldsymbol{\omega}_p + \boldsymbol{\omega}_\theta$$

where

$$\boldsymbol{\omega}_\theta = \nabla \times \boldsymbol{u}_p$$

$$\boldsymbol{\omega}_p = \nabla \times \boldsymbol{u}_\theta.$$

The Navier-Stokes equation may itself be split into aximuthal and poloidal parts. To eliminate the pressure from the poloidal equation we may take its curl and obtain a vorticity advection-diffusion equation. The resulting equations are,

$$\boldsymbol{\omega}_p \times \boldsymbol{u}_p = \nu\nabla^2 \boldsymbol{u}_\theta + \frac{1}{\rho} F_\theta \hat{\boldsymbol{e}}_\theta$$

$$\nabla \times (\boldsymbol{u}_p \times \boldsymbol{\omega}_\theta) + \nu\nabla^2 \boldsymbol{\omega}_\theta = -\frac{\partial}{\partial z}\left(\frac{u_\theta^2}{r}\right)\hat{\boldsymbol{e}}_\theta. \tag{7}$$

The first equation represents the azimuthal force balance. It is through the inertial term $\boldsymbol{\omega}_p \times \boldsymbol{u}_p$ that the azimuthal motion is coupled to the poloidal motion. For large Reynolds numbers the inertial force will be much larger than the shear term, and we expect the primary force balance to be between $\boldsymbol{J} \times \boldsymbol{B}$ and inertia.

We shall show that this inertial term represents a flux of angular momentum out of the forced region.

The second equation is the steady-state advection-diffusion equation for the azimuthal vorticity and corresponding poloidal velocity. It is coupled to the azimuthal velocity through a source term which is the axial gradient of the centripetal acceleration. This additional term derives from $\nabla \times (\boldsymbol{u}_\theta \times \boldsymbol{\omega}_p)$ and represents sweeping of poloidal vorticity through an axial variation in the azimuthal velocity, generating azimuthal vorticity. This is illustrated in Fig. 3. Note that if u_θ is independent of z, then equation (7) implies that no field sweeping of $\boldsymbol{\omega}_p$ occurs. This is to be expected since in such a situation the poloidal vortex lines lie parallel to the axis and each point on a vortex line experiences the same azimuthal velocity.

These equations of motion may be rewritten as scalar transport equations for angular momentum Γ and ω_θ/r.

$$\Gamma = u_\theta r.$$

The azimuthal force balance becomes a transport equation for Γ, and the advection-diffusion equation a transport equation for ω_θ/r.

$$\boldsymbol{u}\cdot\nabla\Gamma = \nu\left\{\nabla^2\Gamma - \frac{2}{r}\frac{\partial\Gamma}{\partial r}\right\} + \frac{r}{\rho}F_\theta \tag{8}$$

$$\boldsymbol{u}\cdot\nabla\left(\frac{\omega_\theta}{r}\right) = \nu\left\{\nabla^2\left(\frac{\omega_\theta}{r}\right) + \frac{2}{r}\frac{\partial}{\partial r}\left(\frac{\omega_\theta}{r}\right)\right\} + \frac{\partial}{\partial z}\left(\frac{\Gamma^2}{r^4}\right). \tag{9}$$

From equation (8) we see that, in the absence of shear and magnetic forces, the angular momentum is advected unchanged along a streamline. This is in accordance with Kelvins circulation theorem applied to a material hoop centred on the z axis.

An examination of equation (9), or a consideration of vortex sweeping, suggests that the secondary poloidal flow must be of the form shown in Fig. 4. The decrease of angular velocity with depth acts as a source of negative azimuthal vorticity. This requires rotation in an axial plane as shown in the figure.

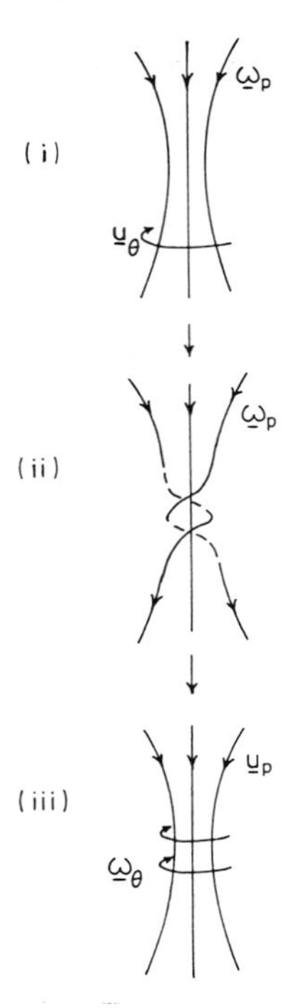

Fig. 3. Sweeping of poloidal vorticity by an axial gradient in azimuthal velocity to produce azimuthal vorticity.

The primary effect of the body force is to spin up the fluid as it passes through the forced region, in accordance with equation (8). The poloidal flow then sweeps this angular momentum into the unforced region.

The viscous terms in equations (8) and (9) are not purely diffusive, they also contain source terms. However, their primary role is to allow the vortex sheet, created at the wall by the no-slip condition, to diffuse into the flow. Since this is a relatively slow process, the Reynolds number being assumed to be large, the diffusion occurs primarily in the unforced region.

The length scales for r and z in the forced region are dictated by the spatial variation of the body force (except in the boundary layer adjacent to the wall), and are of order R. It follows from equation (8) that the magnetic force is locally balanced by inertia, assuming the Reynolds number is large. However, there is also a well known integral requirement on closed streamline flows. It may be derived from the Navier-Stokes equation,

$$\boldsymbol{\omega} \times \boldsymbol{u} = -\nabla(p/\rho + u^2/2) + \nu\nabla^2\boldsymbol{u} + \boldsymbol{F}/\rho.$$

Integrating around a closed streamline we deduce,

$$\oint \boldsymbol{F} \cdot \mathrm{d}\boldsymbol{r} + \rho\nu\oint \nabla^2\boldsymbol{u} \cdot \mathrm{d}\boldsymbol{r} = 0. \tag{10}$$

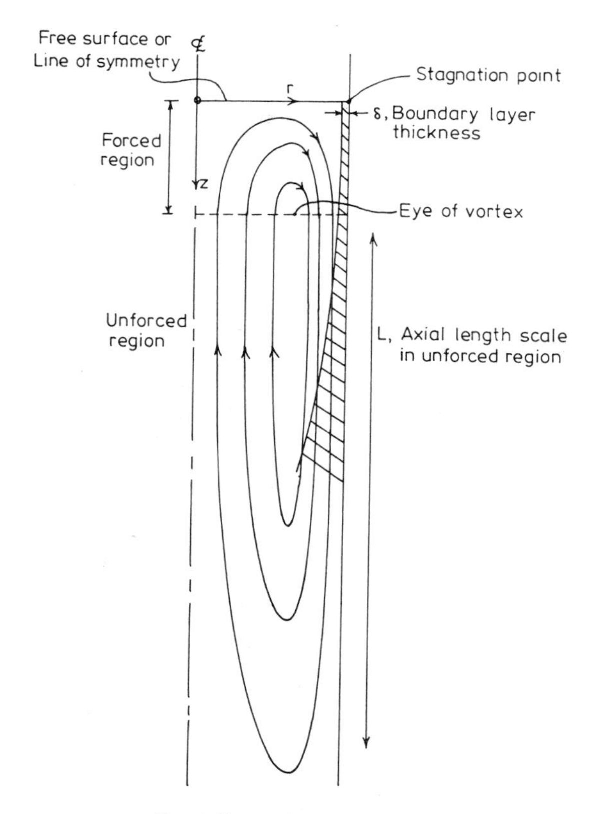

Fig. 4. Shape of poloidal eddy.

This is a form of Bernoulli's equation showing that the work done on a fluid element by the body force must be dissipated by shear. This has implications regarding the length scale for z in the unforced region, and the axial position of the eye of the vortex.

The eye of the vortex cannot lie in the forced region since this would imply that closed streamlines exist entirely within that region, and since the flow there is essentially inviscid, equation (10) could not be satisfied. On the other hand, if the eye of the vortex lay some distance below the forced region, then equation (10) would require the velocity field in the region of the eye of the vortex to satisfy

$$\oint \nabla^2 \boldsymbol{u} \cdot \mathrm{d}\boldsymbol{r} = 0.$$

In general the velocity field will not satisfy this constraint. Instead, the eye of the vortex positions itself at the edge of the forced region, with each streamline passing through both the forced and unforced regions. This is shown in Fig. 4.

The boundary layer on the wall grows at a rate,

$$\delta \sim \sqrt{\frac{\nu z}{u}}\,.$$

Equation (10) requires all streamlines to pass through this boundary layer. This suggests that the axial length scale in the unforced region, L, satisfies,

$$R \sim \sqrt{\frac{\nu L}{u}}$$

$$\Rightarrow L \sim \mathrm{Re} \cdot R.$$

This result will be deduced later on dimensional grounds. It implies that the penetration of the swirl beyond the forced region is extensive.

It is useful to consider the angular momentum integral applied to the problem.

$$\boldsymbol{T} = \oint (\rho \boldsymbol{r} \times \boldsymbol{u}) \boldsymbol{u} \cdot \mathrm{d}\boldsymbol{s}.$$

If the control volume is taken as the flow field as a whole, then the applied magnetic torque must be balanced by shear on the walls.

$$\int_0^\infty \int_0^R F_\theta r^2 \, \mathrm{d}r \, \mathrm{d}z = R^2 \int_0^\infty | \tau_{r\theta} |_R \, \mathrm{d}z.$$

If the momentum integral is applied from the free surface to the bottom of the forced region, and the wall shear in the forced region neglected in comparison with the total wall shear, then,

$$\int_0^\infty \int_0^R F_\theta r^2 \, \mathrm{d}r \, \mathrm{d}z = \rho \int_0^R r^2 \bar{u}_\theta \bar{u}_z \, \mathrm{d}r.$$

The bars over the velocity terms indicate values at the bottom of the forced region. Combining these equations and substituting for F_θ from equation (3) we deduce,

$$\tfrac{1}{8} \overline{V}^2 R^2 \int_0^\infty f(z/R) \, \mathrm{d}z = \int_0^R r^2 \bar{u}_\theta \bar{u}_z \, \mathrm{d}r = R^2 \int_0^\infty \frac{| \tau_{r\theta} |_R}{\rho} \, \mathrm{d}z. \tag{11}$$

$\uparrow$ Total applied magnetic torque = $\uparrow$ Flux of angular momentum out of forced region = $\uparrow$ Torque due to wall shear

It follows from equation (11) that in the forced region,

$$u_\theta u_z \sim \overline{V}^2.$$

By comparing inertial terms in equation (9), and invoking continuity, it may be concluded that,

$$u_r \sim u_z \sim u_\theta.$$

Combining these two estimates we deduce that in the forced region,

$$u_r \sim u_z \sim u_\theta \sim \overline{V}.$$

Thus the secondary poloidal flow is as large as the primary azimuthal flow, all three velocity components being of order $\overline{V}$.

In the unforced region, however, we expect the axial length scale to be large. Comparing the first and last terms in equation (11) we deduce,

$$\overline{V}^2 R^3 \sim R^2 \int_0^\infty \frac{|\tau_{r\theta}|_R}{\rho} \, dz \sim R^2 \nu \frac{u_\theta}{\delta} L.$$

Taking the boundary layer thickness in the unforced region to be of order R we deduce,

$$L \sim \mathrm{Re} \cdot R. \tag{12}$$

This is the same estimate for L that was obtained from a consideration of the force integral $\oint \boldsymbol{F} \cdot d\boldsymbol{r}$. Note that the large axial length scale in the unforced region requires that,

$$u_r \sim \overline{V}/\mathrm{Re} \quad \text{in the unforced region.}$$

In summary, we have the following scaling relationships,
Forced region:

$$u_r \sim u_\theta \sim u_z \sim \overline{V}$$

$$r \sim z \sim R$$

Unforced region:

$$u_\theta \sim u_z \sim \overline{V}, \quad u_r \sim \overline{V}/\mathrm{Re}$$

$$r \sim R, \quad L \sim R \cdot \mathrm{Re}.$$

In the previous section we saw that the core angular velocity is predicted by a one-dimensional model as,

$$\Omega_0 = \frac{\overline{V}}{R}\left\{0.88 \ln \frac{\overline{V}R}{\nu} + 1.0\right\}.$$

In a typical continuous casting plant the Reynolds number is of order 10^5. This equation then gives,

$$\Omega_0 \sim 10 \frac{\overline{V}}{R}.$$

The dimensional analysis given above suggests that in a two-dimensional axisymmetric flow, $\Omega_0 \sim \overline{V}/R$. In fact, we shall see in the next section that typically, $\Omega_0 \sim 2\overline{V}/R$. This is a factor of five smaller than that predicted by the one-dimensional analysis. The reason for the difference is, of course, that in one case the magnetic body force is balanced by shear, while in the other it is balanced by inertia.

V. Numerical experiments of two-dimensional stirring

The arguments presented in the preceding section can be substantiated by means of finite difference computations of two-dimensional axisymmetric stirring. The turbulence is modelled via the widely used $k-\epsilon$ model which entails the solution of transport equations for the kinetic energy of turbulence and its dissipation rate. The Reynolds stresses are related to the rate of strain using an eddy viscosity hypothesis [8].

$$\sigma_{ij} = \rho \nu_t \left(\frac{\partial u_i}{\partial x_j} + \frac{\partial u_j}{\partial x_i} \right) - \tfrac{2}{3} \rho k \delta_{ij}.$$

The eddy viscosity ν_t is then related to the turbulence kinetic energy k and viscous dissipation, ϵ.

$$\nu_t = c_\mu k^2 / \epsilon$$

where c_μ is a constant.

The transport equation for k is based on the turbulence kinetic energy equation, while the transport equation for ϵ is somewhat more empirically based.

We have used the standard form of the transport equations for k and ϵ which may be found in [8]. Some workers in swirling flows suggest that these standard equations need to be altered to take into account the effect of streamline curvature on the turbulence. Typically, by analogy with the correction for buoyancy [8], the source term in the ϵ equation may be altered to take into account the stabilising or destabilising effect of rotation. (We would expect the turbulence to be destabilised for $\partial \Gamma / \partial r < 0$, and stabilised for $\partial \Gamma / \partial r > 0$.) However, there is no universally accepted correction for curvature and we have not used any such correction in the computations.

There is some debate as to the generality of different turbulence models. However, we have shown that the flow in the forced region is largely controlled by the inertia of the mean flow and that the primary effect of shear is to control the length of the diffusive region. We expect, therefore, that many of the broad features of the flow are insensitive to the details of the turbulence modelling.

Computations were performed using a finite difference code employing a power-law differencing scheme. The details of the solution methodology can be found elsewhere [6].

The electromagnetic body force used in the computations is that given by equation (3), with an inverse fourth power law distribution for $f(z/R)$.

$$f(z/R) = 2\left[1 + (z/R)^4\right]^{-1}.$$

This force drops to 5% of its initial value after one diameter depth ($z = 2R$).

Two cases were analysed corresponding to different magnitudes of the body force, and hence $\overline{V}$. The cases considered correspond to $\overline{V} = 1$ cm/s and $\overline{V} = 10$ cm/s. If the analysis presented in Section IV is correct, then u should scale on $\overline{V}$. We shall see that this is indeed the case. The relevant physical properties used in the computations are given in Table 1.

Table 1. Physical properties used in computations.

Case	ρ (kg/m^3)	ν (m^2/s)	R (m)	$\overline{V}$ (m/s)	Roughness parameter E
1	7×10^3	10^{-6}	0.1	0.01	Smooth
2	7×10^3	10^{-6}	0.1	0.1	0.1 (k′ ~ 1/2 mm)

Fig. 5. Computed poloidal flow pattern (Case 2).

In order to investigate the effect of wall roughness on the flow, a rough wall was used in the second case. The roughness is specified in terms of a roughness parameter, E. This appears in a 'wall function' used in the computations to relate the wall slip velocity, u_s, to the local shear velocity, V_* [8].

$$u_s/V_* = 2.5 \ln(EV_* y/\nu).$$

For a smooth wall $E = 9$, and for a rough wall $E = \nu/V_* k'$, where k' is the roughness height. We chose $E = 0.1$ for the rough wall, which, for the Reynolds number used, corresponds approximately to $k' = 1/2$ mm.

Figure 5 shows the computed poloidal flow for the rough wall case. The difference between the axial and radial length scales is clear. The eye of the poloidal vortex lies at $z = 2.0\ R$, which corresponds, approximately, to the bottom of the forced region.

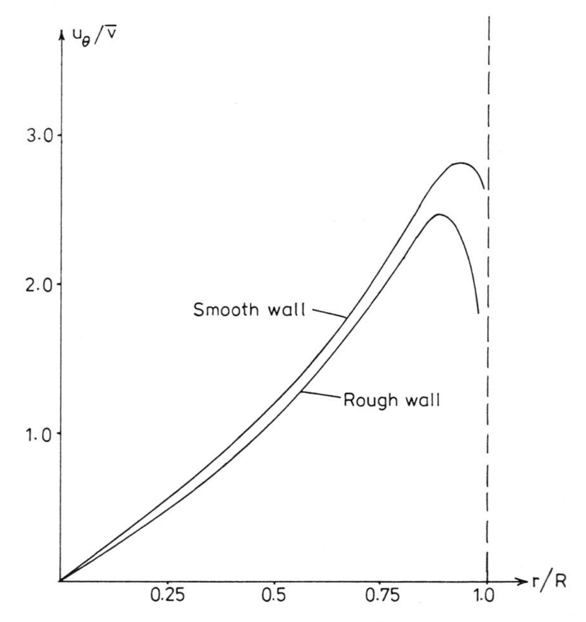

Fig. 6. Computed azimuthal surface velocity.

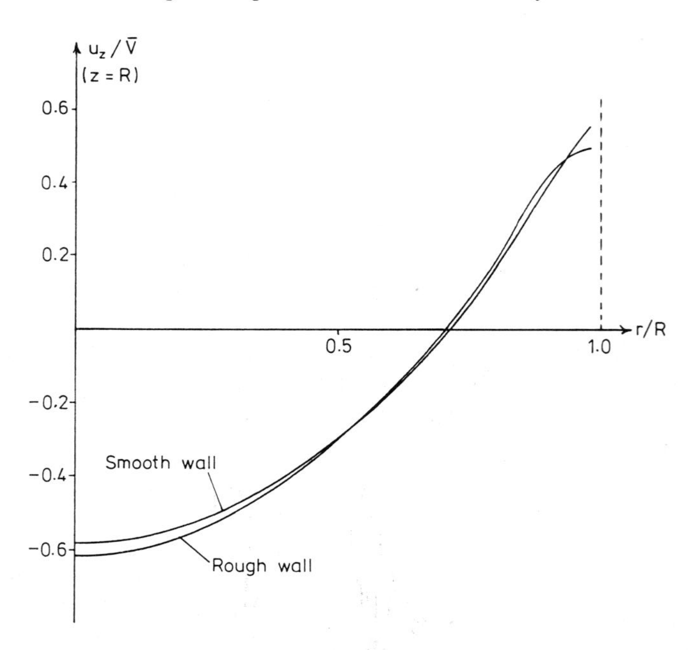

Fig. 7. Computed axial velocity ($z = R$).

Figures 6 and 7 show the computed azimuthal velocity profile at the free

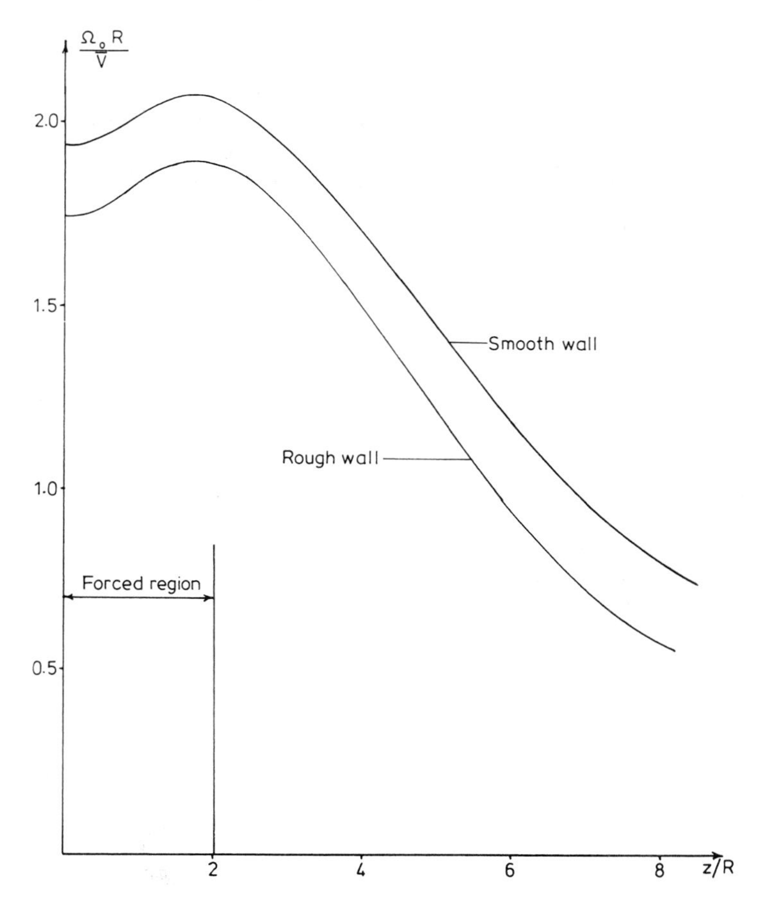

Fig. 8. Variation of core angular velocity with depth.

surface ($z = 0$) and axial velocity profile at $z = R$. Both the smooth and rough wall results are given. It is clear that the axial and azimuthal velocities are of the same order, and are insensitive to the wall shear.

Figure 8 shows the variation of core angular velocity with depth. The swirl penetrates well out of the forced region, being carried by the secondary flow.

The computed eddy viscosity varies throughout the flow field. Typically the effective Reynolds number $\mathrm{Re}_t = \overline{V}R/\nu_t$, based on the eddy viscosity, has a value of ~ 30 in both cases. The similarity in Re_t for the smooth and rough wall flows explains why the rates of decay of Ω, shown in Fig. 8, are similar.

It is clear that the results of the computations are in accordance with the discussion of Section IV.

VI. Conclusions

On ignoring the axial variation of the electromagnetic body force, a simple, one-dimensional, swirl flow is produced. There are no inertial effects except for the centripetal acceleration, which is balanced by a radial pressure gradient. It follows that the only forces available to balance the magnetic torque are shear stresses. Information about the relationship between shear stress and velocity gradients in the fluid is then sufficient to allow the velocity profile to be calculated.

These one dimensional models are misleading in the context of magnetic stirring. In a real stirrer, the magnetic forcing occurs only over a relatively short length of the cylinder. This results in differential rotation between forced and unforced regions. This differential rotation drives a secondary poloidal flow, where fluid particles now follow helical paths.

Dimensional analysis and numerical experiments suggest that the secondary flow is as large as the primary swirl flow. This secondary flow sweeps angular momentum out of the forced region so that the forced vortex ultimately penetrates some distance beyond the stirrer. Since inertial forces are now non-zero, the magnetic body force is locally balanced by inertia, rather than shear. This results in a quite different magnitude of swirl than that predicted by the one-dimensional model.

The primary role of shear is to dissipate the energy created by $\boldsymbol{J} \times \boldsymbol{B}$. This occurs predominantly in the unforced region, in which the axial length scale is controlled by the shear.

The existence of strong secondary flows during rotary magnetic stirring has been largely ignored in the literature. Their effect in extending the forced vortex well beyond the immediate vicinity of the magnetic stirrer is clearly important with regard to interpreting the metallurgical influence of stirring.

Acknowledgements

The authors would like to thank TI Research Labs., and Dr J.C.R. Hunt of Cambridge University for their advice and support in this study. The computer code was a modified version of FLUENT made available by Creare R&D.

References

1. P.A. Davidson: Estimation of turbulent velocities induced by magnetic stirring during continuous casting, *Materials Sc. and Technology* 1 (1985) 994–999
2. L.D. Landau and E.M. Lifshitz: *Electrodynamics of Continuous Media.* Pergamon Press (1960) Chap. VII.
3. H.S. Marr: Electromagnetic stirring in continuous casting of steel. In: M.R.E. Proctor and H.K. Moffatt (eds) *Proceedings of Metallurgical Applications of MHD.* IUTAM, Cambridge (1982) pp 143–153.

4. D.A. Melford and P.A. Davidson; The solidification of continuously cast steel under conditions of forced flow. In: Conf. *Modelling of Casting and Welding Processes*. New England College, USA (1983) pp 101–104.
5. H.K. Moffatt: Rotation of a liquid metal under the action of a rotating magnetic field. In: H. Branover and A. Yakhot (eds) *MHD - Flows and Turbulence II*. Israel University Press (1978) pp45–62.
6. S.V. Patankar; *Numerical Heat Transfer and Fluid Flow*. Hemisphere (1980).
7. T. Robinson and K. Larsson; An experimental investigation of a magnetically driven rotating liquid-metal flow. *J. Fluid Mech.* 60 (1973) 641–664
8. W. Rodi: Turbulence models and their applications in hydraulics. State-of-the-art paper IAHR (1984).
9. K.H. Tacke and K. Schwerdtfeger: Stirring velocities in continuously cast round billets as induced by rotating electromagnetic fields. *Stahl und Eisen* 99 (1979) 7–12.

Applied Scientific Research 44: 261–275 (1987)

On determining the shape of weld pools

R.E. CRAINE
Faculty of Mathematical Studies, The University, Southampton S09 5NH, UK

Abstract. The equations governing heat and fluid flow in weld pools for the TIG fusion welding process are presented and this coupled system is solved numerically using finite differences. Electromagnetic forcing terms, buoyancy forces, shear forces on the pool surface due to the variation in surface tension with temperature and an additional uniform magnetic field applied normal to the workpiece are all included in our model and results are displayed indicating the relative importance of these four mechanisms.

1. Introduction

Until comparatively recently little was known about the physical and chemical processes involved in welding and satisfactory welds were only produced by skilled practitioners using 'rule of thumb' techniques based on experience. Since the modern trend is towards the use of automatic welding procedures, there has been considerable effort expended in recent years in trying to understand more fully the underlying physical mechanisms.

In this paper attention is concentrated on a commonly-used fusion welding process known as TIG (Tungsten electrode with an Inert Gas shroud). In the TIG process a high current electric arc is moved along the line of contact between the two components to be joined. Sufficient heat is deposited on the surface of the materials by the arc to cause the components to melt. Then, as the arc moves away, the molten material solidifies and the components are fused together.

The first significant mathematical attempt to model the fusion welding process was made by Rosenthal [1], who derived a number of solutions of the steady state heat conduction equation $\kappa\nabla^2 T = \boldsymbol{v}\cdot\nabla T$, where κ is the thermal diffusivity of the material, T is the temperature at any point inside the material and $\boldsymbol{v}$ is the velocity of the arc relative to the workpiece. Rosenthal's model has proved popular with metallurgists for it reveals, for instance, the characteristic tail of the weld pool behind the heat source and it can also be relatively easily adapted to take account of different geometrical arrangements, distributed heat sources, varying thermal properties of the material and so on (see [2]). The model does, however, neglect fluid motion in the weld pool, and the existence of such motion has been confirmed experimentally [3,4].

In the TIG process the electrode is non-consumable and the major reasons for convection in the pool seem to be:
(a) electromagnetic body forces inside the weld pool,
(b) buoyancy effects inside the pool
and
(c) shear forces on the free surface of the pool due to both
(i) the gas jet and (ii) the variation in surface tension with temperature (the Marangoni effect).

Until the last few years only (a) has received much attention. Some experiments were performed with liquid mercury [5], which could be forced to flow in different directions according to how electric current entered or left the pool. A little later the first theoretical attacks on the effects of (a) were made [6–8], by considering the flow of an electrically conducting fluid in a semi-infinite region when the current in the fluid diverged from a stationary point source on its surface. The flow induced by a similar current source on the surface of fluid confined to a hemispherical container was investigated in [9]. Unfortunately, the solutions in [6–9] all suffered from the inherent weakness that the velocity field became non-physical on the axis of symmetry for the values of the applied current typically used in welding. Replacing the point source of current by a distributed one improved the situation and, after initial work on the linear problem (valid at low currents) in [10], Atthey [11] used a distributed source model in solving the nonlinear problem in a hemispherical weld pool in a plate at realistic currents by numerical finite difference methods.

At this stage in the theoretical development of the area, therefore, there existed the two separate, but incomplete, approaches – Rosenthal's solution of the heat conduction equation for a moving heat source but a stagnant pool and, secondly, the solutions in [6–11] (and other related ones, mainly by Sozou and his co-workers) for the flow of electrically conducting fluids, created by stationary current sources, in regions of simple predetermined shape with thermal effects ignored.

Within the past few years attention has turned inevitably to combining the fluid and heat flow effects. The shape of the solid/liquid boundary in a semi-infinite material has been determined in [12,13] by finding a numerical solution to the coupled fluid and heat flow equations, but only the electromagnetic forcing terms (effects (a)) were included. The results in [12,13] revealed that at realistic welding currents the electromagnetic forces produced by a stationary, axisymmetric, distributed current source were sufficiently large to have a significant influence on the position of the solid/liquid interface. The shape of this boundary was found to be particularly sensitive to the magnitude and concentration of the applied current source.

Quite independently of the above work, investigations of a similar type were being carried out at M.I.T. and results for the development of the shape of the solid/liquid boundary with time and for the associated motion in the molten metal due to an applied, stationary, axisymmetric, distributed source of current on the surface of a thick plate have been published [14]. Most of the results discussed in [14] concern the relative importance of the forcing mechanisms (a) and (b), but some preliminary work was included on the effects of also introducing (c)(ii).

Further numerical solutions of the coupled equations for a plate for typical welding currents have appeared in [15]; the forcing mechanisms (a) and (c)(ii) being included but buoyancy ((b)) omitted. Also of relevance are the results obtained for surface tension driven convective flow in laser melted pools [16]. Additional results for the heat and fluid flow in stationary arc welds, with forcing terms (a), (b) and (c)(ii) present, can be found in [17] for various surface tension temperature coefficients and applied current density distributions. Particularly noteworthy is the very recent paper of Kuo and Wang [18] in which some numerical results are obtained in three-dimensions for a moving source of current and heat on the surface of a plate, with the driving mechanisms (a), (b) and (c)(ii) all incorporated.

The absence from all the papers mentioned above of the shearing force due to the gas jet ((c)(i)) is not important since it has been shown in [19] that arc pressure effects are only significant at extremely high currents ($\geqslant 500$ A).

The results in [12–18] clearly indicate that the electromagnetic forcing terms, buoyancy and the shear forces generated on the free surface of the pool by the variation in surface tension with temperature can each lead to major changes in the position and shape of the solid/liquid boundary, but that it is normally the variation in surface tension that is the dominant mechanism. The change in surface tension with temperature is numerically quite small but in weld pools the temperature differs by hundreds of degrees over distances of a few millimetres and the cumulative effect of this variation on the position of the solid/liquid interface is found to be considerable. The importance of the temperature coefficient of surface tension was proposed in [20] and subsequent experiments (see, for example, [21–25]) support their suggestion.

In this paper additional results are presented that reveal further useful information on the relative importance of the three forcing mechanisms (a), (b) and (c)(ii) and particularly on the effects of applying to the workpiece an additional uniform magnetic field. The reason for including such a magnetic field is of interest. It has long been recognized that it is very difficult to design reliable welding processes, even after taking elaborate care to control all of the main process variables. This lack of reproducibility is of particular concern in automatic processes where tolerances may be critical. One practical solution to the problem was found to be the application of an external magnetic field that was sufficiently large to swamp the effects of the self-induced magnetic field and lead to bulk motion of the fluid in an azimuthal direction [26]. This additional field was found experimentally to decrease the depth of the weld pool but, most importantly, also to significantly improve the overall reproducibility of the TIG weld [27]. Some theoretical work on the fluid flows that occur in the TIG process when a uniform externally applied magnetic field is introduced, but thermal effects are excluded, can be found in [28].

In this paper we consider a stationary, axisymmetric distributed source of current (and heat) on the surface of a semi-infinite block of metal which is also subjected to an additional uniform magnetic field that is applied normal to the surface. Electromagnetic forces, buoyancy and shear forces due to the variation of surface tension are all included. The governing equations for the problem are stated in section 2, in stream function-vorticity form. In section 3 a numerical solution of the coupled equations is sought using finite differences. Many features of the numerical procedure are similar to those discussed in the simpler problem considered in [13] and, consequently, only a brief outline of the method is presented here. Some numerical results are discussed in section 4 and the main conclusions to be drawn from our work are stated in section 5.

2. Basic equations

In practical welding situations the magnetic Reynolds number is always much less than unity and, consequently, back e.m.f. effects can be neglected. The current and magnetic field distributions can therefore be determined independently of considerations of the fluid flow.

Within the weld pool the continuity, momentum and heat convection equations

$$\nabla \cdot \boldsymbol{v} = 0, \tag{2.1}$$

$$\rho\left(\frac{\partial \boldsymbol{v}}{\partial t} + \nabla\left(\tfrac{1}{2}v^2\right) - \boldsymbol{v}\times\boldsymbol{\omega}\right) = -\nabla p - \eta\nabla\times\boldsymbol{\omega} + \boldsymbol{j}\times\boldsymbol{B} + \boldsymbol{F} \tag{2.2}$$

and

$$\kappa\nabla^2 T = \boldsymbol{v}\cdot\nabla T \tag{2.3}$$

must be satisfied, where $\boldsymbol{v}$ denotes the velocity of the fluid of density ρ and coefficient of viscosity η and $\boldsymbol{\omega} = \nabla\times\boldsymbol{v}$ is its vorticity, p denotes the pressure, $\boldsymbol{j}$ and $\boldsymbol{B}$ represent the current density and magnetic induction respectively, $\boldsymbol{F}$ denotes the buoyancy force, κ is the thermal diffusivity and T is the temperature. In the solid $\boldsymbol{v} = \boldsymbol{0}$ and equation (2.3) clearly reduces to Laplace's equation.

In this paper the resultant value for $\boldsymbol{B}$ is the sum of $\boldsymbol{B}_0$, the externally applied uniform magnetic field, and $\boldsymbol{B}_s$, the self-field, a term arising directly from the current density induced by the applied source and related to it through Maxwell's equation $\nabla\times\boldsymbol{B} = \mu_0\boldsymbol{j}$, where μ_0 is the permeability. Introduce spherical polar coordinates (r, θ, ϕ) (see Fig. 1) then, assuming the applied magnetic field is parallel to $\theta = 0$ and acts into the semi-infinite block of material, the appropriate form for $\boldsymbol{B}_0$ is

$$\boldsymbol{B}_0 = B_0(\cos\theta, -\sin\theta, 0). \tag{2.4}$$

The uniformity of $\boldsymbol{B}_0$ ensures that the resultant value of $\boldsymbol{j}$ arises solely from the applied current source. Welding arcs are either relatively steady or in a

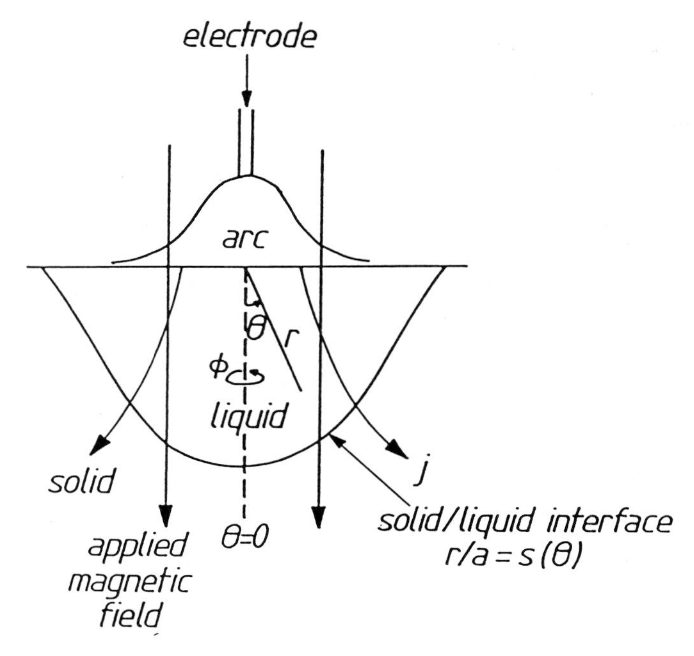

Fig. 1. Schematic section of axisymmetric weld pool produced by TIG welding process.

state of high frequency random motion but, in either case, it is reasonable to assume that the current density normal to the surface $\theta = \frac{1}{2}\pi$ is represented by the Gaussian distribution

$$j_\theta = -\frac{I}{\pi r_0^2} \, e^{-r^2/r_0^2} \quad \text{on } \theta = \tfrac{1}{2}\pi, \tag{2.5}$$

where I is the total current entering the material and r_0 denotes a characteristic radius for the decay of the current source. For our axisymmetric problem the quantities $\boldsymbol{j}$ and $\boldsymbol{B}_s$ take the forms

$$\boldsymbol{j} = (j_r, j_\theta, 0), \quad \boldsymbol{B}_s = (0, 0, B_{s\phi}), \tag{2.6}$$

where the non-zero components j_r, j_θ and $B_{s\phi}$ are determined explicitly from the electric potential, which can itself be calculated by solving Laplace's equation using Hankel transforms and the boundary condition (2.5) (see [13] for further details).

It is convenient to introduce a stream function ψ and write

$$\boldsymbol{v} = \left(\frac{1}{r^2 \sin\theta} \frac{\partial \psi}{\partial \theta}, \; -\frac{1}{r \sin\theta} \frac{\partial \psi}{\partial r}, \; w \right), \tag{2.7}$$

where both ψ and w are independent of ϕ. Equation (2.1) is then satisfied identically. Let a define a typical weld pool dimension, then we can introduce the non-dimensional variables R, τ, Ψ, V, Ω and Φ through the definitions

$$R = r/a, \quad \tau = \nu t/a^2, \quad \Psi = 2\psi/(a\nu), \tag{2.8}$$

$$V = 2a v/\nu, \quad \Omega = 2a^2\omega/\nu, \quad \Phi = (T - T_0)/(T_m - T_0), \tag{2.9}$$

where ν $(=\mu/\rho)$ is the kinematic viscosity, T_0 is the ambient temperature and T_m the melting temperature.

The buoyancy force $\boldsymbol{F}$ is calculated with the aid of the Boussinesq approximation and, recalling the axisymmetric nature of our problem and making use of the definitions (2.7) to (2.9), the vector equation (2.2) is equivalent to the three scalar equations

$$\frac{\partial Q}{\partial \tau} + \tfrac{1}{2}R^2 \sin^2\theta \,\mathrm{Div}\{(QV - W\Omega)/(R^2 \sin^2\theta)\}$$
$$= DQ + 2Gr\, R \sin\theta \left(\sin\theta \frac{\partial \Phi}{\partial R} + \frac{\cos\theta}{R}\frac{\partial \Phi}{\partial \theta}\right) + 2KI_1I_2, \tag{2.10}$$

$$\frac{\partial W}{\partial \tau} + \tfrac{1}{2}\,\mathrm{Div}(WV) = DW + NR \sin\theta\, I_1, \tag{2.11}$$

$$D\Psi = -Q, \tag{2.12}$$

where Div denotes the usual divergence operator in the variables (R, θ, ϕ), D is the operator

$$D \equiv \frac{\partial^2}{\partial R^2} + \frac{1}{R^2}\left(\frac{\partial^2}{\partial \theta^2} - \cot\theta \frac{\partial}{\partial \theta}\right) \tag{2.13}$$

and the variables Q and W are defined in terms of the azimuthal components of Ω and V by

$$Q = R \sin\theta\, \Omega_\phi, \quad W = R \sin\theta\, V_\phi, \tag{2.14}$$

respectively. In addition, in equations (2.10) to (2.12) the constants Gr (the Grashof number), K and N are defined through

$$Gr = \frac{\beta g (T_m - T_0) a^3}{\nu^2}, \quad K = \frac{\mu_0 I^2}{2\pi^2 \rho \nu^2}, \quad N = \frac{I B_0 a}{\pi \rho \nu^2}, \tag{2.15}$$

where β is the coefficient of volumetric expansion of the molten material and g is the acceleration due to gravity, and I_1 and I_2 denote the integrals

$$I_1 = \int_0^\infty \mathrm{e}^{(-0.25R_0^2x^2 - xR\cos\theta)} x J_1(xR\sin\theta)\,\mathrm{d}x, \tag{2.16}$$

$$I_2 = \int_0^\infty \mathrm{e}^{(-0.25R_0^2x^2 - xR\cos\theta)} J_1(xR\sin\theta)\,\mathrm{d}x, \tag{2.17}$$

which are both related to the components of the electromagnetic quantities $\boldsymbol{j}$ and $\boldsymbol{B}$, J_1 being the first order Bessel function of the first kind.

In a similar way the heat convection equation (2.3) becomes, in non-dimensional variables,

$$\mathrm{Div\,Grad}\,\Phi = \tfrac{1}{2}Pr\, V \cdot \mathrm{Grad}\,\Phi, \tag{2.18}$$

where Grad denotes the usual gradient operator with respect to the variables (R, θ, ϕ) and the Prandtl number Pr is given by

$$Pr = \nu/\kappa. \tag{2.19}$$

Equations (2.10) to (2.12) and (2.18) form a coupled set of equations for the independent variables Q, W, Ψ and Φ. A numerical solution of this system is required subject to the boundary conditions

$$\Psi = Q = \frac{\partial Q}{\partial \theta} = 0, \quad W = 0, \quad \frac{\partial \Phi}{\partial \theta} = 0, \quad \text{all on } \theta = 0, \tag{2.20}$$

$$\Psi = 0, \quad Q = S_T R \frac{\partial \Phi}{\partial R}, \quad \frac{\partial W}{\partial \theta} = 0, \quad \frac{\partial \Phi}{\partial \theta} = \left(\frac{Q^*}{R_0^2}\right) \mathrm{Re}^{-R^2/R_0^2},$$

$$\text{all on } \theta = \tfrac{1}{2}\pi, \tag{2.21}$$

$$\Psi = 0, \quad Q = -\frac{\partial^2 \Psi}{\partial R^2}, \quad W = 0, \quad \Phi = 1, \tag{2.22}$$

all on solid/liquid interface,

$$\Phi = \frac{Q^*}{2R_l} \quad \text{on the hemisphere } R = R_l, \tag{2.23}$$

where

$$R_0 = \frac{r_0}{a}, \quad Q^* = \frac{\overline{Q}}{\pi k a (T_m - T_0)}, \quad S_T = -\frac{2a(T_m - T_0)(\partial \gamma/\partial T)}{\rho \nu^2}, \tag{2.24}$$

k being the coefficient of thermal conductivity, $\overline{Q}$ the total power dissipated by the arc on the surface and γ the coefficient of surface tension. Most of the above boundary conditions are the same as those used in [13]. Of the extra conditions $(2.20)_4$ follows from the definition $(2.14)_2$ and the finiteness of V_ϕ, $(2.21)_3$ arises from zero shear force on the top surface in the azimuthal direction and $(2.22)_3$ is the usual no-slip condition. Allowing the surface tension to vary with temperature leads to condition $(2.21)_2$, where S_T is defined through $(2.24)_3$. It should be pointed out condition (2.23), which essentially states that sufficiently far from $R = 0$ the distributed source could be replaced by a point source, is used instead of applying a condition at infinity. Also, condition $(2.21)_4$ is formulated by assuming that the distribution of heat flux on $\theta = \frac{1}{2}\pi$ is of an identical form to that for the current source: a result confirmed for high current inert gas arcs by Nestor [29].

3. Numerical solution

A numerical finite difference solution of the governing equations is now sought. In order to deduce the position of the solid/liquid interface, $R = s(\theta)$, it is necessary at earlier stages of the procedure to iterate the shape of this interface. A fixed mesh could lead to inaccuracies, so a change in coordinates from (R, θ) to (ξ, θ) through the transformation $\xi = R/s(\theta)$ is introduced. This change ensures that the weld pool is mapped onto the hemisphere $0 \leqslant \xi \leqslant 1$ and therefore makes it easy for us to introduce mesh points that always coincide with the current position of the solid/liquid interface. The transformation does make the governing differential equation more complicated, but the disadvantage of a slight increase in computing time is thought to be outweighed by the improvement in accuracy.

In (ξ, θ) space the governing equations (2.10) to (2.12) and (2.18) become

$$\frac{\partial Q}{\partial \tau} + \frac{\xi^2 \sin^2\theta}{2s} \operatorname{Div}^{(1)}\left\{\left(QV^{(1)} - W\Omega^{(1)}\right)/(s\xi \sin\theta)^2\right\} = D^{(1)}Q + 2Gr\xi \sin\theta\left\{\left(\sin\theta - \frac{s' \cos\theta}{s}\right)\frac{\partial \Phi}{\partial \xi} + \frac{\cos\theta}{\xi}\frac{\partial \Phi}{\partial \theta}\right\} + 2KI_1I_2, \tag{3.1}$$

$$\frac{\partial W}{\partial \tau} + \frac{1}{2s^3}\operatorname{Div}^{(1)}\left(WV^{(1)}\right) = D^{(1)}W + Ns\xi \sin\theta\, I_1, \tag{3.2}$$

$$D^{(1)}\Psi = -Q \tag{3.3}$$

and

$$D^{(2)}\Phi = \beta\left(\frac{\partial \Psi}{\partial \theta}\frac{\partial \Phi}{\partial \xi} - \frac{\partial \Psi}{\partial \xi}\frac{\partial \Phi}{\partial \theta}\right), \tag{3.4}$$

in which the variables Q, W, Ψ and Φ are now functions of ξ, θ and τ, $\operatorname{Div}^{(1)}$ denotes the divergence operator in (ξ, θ) coordinates, a $'$ denotes differentiation with respect to θ, $V^{(1)}$ and $\Omega^{(1)}$ are defined by

$$V^{(1)} = \left(\frac{1}{\xi^2 \sin\theta}\frac{\partial \Psi}{\partial \theta}, -\frac{1}{\xi \sin\theta}\frac{\partial \Psi}{\partial \xi}, 0\right), \tag{3.5}$$

$$\Omega^{(1)} = \left(\frac{1}{\xi^2 \sin\theta}\frac{\partial W}{\partial \theta}, -\frac{1}{\xi \sin\theta}\frac{\partial W}{\partial \xi}, 0\right), \tag{3.6}$$

and the operators $D^{(1)}$ and $D^{(2)}$ denote

$$D^{(1)} = \alpha_1\frac{\partial^2}{\partial \xi^2} + \alpha_2\frac{\partial^2}{\partial \xi \partial \theta} + \alpha_3\frac{\partial^2}{\partial \theta^2} + \alpha_4\frac{\partial}{\partial \xi} + \alpha_5\frac{\partial}{\partial \theta}, \tag{3.7}$$

$$D^{(2)} = \alpha_1\frac{\partial^2}{\partial \xi^2} + \alpha_2\frac{\partial^2}{\partial \xi \partial \theta} + \alpha_3\frac{\partial^2}{\partial \theta^2} + \alpha_6\frac{\partial}{\partial \xi} - \alpha_5\frac{\partial}{\partial \theta}, \tag{3.8}$$

where

$$\alpha_1 = \left(s^2 + (s')^2\right)/s^4, \quad \alpha_2 = -2s'/(\xi s^3), \quad \alpha_3 = 1/(\xi s)^2, \tag{3.9}$$

$$\alpha_4 = \left(-s'' + 2(s')^2 s^{-1} + s' \cot\theta\right)/(\xi s^3), \quad \alpha_5 = -\cot\theta/(\xi s)^2, \tag{3.10}$$

$$\alpha_6 = \left(-s'' + 2(s')^2 s^{-1} - s' \cot\theta + 2s\right)/(\xi s^3),$$

$$\beta = Pr/\left(2\xi^2 s^3 \sin\theta\right). \tag{3.11}$$

The boundary conditions (2.20) to (2.22) remain formally the same in (ξ, θ) space except that $(2.21)_{2,4}$, $(2.22)_2$ and (2.23) are replaced, respectively, by

$$Q = S_T \xi \frac{\partial \Phi}{\partial \xi}, \quad \frac{1}{\xi}\frac{\partial \Phi}{\partial \theta} = \frac{s'}{s}\frac{\partial \Phi}{\partial \xi} + \frac{Q^*}{R_0^2} s\, \mathrm{e}^{-(\xi s/R_0)^2}, \quad \text{on } \theta = \tfrac{1}{2}\pi, \tag{3.12}$$

$$Q = -\alpha_1 \frac{\partial^2 \Psi}{\partial \xi^2} \quad \text{on } \xi = 1 \text{ (the solid/liquid interface)}, \tag{3.13}$$

and

$$\Phi = Q^*/(2s\xi_l). \tag{3.14}$$

The final boundary condition (3.14) is applied on the surface in (ξ, θ) space that corresponds to the large hemisphere of radius R_l in (R, θ) space. The surface in (ξ, θ) space will not generally intersect the mesh points so some interpolation is required. However, as the temperature Φ is only slowly varying in this region the errors introduced by the interpolation are extremely small.

The system considered in this paper extends that investigated in [13] through the inclusion of the additional variable W, the nonzero shear stress on the free surface and the buoyancy terms, but a similar method of solution to that given in [13] can be adopted. In fact, since much of the discretization is identical to that presented in [13] and the additional terms are considered in an entirely analogous way, it is felt unnecessary to display in this paper the lengthy finite difference forms of the equations and boundary conditions. Only a brief outline of the numerical procedure will therefore be presented.

Suppose the position of the solid/liquid interface is known at some stage in the iterative process. Then, using the simple transformation of variables discussed earlier, the interior of the weld pool is mapped on to the hemisphere $0 \leqslant \xi \leqslant 1$ and the next stage in the procedure requires the solution of equations (3.1) to (3.3) within this hemisphere. The updated values of Q and W are calculated from the finite difference forms of (3.1) and (3.2), derived using the Dufort-Frankel leap-frog method, and the corresponding values of Ψ are then determined from the discretized form of (3.3) using successive over-relaxation. The calculation of Q, W and Ψ is then repeated at successive time steps until convergence is achieved. Once the fluid flow is known the solution for Ψ is substituted into the finite difference form of (3.4). The Péclet number associated with (3.4) seems likely to lie between 10 and 100 in typical welding situations and so to improve the stability of our numerical scheme a further transformation of variables is carried out before the equation is discretized (see [13] for full details). The temperature Φ is then found from (3.4) using a relaxation method and from this solution the new position of the melting isotherm (the solid/liquid boundary) is calculated. The whole procedure is repeated until the position of the solid/liquid boundary is unchanged, to a specified accuracy, from one iteration to the next.

4. Results and discussion

All the results presented in this section were obtained with a 11×11 mesh inside the weld pool. The radius of the outer hemisphere, R_l, was taken to be three times the maximum radius of the weld pool, but the final results were found to be insensitive to an increase in this radius, and also to the particular initial conditions chosen for our solutions. For all results we assumed $Q^* = 2$, which effectively fixes the size of the weld pool, and the typical weld pool dimension, a, was chosen to be 2×10^{-3}. Also introduced were the values $\rho = 6.48 \times 10^3$, $\nu = 4.475 \times 10^{-7}$, $Pr = 0.1$, $T_m = 1809^0 K$ and $T_0 = 293^0 K$, all appropriate for the welding of steel.

Figure 2 shows results obtained for the shape of the solid/liquid interface for an applied current of 50 Amps and for a decay radius $R_0 = 1.0$ in the absence of an externally applied magnetic field ($B_0 = 0$). Results are displayed for conduction only, for convection created by the electromagnetic forces only (the situation considered in [12,13]) and also for models in which the additional mechanisms of buoyancy and variation of surface tension with temperature are included. As remarked in [12,13], the electromagnetic forces on their own considerably increase the depth/width ratio of the weld pool, but Fig. 2 demonstrates that this increase is considerably reduced when realistic buoyancy terms (with $\beta = 1.36 \times 10^{-4}$) are added. Positive values for S_T, the constant governing the surface tension effects, are also seen to significantly reduce the size of the weld pool. Figures 3a,b show the streamlines and isotherms for two cases where buoyancy is included for different values of S_T.

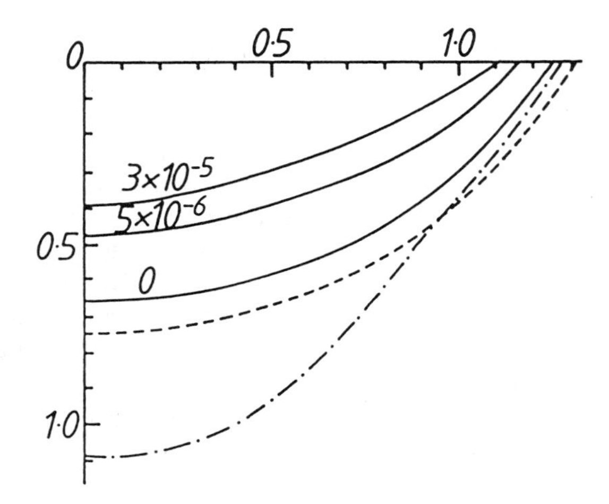

Fig. 2. Shape of solid/liquid boundary when $I = 50$, $R_0 = 1.0$, $B_0 = 0$ (i.e. $K = 1.2 \times 10^5$, $N = 0$): — — — — heat conduction only; –·–· electromagnetic forcing terms only ($Gr = S_T = 0$); ——— electromagnetic, buoyancy ($Gr = 0.81 \times 10^5$) and variation in surface tension included at stated values of $-\partial\gamma/\partial T$ ($\partial\gamma/\partial T = -5 \times 10^{-5}$ equiv. to $S_T = 0.23 \times 10^5$, $\partial\gamma/\partial T = -3 \times 10^{-5}$ equiv. to $S_T = 1.4 \times 10^5$).

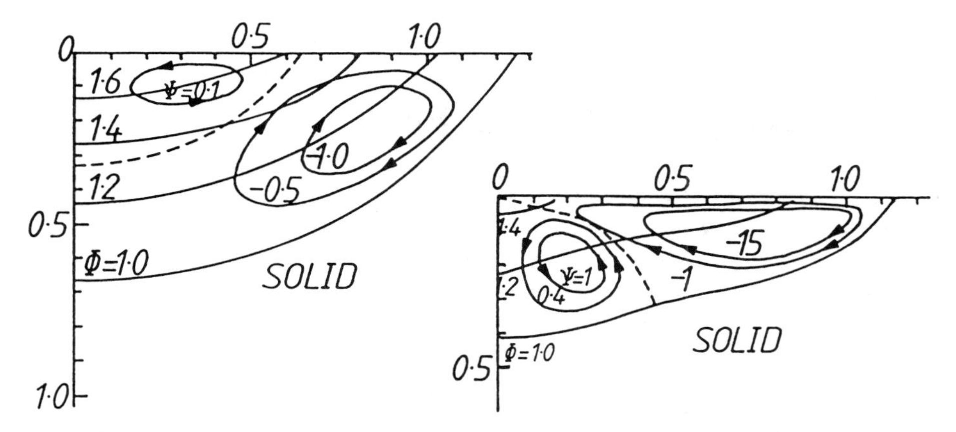

Fig. 3. Streamlines and isotherms when $I = 50$, $R_0 = 1.0$, $B_0 = 0$, $\beta = 1.36 \times 10^{-4}$ (i.e. $K = 1.2 \times 10^5$, $Gr = 0.81 \times 10^5$, $N = 0$): (a) $\partial\gamma/\partial T = 0$ (i.e. $S_T = 0$); (b) $\partial\gamma/\partial T = -3 \times 10^{-5}$ (i.e. $S_T = 1.4 \times 10^5$).

Although both these figures display counter-rotating loops for the poloidal flow other results that have been obtained indicate that single loops can arise for certain values of the parameters.

When an additional uniform magnetic field is applied to the material our numerical results reveal some qualitative changes in flow patterns and weld pool shape. Plots of the pool shape for $B_0 = 5 \times 10^{-4}T$, again for a current source with $I = 50A$ and $R_0 = 1.0$, are given in Fig. 4 for a number of the cases considered in the previous paragraph. Comparison of Figs. 2 and 4 shows that, when B_0 is non-zero, the depth/width ratio of the weld pool is reduced by varying amounts: the reduction being least when $S_T > 0$. The isotherms, streamlines and contours of constant W, for one particular choice

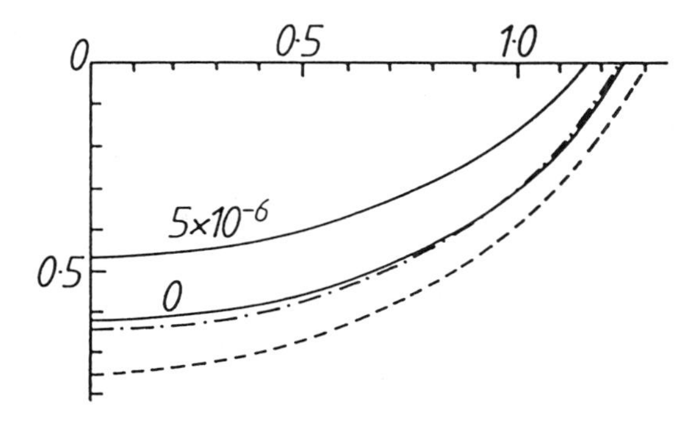

Fig. 4. Shape of solid/liquid boundary when $I = 50$, $R_0 = 1.0$, $B_0 = 5 \times 10^{-4}$ (i.e. $K = 1.2 \times 10^5$, $N = 1.2 \times 10^4$): — — — — heat conduction only; –·–· electromagnetic forcing terms only ($Gr = S_T = 0$); ——— electromagnetic, buoyancy ($Gr = 0.81 \times 10^5$) and variation in surface tension included at stated values of $-\partial\gamma/\partial T$ ($\partial\gamma/\partial T = -5 \times 10^{-6}$ equiv. to $S_T = 0.23 \times 10^5$).

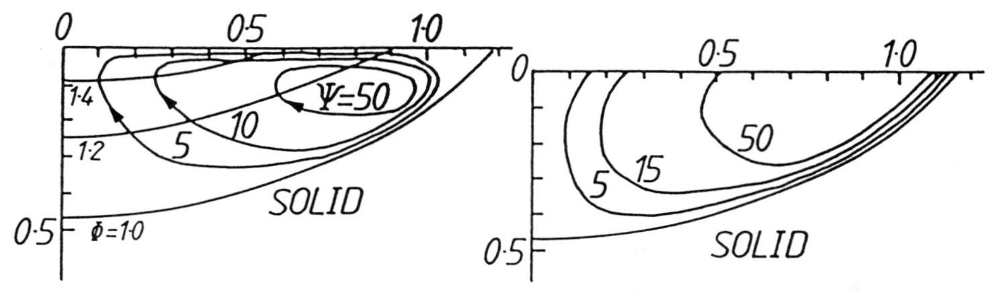

Fig. 5. Plots for $I=50$, $R_0=1.0$, $B_0=5\times10^{-4}$, $\beta=1.36\times10^{-4}$, $\partial\gamma/\partial T=-5\times10^{-6}$ (i.e. $K=1.2\times10^5$, $Gr=0.81\times10^5$, $N=1.2\times10^4$, $S_T=0.23\times10^5$): (a) streamlines and isotherms; (b) contours of constant W.

of parameters, are shown on Figs. 5a,b. Comparison of Fig. 5a with the corresponding results when $B_0=0$ (closely similar to Fig. 3b) reveals little variation in weld pool shape although the flow patterns close to the axis are distinctly different.

The velocities corresponding to Figs. 3a and 3b are not displayed explicitly but a simple calculation from the numerical results for Ψ shows that for steel the maximum poloidal velocities are about 1mm s^{-1} when $S_T=0$ but about 50mm s^{-1} when $S_T=1.4\times10^5$ Moreover, choosing the values for I, R_0, B_0 and β corresponding to Fig. 5 we find that the maximum azimuthal and poloidal velocities are about 20mm s^{-1} and 3mm s^{-1} respectively when $S_T=0$ and 12mm s^{-1} and 50mm s^{-1} respectively for $S_T=0.23\times10^5$.

Estimates of the effects and relative importance of the three forcing mechanisms (a), (b) and (c)(ii) can be found from an order of magnitude analysis. For the current source considered in this paper the integrals I_1 and I_2 (defined by (2.16) and (2.17)) are both 0(1). Suppose first that there is no externally applied magnetic field ($B_0=0$) in which case $N=0$ and there is no azimuthal motion. In this situation it is evident from (2.10), the differential equation governing the poloidal motion, that the competing forcing mechanisms are buoyancy and electromagnetism. Since the values of Gr and K corresponding to Figs. 2 and 3 are of the same order it is not surprising that buoyancy effects are comparable to electromagnetic ones when both are included. Clearly the electromagnetic forcing terms would take increasing importance if the applied current source were more concentrated (increasing I_1 and I_2) and/or the total applied current were increased. Calculations based on (2.10) suggest that in the absence of surface tension effects poloidal velocities for steel in the range 1-10mm s^{-1} could be expected. The inclusion of the variation of surface tension with temperature is crucial, however, since from $(2.21)_2$ it is immediately deduced that for steel velocities of $0(10^{-4}S_T)\mathrm{ms}^{-1}$could occur. For the values of S_T used in our numerical results this suggests velocities of $0(1)\mathrm{ms}^{-1}$ and in practical situations S_T could be even 10 times the maximum value used in this paper. Variation in surface tension with temperature, if present, is generally found to be the dominant forcing mechanism when $N = 0$ therefore, but since it is a condition to be applied only on the free surface and not throughout the pool one would expect that the approximate velocities stated above are likely to be over-estimates.

When the externally applied magnetic field is non-zero (i.e. $N \neq 0$) azimuthal motion occurs and (2.11) suggests that for steel the azimuthal speed is $0(10^{-5}N)\mathrm{ms}^{-1}$. For the values of N used in the displayed results azimuthal speeds of $0(10^{-1})\mathrm{ms}^{-1}$ might consequently be expected. A careful inspection of equation (2.10) then reveals that the terms arising from this azimuthal motion are greater than the electromagnetic and buoyancy terms, resulting in poloidal flows with speeds also of $0(10^{-1})\mathrm{ms}^{-1}$. Once again the surface forces arising from the change in surface tension with temperature, if present, are likely to dominate the other forcing mechanisms but less strongly than when $N = 0$.

In this paper we have determined the shape of the weld pool produced in the steady state by a stationary current source so our problem is independent of time. However, the unsteady equations (2.10) and (2.11) can be used to estimate the time scales that would be appropriate if some variation with time were allowed. Order of magnitude analyses suggest that for the particular non-zero values for K, Gr and S_T chosen in our displayed results for steel the time scale is $0(1)$s when $N \neq 0$ but reduces to $0(10^{-1})$s when $N = 0$.

It is easily confirmed that the predictions made above from our order of magnitude analyses are consistent with the numerical solutions displayed in Figs. 2 to 5.

5. Conclusion

The numerical results presented in the preceding section, and others obtained with our programme and by previous authors, clearly show that the shape of the solid/liquid interface of a weld pool in steel under typical welding conditions is strongly dependent on convection effects in the pool. The electromagnetic forcing terms, buoyancy effects and the variation of surface tension with temperature must all be included in a satisfactory model since each mechanism can have significant influence on weld pool shape for certain values of the main parameters.

Application of an additional uniform magnetic field creates a strong swirling motion in the pool and leads to changes in weld pool shape, especially when $S_T = 0$. For the non-zero values of S_T considered in this paper the weld pool shape is comparatively unaltered but the applied field causes considerable variation in fluid flow within the pool.

Useful information on the relative importance of the forcing mechanisms has been determined in section 4 from an order of magnitude analysis but our numerical results reveal that considerable differences in these forces occur across the weld pool and some mechanisms can become important only in small sections of the pool leading to quite complicated flow patterns. In most situations, however, the dominant forcing term is that due to the variation of surface tension with temperature.

Figure 4 indicates that the depth to width ratio of a weld pool is much more uniform when the external magnetic field is present and this fact could be of importance in designing an efficient control process. Further work on this problem is being carried out.

References

1. D. Rosenthal: The theory of moving sources of heat and its application to metal treatments. *Trans. ASME.* 68 (1946) 849–866.
2. H.S. Carslaw and J.C. Jaeger: *Conduction of heat in solids*, 2nd edn. Oxford: Clarendon Press (1959).
3. V. Kublanov and A. Erokhin: On metal motion in a stationary weld pool under the action of electromagnetic forces and gas flow velocity head in arc welding. *Int. Inst. Weld. Doc.* No. 212-318-74 (1974) 1–12.
4. I.E. Butsenieks, D.E. Peterson, V.I. Sharamkin and E.V. Shcherbinin: Magnetohydrodynamic fluid flows in a closed space with a nonuniform electric current. *Magnetohydrodyn.* 12 (1976) 70–74.
5. R.A. Woods and D.R. Milner: Motion in the weld pool in arc welding. *Weld. J.* 50 (1971); *Res. Suppl.* 163-s to 171-s.
6. S. Lundquist: On the hydromagnetic viscous flow generated by a diverging electric current. *Ark. Fys.* 40 (1969) 89–95.
7. J.A. Shercliff: Fluid motions due to an electric current source. *J. Fluid Mech.* 40 (1970) 241–250.
8. C. Sozou: On fluid motions induced by an electric current source. *J. Fluid Mech.* 46 (1971) 25–32.
9. C. Sozou and W.M. Pickering: Magnetohydrodynamic flow due to the discharge of an electric current in a hemispherical container. *J. Fluid Mech.* 73 (1976) 641–650.
10. J.G. Andrews and R.E. Craine: Fluid flow in a hemisphere induced by a distributed source of current. *J. Fluid Mech.* 84 (1978) 281–290.
11. D.R. Atthey: A mathematical model for fluid flow in a weld pool at high currents. *J. Fluid Mech.* 98 (1980) 787–801.
12. R.E. Craine and J.G. Andrews: The shape of the fusion boundary in an electromagnetically stirred weld pool. In: *Proc. IUTAM Symposium on Metallurgical applications of Magnetohydrodynamics*, The Metals Society (1984) pp. 301–313.
13. R.E. Craine and J.G. Andrews: The shape of the fusion boundary in an electromagnetically stirred weld pool. *C.E.G.B. Lab. Note* TPRD/M/1355/N84 (1984) 1–16.
14. G.M. Oreper and J. Szekely: Heat- and fluid-flow phenomena in weld pools. *J. Fluid Mech.* 147 (1984) 53–79.
15. Y. Fautrelle: Motion and pool profiles in arc welding. *Actes du colloque módelisation et calcul en électromagnétisme et applications*. A. Bossavit (1984) pp. 152–162.
16. C. Chan, J. Mazumder and M.M. Chen: A two-dimensional transient model for convection in laser melted pool. *Met. Trans A* 15 (1984) 2175–2184.
17. S. Kou and D.K. Sun: Fluid flow and weld penetration in stationary arc welds. *Met. Trans A* 16 (1985) 203–313.
18. S. Kou and Y.H. Wang: Weld pool convection and its effect. *Weld. J.* 65 (1986); *Res. Suppl.* 63-s to 70-s.
19. M.L. Lin and T.W. Eager: Influence of arc pressure on weld pool geometry. *Weld. J.* 64 (1985); *Res. Suppl.* 163-s to 169-s.

20. C.R. Heiple and J.R. Roper: Effect of Selenium on GTAW fusion zone geometry. *Weld. J.* 60 (1981); *Res. Suppl.* 143-s to 145-s.
21. C.R. Heiple and J.R. Roper: Mechanism for minor element effect on GTA fusion zone geometry. *Weld. J.* 61 (1982); *Res. Suppl.* 97-s to 102-s.
22. C.R. Heiple, J.R. Roper, R.T. Stagner and R.J. Aden: Surface active element effects on the shape of GTA, laser and electron beam welds. *Weld. J.* 62 (1983); *Res. Suppl.* 72-s to 77-s.
23. P. Burgardt and C.R. Heiple: Interaction between impurities and welding variables in determining GTA weld shape. *Weld. J.* 65 (1986); *Res. Suppl.* 150-s to 155-s.
24. K.C. Mills, B.J. Keene, R.F. Brooks and A. Olusanya: The surface tensions of 304 and 316 type stainless steels and their effects on weld penetration. In: *Proc. Cent. Conf., Metallurgy Dept., Univ. of Strathclyde*, June 1984, R1-R11.
25. B.J. Keene, K.C. Mills and R.F. Brooks: Surface properties of liquid metals and their effects on weldability. *Mat. Sci. and Tech.* 1 (1985) 568–571.
26. A.M. Boldyrev, N.P. Tolokonikov and E.B. Doroveev: Concerning the possibility of obtaining stable reversible motion of the molten metal in the weldpool with electromagnetic stirring. *Weld. Prod.* 22 (1975) 4–7.
27. R.A. Willgoss: Electromagnetic stirring: a potential aid to the close control of welding processes. *C.E.G.B. Lab. Note* RD/M/N1138 (1980) 1–12.
28. R.E. Craine and N.P. Weatherill: Fluid flow in a hemispherical container induced by a distributed source of current and a superimposed uniform magnetic field. *J. Fluid Mech.* 99 (1980) 1–11.
29. O.H. Nestor: High intensity and current density distributions at the anode of high current, inert gas arcs. *J. Appl. Phys.* 33 (1962) 1638–1648.

T.J. Smith (ed.) Modelling the Flow and Solidification of Metals

Recent developments in modelling metal flow and solidification

T.J. SMITH[1] & A.F.A. HOADLEY[2]
[1]*GST Professional Services Limited, Cambridge, and Department of Chemical Engineering, University of Cambridge, UK;* [2]*Department of Chemical Engineering, University of Cambridge, UK*

Received February 1987

Abstract. This paper discusses the modelling of a number of metallurgical systems involving flow and solidification. The discussion is neither definitive nor critical, but is an attempt to provide information on important processes not covered elsewhere in this volume. The subjects covered include rapid solidification, semi-solid casting, natural convection and alternative formulations for solidification modelling. In addition, some new results for hyperbolic heat transfer and heat losses during mould filling are presented. Throughout the discussion, the practical application of the various models is stressed and wherever possible use is made of industrial examples. Finally, a substantial, although by no means exhaustive bibliography is provided to indicate where further details of the work may be found.

Introduction

The papers presented in this volume describe a number of advances in modelling the flow and solidification of metals. The subjects cover a broad range from the structure of dendrite growth to the simulation of complete metal processing operations. Part of the value of this volume lies in the spectrum of topics covered. The individual papers, together with the bibliographies they contain, cover a substantial proportion of the problems encountered in modelling the processing of metals. Even so there are still some areas which have received little or no coverage in this volume despite some important recent work in these fields. Therefore, it is the purpose of this paper to draw attention to these developments, to comment on their significance and to provide details of where further information may be obtained. It is not intended that this should be a definitive or even critical review of this topic. It is simply a guide to important work not included in the preceding papers.

The scope of the paper closely follows its theme. The next section deals with solidification processes and comments on the global models of dendrite growth developed by Huppert and his co-workers [46, 48, 103]. Also the important areas of rapid solidification and semi-solid casting are considered. There then follows a section on the numerical modelling of solidification. Aspects of this subject have been covered by Flood and Hunt [28] and by Lewis and Roberts [63], but there are a number of other techniques which are widely used and must be included for the sake of completeness. Also described is the intriguing proposal of a non-Fourier, hyperbolic Stefan problem, the concept of which is still controversial and the mathematics of which are still under development.

This completes the discussion of solidification with the remainder of the paper concentrating on metal flow. The subject is split into forced convection and free convection problems. Forced convection covers any flow of metal driven by externally applied forces. As such it covers the whole range of problems encountered during the filling of a mould cavity or during the transfer or stirring of metal. Certain aspects of the problem are covered in the papers by Smith and Welbourn [87], Johansen et al. [51], Kunda and Poots [61] and Rogers et al. [74], but there remain a number of significant works not covered in these papers, particularly relating to the calculation of free surface flows and heat losses during mould filling. Consequently, further details of these works are presented.

Free or natural convection is driven in liquid metals by density gradients within the melt set up by spacial variations in temperature or composition. The presence of these motions can significantly effect heat and mass transfer at the solidification interface thereby altering the propagation of solidification through a casting. Free convection also changes the heat transfer between the melt and its container which influences the subsequent behaviour of the metal. Here the geometry of the container is particularly important. A number of different models for the effect of natural convection on solidification have been developed, some of which are discussed.

Following some concluding remarks the paper is completed by a comprehensive bibliography designed to complement those appearing in the remainder of the volume.

Solidification processes

Dendrite growth

A number of papers in this volume [20, 28, 44, 63] and elsewhere [13, 38, 48, 57, 91, 101–103] consider the problem of modelling the physical process of solidification in a multicomponent melt. The common theme of this work is a concentration on modelling the physics of the solidification process on the scale of the dendrite which can then be integrated over a large volume to allow solidification calculations to be performed. A similar approach was applied in a model of horizontal continuous casting developed by Schneider [80]. Here, analytical solutions of the model equation were possible due to the quasi-steady nature of the problem. However, the accuracy of such models is doubtful as there is considerable controversy surrounding the modelling of the shape of a single dendrite and its inherent instability [45]. A model for the stability to sidebranching leads to the result that the velocity corresponding to marginal stability is close to that observed for dendrite growth. Unfortunately, this does not explain why a dendrite grows at this speed. An alternative approach to modelling dendrite growth has been to treat a dendrite as a slender body and match solutions for the far field and the near field. When surface tension is included, the model predicts that dendrites can grow only at a velocity less than a critical value although the relationship between this critical velocity and that for marginal stability is not yet known [45].

Another recent development has been the generalisation, in a statistical sense, of the development of undercooling due to local curvature effects to predict the mean undercooling. This is found to be proportional to the mean curvature which theoretical analysis reveals decreases as the inverse cube-root of time. These model studies have now been verified experimentally for a number of systems having different solidification morphologies [45].

An alternative approach is to argue that the details of dendritic growth are irrelevant to the modelling of large scale solidification development and only concern the grain structure produced. Further, it is proposed that a model of solidification in multicomponent systems can be derived merely by satisfying global energy and mass conservation. An intriguing insight into some of the mathematical problems associated with the various approaches is given by Crowley and Ockendon [20].

One of the most significant developments in the global modelling of multicomponent systems has been provided recently by Huppert and Worster [48]. These authors develop equations which relate the bulk properties of the mushy region averaged over a length scale much larger than a typical dendrite spacing. This model has then been applied to the crystallisation of aqueous solutions having simple phase diagrams. These solutions are experimentally convenient yet mimic most of the important phenomena observed in the solidification of more general binary melts. Within such a simple system there are six different situations depending on whether the cooling is from above or below and whether the initial composition is less than, equal to or greater than the eutectic value. Melts cooled from below are thermally stable while those cooled from above are thermally unstable.

In aqueous salt solutions there is no longer a point of maximum density even at concentrations much less than the eutectic value. If the salt concentration is less than the eutectic value heavy fluid is released on solidification while if the concentration is greater than the eutectic value light fluid is released. There is only one combination which is both thermally and compositionally stable: a hypoeutectic solution cooled from below. In any binary system, the growth of the solidification interface depends on the molecular transfer of heat and mass. However, for aqueous solutions the molecular diffusivity of the solute is much less than the thermal diffusivity. For the situation in which the interface is morphologically stable, two regimes occur. If the solute volume concentration in the melt is much greater than the square of the ratio of the mass and thermal diffusivities then solidification is controlled by the diffusion of solute away from the solidification interface. However, if the initial concentration in the melt is sufficiently small of the order of the ratio of the diffusivities then interface growth is controlled by the rate at which heat can diffuse away from the interface into the solid.

Unfortunately, the condition on the undercooling to ensure morphological stability is sufficiently restrictive to result in most practical applications having a morphologically unstable interface. Under these circumstances Huppert and Worster [48] show that the growth of the block is governed by thermal diffusion. It is also demonstrated that the porosity within the solid increases significantly with increasing initial solute concentration. This has important implications for predicting the occurrence of porosity defects in castings [49]. Initially, the solid fraction was assumed to be constant everywhere within the solid. This restriction has now been removed and a model developed which allows the prediction of local variations in solid fraction [103].

Rapid solidification

In recent years, the generation of metastable metallurgical structures by rapid solidification has received considerable attention [55, 59, 78] due to the possibility of manufacturing many new materials with improved physical properties. The potential for the commercial exploitation of these new materials is considerable, with production successes already reported for high speed tool steels, corrosion resistant alloys, brazing alloys and austenitic, stainless, seamless tubing [33]. The fundamental principle behind rapid solidification is that heat is removed as fast as possible from a mass of molten metal so that the liquid can cool to a temperature well below its equilibrium freezing point. This then allows solidification to take place under non-equilibrium conditions. This large liquid undercooling with its consequent high solidification front velocity leads to a variety of desirable microstructural effects [9] including:

(i) small grain size
(ii) fine scale or zero microsegregation
(iii) extended solubility of alloying elements
(iv) the formation of metastable crystalline and amorphous phases.

In order to achieve a high rate of heat removal the molten metal mass must be thin in at least one dimension to allow heat to diffuse rapidly from the interior to the boundary. The metal must also be in good thermal contact with an efficient heat sink. A number of ways of achieving these two essential criteria have been derived which have led to the development of a wide variety of rapid solidification processes. Amongst these are:

(i) gun and piston splat quenching
(ii) melt spinning and planar flow casting
(iii) melt extraction
(iv) inert gas or water atomisation
(v) plasma and atomised spray deposition
(vi) laser and electron beam surface melting.

Each of these processes is characterized by a relatively simple geometry as this is desirable for optimised heat flow. Atomisation and splat quenching are the most common for batch processing while continuous processing tends to rely on variations of the melt spinning theme using a rotating cylinder as the substrate.

During rapid solidification, the propagation velocity of the solidification front is largely controlled by interfacial undercooling. This is governed by the nucleation temperature and hence on the competition between latent heat release and heat loss from the metal. In general, large nucleation undercoolings associated with strong temperature inversions at the growing interface allow latent heat to be dissipated in both directions during the initial phase. This promotes very high freezing rates. At later times this can result in reheating to the equilibrium solidification temperature leading to a marked reduction in the growth rate of the freezing front. This process is known as recalescence and can give rise to structural changes such as microsegregation, reduced supersaturation and phase segregation. Hence it is desirable that recalescence is retarded or eliminated by suitable external cooling. A particularly neat derivation of the criterion required for external heat transfer if this is to be achieved was given by Clyne [14]. This can then be used in a rationale to enable the operational regimes for arrested recalescence to be established, significantly reducing the modelling effort required during the design process.

Generally, the requirements of any model of the rapid solidification process are first to predict the cooling rate, liquid undercooling and solidification front velocity. These parameters can then be used to predict the final microstructural effects including grain size, segregation solubility and post solidification phase transitions. A number of models have been proposed which include some or all of these processes [14, 16, 57–59]. The majority of models concentrate on melt spinning as the geometry is particularly suitable to mathematical analysis. Gas atomisation is also particularly amenable to modelling due to the spherical geometry of the particles.

A number of gas atomisation procedures have been proposed, among which that of ultrasonic gas atomisation has attracted a lot of interest [16, 17] owing to the attractive combination of processing conditions which can be obtained. In this process, a jet of the melt passes from the base to the apex of a cone-shaped jet of an inert gas such as helium or argon. This atomises the melt stream which is discharged through a nozzle. Mach numbers in the discharged gas stream are of the order of unity so that the flow is compressible. However, after the initial acceleration, droplet Reynolds numbers, $\mathrm{Re} = [\frac{8}{3}(a^3 g \varrho_p)/(C_D v^2 \varrho_g)]^{1/2}$, based on particle diameter, a, and velocity relative to the gas stream are ~ 10–100 [16]. Here, g denotes the acceleration due to gravity, ϱ_p denotes the particle density, ϱ_g denotes the gas density, v denotes the kinematic viscosity of the gas and C_D is the drag coefficient. This reduces the complexity of the problem provided the timescale for the particle to accelerate to a steady velocity relative to the gas stream and the timescale for the heat transfer to adjust to this equilibrium are small in comparison with the time required for complete solidification. The timescale for a particle to accelerate to its equilibrium velocity can be obtained from the equation of motion for the particle [3]. It is the order of $\varrho_p a/\varrho_g V_g$ in which V_g is the gas velocity. For operating conditions typical of an aluminium alloy in helium this is the order of 10 μs for a 1 μm diameter droplet. As this is comparable to the solidification time for droplets of this size [16], the assumption of a steady flow field around the particle is questionable. Dimensional analysis reveals that the timescale for thermal response is of the order of a^2/α in which α is the thermal diffusivity of the melt. For an aluminium melt this is ~ 50 ns and hence is much less than the solidification timescale. It is interesting to note that a transient numerical solution of the energy equation led to a relaxation time of $0.9a^2/\alpha$ [16] which is very close to that given by simple dimensional analysis.

Using a steady flow assumption, Clyne et al. [16] were able to develop a model for the rapid solidification of aluminium alloy droplets in both helium and argon gas streams based on Clyne's earlier model for melt spinning [14]. It was also assumed that the droplet concentration was sufficiently small that the flow around each droplet is unperturbed by the flow around adjacent drops. The authors were then able to predict that aluminium droplets of less than 2 μm diameter entrained in a fast moving stream of helium gas should experience sufficiently high heat extraction for post nucleation recalescence to be arrested. This should then lead to partitionless solidification and hence a material completely free from segregation. Microscopic and metallurgical examination of an Al–Li alloy powder produced in this way appears to give a structure consistent with these predictions [17]. Here it is worth noting that during particle acceleration to the steady state, heat transfer is likely to be greater than in the steady state. Hence, the estimation of solidification time given by Clyne et al. is likely to be conservative. Further work is required to quantify the effects of an unsteady flow field around the particles and also the effect of particle doublet interactions using an extension of Batchelor's model [5] for mass transfer from particles.

Semi-solid casting

When an alloy is held at a temperature between its liquidus and solidus it exists in a semi-solid state having useful rheological properties. At temperatures well in excess of the liquidus, molten metal is a Newtonian fluid in which the shear stress is directly proportional to the rate at which it is being sheared. However, at temperatures between the liquidus and solidus the metal is neither liquid nor solid and exhibits either pseudo-plastic or Bingham plastic rheological properties. The advantage of these properties is that for an applied stress less than the lower yield stress or Bingham yield stress the material behaves as a solid whereas at an applied stress greater than these values, the material will flow. This behaviour can be utilised in a number of metal forming processes including pressure die-casting, billet extrusion and continuous casting of strip [15]. Also it has been proposed as a viable means for producing metal–matrix composites.

Much of the early work on semi-solid casting was carried out by Flemings and his co-workers at MIT under the name of rheocasting. A considerable amount of work was undertaken to characterise the structure and fluid flow behaviour of the semi-solid metal although a complex mixing technique was required to prepare the material in this form. Recent work at the Fulmer Research Institute [89] has overcome many of these operational problems allowing semi-solid pressure die-casting on a production basis. The advantage to such a process is the production of pore free castings. Considerable energy savings are also possible as the solid–liquid slurry is produced without the necessity of passing through the fully liquid state. Similar energy savings are also possible in the continuous rheocasting of strip while the special microstructural features of rheocast material could be desirable for re-roll gauge strip.

Modelling rheocasting processes is complicated by the non-Newtonian behaviour of the material in which the apparent viscosity is strongly coupled to the thermal and strain rate fields. If U_i denotes velocity in the direction x_i then the elements d_{ij} of the rate of deformation tensor are given by:

$$d_{ij} = \tfrac{1}{2}(U_{i,j} + U_{j,i}) \tag{1}$$

such that the shear rate $\dot{\gamma}$ becomes:

$$\dot{\gamma} = \sqrt{2d_{ij}d_{ij}}\,. \tag{2}$$

The apparent viscosity μ then takes the form [70]

$$\mu(T) = \mu_0\dot{\gamma}^{n-1}f(T) \tag{3}$$

in which T denotes temperature, μ_0 is a reference viscosity, $f(T)$ is some function of temperature and n is an index which depends on the solid fraction of the semi-solid slurry and hence on temperature. The most common form for the function f is:

$$f = \exp\{-b(T - T_0)\} \tag{4}$$

in which T_0 is a reference temperature and b is a material constant.

The above constitutive relationship may then be included in a generalised Newtonian flow model:

$$s_{ij} = 2\mu d_{ij} \tag{5}$$

which together with the equations for the conservation of linear momentum and energy:

$$\varrho[\dot{U}_i + (U_j U_i)_{,j}] = \sigma_{ij,j} + \varrho g_i \tag{6a}$$

$$\varrho[(\dot{CT}) + U_k(CT)_{,k}] = (k_{ij}T_{,i})_{,j} + 2\mu d_{ij}d_{ij} \tag{6b}$$

$$s_{ij} = \sigma_{ij} + p\delta_{ij} \tag{6c}$$

and suitable boundary and initial conditions completely specify the flow problem. Here $\boldsymbol{\sigma}$ denotes the stress tensor, ϱ denotes density, $\mathbf{g}$ is the body force vector, C is the heat capacity and $\mathbf{k}$ is the thermal conductivity tensor.

The flow formulation can be used to describe the whole of a rheocasting process provided suitable modifications are made to the constitutive relationship (3) to allow for post-solidification or viscoplastic effects [105]. However, even for simple geometries such as roll-casting, the above model is sufficiently complex that analytical solutions are possible only under severely restricted assumptions [65]. Numerical solutions for similar polymer processing operations are common [66, 70] although there have been few applications to rheocasting. This is primarily due to uncertainties in the various material constants within the constitutive relationships. A comprehensive discussion of the numerical analysis of these problems is given by Zienkiewicz [105].

Solidification modelling

Alternatives to the enthalpy method

In recent years a number of techniques have been proposed for the numerical solution of thermal problems with a change of phase. Of these, the enthalpy method has become the most popular. Primarily this is due to its ease of application which doesn't require the location of the solidification front to be tracked. This allows a fixed computational mesh to be used with efficient numerical algorithms. Details of the enthalpy method are given by Lewis and Roberts [63] and Smith et al. [85] in this volume and elsewhere [94, 95, 97, 100]. Unfortunately, the enthalpy method suffers from a number of deficiencies [6, 82] particularly when used with non-eutectic alloys [20] or geometries containing boundary singularities [8].

Of the alternatives [29, 40], one is to treat latent heat release as a source term within the elements containing the solidification front [96]. The magnitude of the rate of latent heat release depends on the growth speed of the solidification front and hence this must be determined. This technique is particularly appropriate for the calculation of rapid solidification processing in which the effective growth speed is determined uniquely by the instantaneous undercooling and is not directly influenced by the heat flow [14].

Boundary discontinuities, as may occur at the junction of isothermal and non-conducting parts of a boundary, produce formidable computational difficulties for conventional numerical schemes [8] due to the singularity in the boundary conditions. One way in which the difficulties may be overcome is to employ an inverse formulation in which the space co-ordinates become the dependent variables [7]. The independent variables are then the temperature and a flow function related to the heat flux normal to isothermal surfaces. An added advantage is that the problem of grid generation in arbitrarily shaped domains is combined with the solution procedure as the domain transforms onto a square during inversion. At present the method has only been applied to steady state situations as major difficulties still exist in adapting the technique to transient problems. The most severe is the choice of the second independent variable as the flow function used in steady state problems is inappropriate. Currently, the only suggestion has been to choose some appropriate yet arbitrary analytic function suggested by the geometry of the problem as there is no requirement for the inverse co-ordinates to be orthogonal. However, this can lead to a highly non-linear inverse problem. It should also be noted that the effects of the boundary singularity do not disappear in the transformed system, they simply become more amenable to numerical treatment. In transient problems a starting algorithm is required to minimise the effect of the initial temperature discontinuity. No mathematical exact solution exists for boundary singularities. As these persist into the inverse problem empirical techniques must be devised. Despite these difficulties, inverse formulation techniques appear to be a promising approach for problems involving boundary singularities and there is considerable scope for further work in this area.

Hyperbolic heat transfer

Fourier's law of heat transfer assumes that the conductive flux of heat $\mathbf{q}$ at position $\mathbf{x}$ and time t is proportional to the temperature gradient ∇T at the same position and time. The constant of proportionality, k, is the thermal conductivity such that:

$$\mathbf{q}(\mathbf{x};\, t) \quad = \quad -k\nabla T(\mathbf{x};\, t). \tag{7}$$

Unfortunately, equation (1) implies the propagation of information at infinite speed as a discontinuous spatial temperature distribution leads to an infinite flux. If the heat capacity is non-zero as it must be in all practical cases, then the propagation speed is infinite.

An alternative hypothesis, which overcomes any philosophical objections to equation (7), is based on the assumption that there is a time lag τ between the onset of a temperature gradient and the conductive heat flux it drives. Thus:

$$\mathbf{q}(\mathbf{x};\, t + \tau) \quad = \quad -k\nabla T(\mathbf{x};\, t). \tag{8}$$

Use of equation (8) in the energy conservation equation:

$$\varrho C T_{,t} + \nabla \cdot \mathbf{q} = 0 \tag{9}$$

leads to a hyperbolic equation of the form:

$$\varrho C(\tau T_{,tt} + T_{,t}) = \nabla \cdot (k \nabla T) \tag{10}$$

in which ϱ denotes density, C denotes specific heat and $T_{,t} = (\partial T/\partial t)$, etc. It will be seen that equation (10) is a generalised form of the Telegrapher's equation which is well known in electrical theory.

Equation (10) has the mathematical form of a dissipative wave equation with a characteristic finite wave speed given by $(k/\varrho C\tau)^{1/2}$ and reduces to the more conventional diffusion equation in the limit $\tau \to 0$. However, even after extensive literature on the subject (see [25, 27]), it is still not clear whether equation (10) has any physical relevance or indeed what relationship the variable T governed by equation (10) has to temperature. Despite these objections, Telegrapher's equation has some useful mathematical properties which make it attractive as an analogue to the diffusion equation although some disadvantages appear when it is applied to Stefan problems.

Consider the application of equation (10) to the solidification of a one-dimensional, semi-infinite melt ($x > 0$) having constant physical properties initially at its fusion temperature, T_f. Solidification is initiated by reducing the boundary temperature of the melt to $T_0 < T_f$ at which it is held constant for all time. Let θ be the time scale over which interest is focussed on the experiment such that $\tau/\theta = \varepsilon \ll 1$. A suitable scaling of equation (10) is then given by:

$$\phi = \frac{T - T_0}{T_f - T_0}; \quad s = \frac{t}{\theta}; \quad y = x(k\theta/\varrho C)^{-1/2} \tag{11a, b, c}$$

which leads to:

$$\varepsilon\phi_{,ss} + \phi_{,s} = \phi_{,yy}. \tag{12}$$

In the limit $\varepsilon \to 0$, that is at times much larger than the delay time τ, the solution of equation (12) will be diffusion-like. However, it is also necessary to know the behaviour of equation (10) at times of the order of the delay timescale τ. Therefore, consider the rescaling given by:

$$r = s\varepsilon^{-1}; \quad z = y\varepsilon^{-1/2}. \tag{13a, b}$$

Table 1. Difference between solutions for hyperbolic and parabolic heat transfer at $\varepsilon = 10^{-2}$

s	0.0025	0.005	0.0075	0.01	0.02	0.03
y						
0.05	0.118	0.006	0.005	0.003	0.001	0.001
0.10	−0.078	0.026	0.007	0.005	0.003	0.003
0.20	−0.194	0.016	0.012	0.008	0.005	0.004
0.30	−0.057	0.000	0.001	0.006	0.007	0.006
0.40	−0.013	−0.103	−0.007	0.002	0.009	0.007

Substitution of (13a, b) into (10) leads to:

$$\phi_{,rr} + \phi_{,r} = \phi_{,zz} \tag{14}$$

which shows that at times of $O(\tau)$ the solution is wave-like and hence substantially different to that for the conventional Stefan problem. Therefore, the conclusion is that well outside the domain $(x; t) \in [0, \varepsilon^{1/2}] \times [0, \varepsilon]$ solutions based on (10) will be indistinguishable from those of the diffusion equation. Numerical experiments for a single phase problem indicate that provided $\varepsilon < 10^{-2}$ there is no significant difference between solutions obtained using equation (10) from those obtained from the diffusion equation [88]. This is due to the decay of wave-like solutions over a timescale of $O(\varepsilon)$ as shown in Table 1.

These features have been used to advantage by Johns [52] to construct an unconditionally stable explicit numerical scheme based on the transmission line matrix method for the solution of equation (8). This has been applied with some success to the solution of diffusion problems [53] although the mathematical consequences of the requirement for additional boundary conditions due to the higher order of (10) were not addressed. There are also advantages with conventional discretisations. If second order central differences are used for both the spatial and temporal derivatives the resulting explicit algebraic analogue of (10) is conditionally stable, requiring $(\alpha/\tau)^{1/2}\Delta t/\Delta x \leqslant 1$. It is interesting to compare this stability requirement with that for the diffusion equation under the same conditions, that is $2\alpha\Delta t/\Delta x^2 \leqslant 1$. Noting that $x = \lambda/n$ with n an integer and recalling that $\lambda = (\alpha\theta)^{1/2}$, the ratio of the maximum time step for the hyperbolic problem to that for the parabolic diffusion equation becomes $2n\varepsilon^{1/2}$. Thus, a larger time step can be used for equation (10) than for the diffusion equation provided $\varepsilon > (2n)^{-2}$. For a typical value of $n = 20$ this requires $\varepsilon > 6.25 \times 10^{-4}$. Thus, use of equation (10) considerably extends the stability domain of explicit schemes without introducing a significant departure from a solution to the equivalent diffusion equation.

While in these applications, the time lag τ is arbitrary, it has yet to be defined in a physical sense. Here, it is proposed that τ is of the order of the timescale kh/α common to all practical solidification problems in which h is the metal-mould heat transfer coefficient, k is the thermal conductivity and α is the thermal diffusivity. Work is in progress to further investigate the implications of this hypothesis.

Various authors have considered the effect of equation (10) on the propagation of the solidification front [15, 77, 88] albeit with an incorrect Stefan condition at the interface in [25, 77]. Solomon et al. [88] corrected this error and also discussed the relevance of the Telegrapher's equation to phase change problems. A number of difficulties were highlighted, in particular the existence of solutions which violate the condition of positive entropy production. In these solutions, the speed of the phase change front is greater than the wave speed $(\alpha/\tau)^{1/2}$ such that the front becomes a space-like wave and the Stefan conditions are too few for the problem to be well posed. To eliminate these unrealistic solutions it is necessary to impose the additional constraint that entropy production must be positive. However, if solidification is driven by an imposed heat flux at $x = 0$ then this constraint requires that the boundary flux must not decay faster than $\exp(-t/\tau)$. It is worrying that certain Stefan problems can be posed for which equation [10] provides no realistic solution. Fortunately, from a practical point of view these restrictions are minimal and are certainly outweighed by other potential advantages, particularly enhanced numerical stability. Even so, special care must still be taken in the vicinity of the solidification front where the enthalpy varies rapidly. As discussed above the Telegrapher's equation permits discontinuous solutions which can appear as a discontinuous enthalpy wave travelling at speed $(\alpha/\tau)^{1/2}$ away from the solidification front. These waves are attenuated over distances of the order $\varepsilon^{1/2}$. However, to take maximum benefit from the enhanced stability of Telegrapher's equation, $\varepsilon \sim 10^{-2}$ such that the attenuation distance of the parasitic waves is ~ 0.1. This is a significant proportion of the solution domain. Thus further work is still required before the full potential of hyperbolic heat transfer as a model for solidification processes can be realised.

Forced convection

Heat losses in runner systems and thin walled castings

If accurate predictions of solidification development in metal castings are to be made using models of the type discussed in this volume, it is necessary that the initial temperature distribution is known. This distribution is the result of heat losses in runner systems [23a, 54] and during filling of the mould cavity [22, 64, 69]. Thus, it is necessary that these heat losses can be modelled. Also in many thin-walled castings, misruns are a major source of scrap. These occur when the metal freezes before the mould cavity is completely filled, shutting off part of the cavity.

Some important results can be obtained simply by dimensional analysis. These are particularly relevant to runner systems but also apply to thin-walled castings [84]. Specifically, there are two regimes of heat transfer depending on whether the major resistance to heat flow is the mould–metal interface or the diffusivity of the mould. The transition between one regime and the other is controlled by the Peclet number, Pe, of the flow and the dimensionless superheat, S, such that for:

$$S\mathrm{Pe}^{1/2} \gg \frac{(\varrho C\alpha^{1/2})_{\mathrm{mould}}}{(\varrho C\alpha^{1/2})_{\mathrm{metal}}}\left(\frac{l}{b}\right)^{1/2} = R$$

the cooling rate, Q at the leading edge of the metal, is constant and of magnitude:

$$Q \sim \frac{h(T_f - T_0)}{(\varrho C)_{\mathrm{metal}} b}$$

in which h is the mould metal heat transfer coefficient [39, 50], T_f is the fusion temperature of the metal, T_0 is the initial temperature of the mould, b is the runner half width, l is the runner length, ϱ denotes density and C denotes heat capacity. Alternatively, for $S\mathrm{Pe}^{1/2} \ll R$:

$$Q \sim \frac{h(T_* - T_f)}{(\varrho C)_{\mathrm{mould}}}\left(\frac{U}{\alpha_{\mathrm{mould}} l}\right)^{1/2}$$

in which U is the mean flow velocity and T_* is the pouring temperature.

It is usual for the pouring time of each mould of a given casting to be constant. Consequently, the only variable between moulds of the same casting is the superheat which depends on the pouring temperature. For white iron in a green sand mould $SPe^{1/2}/R$ is typically three to four. However, a reduction in pouring temperature and hence S can result in this being less than one. As the ratio of the two cooling rates is equal to $SPe^{1/2}/R$ the rate of heat loss incurred by this reduction in superheat is increased by a factor $R/SPe^{1/2} > 1$. Under these circumstances the adverse effects are cumulative as there is less initial superheat being lost more quickly, a situation which can easily lead to misruns.

The above analysis has taken no account of the increase in viscosity and hence decrease in fluidity of the metal as it cools [60]. The magnitude of this effect is difficult to estimate for castings having a realistic geometry. However, approximate quantitative results can be derived for the fluidity test spiral [27]. For white iron in a green sand mould the reduction in flow rate and hence Pe between the first and last casting poured from the same ladle could be as much as 60%. Again the effect is cumulative with the other effects discussed above. Consequently, these results suggest that all these processes must be included in any predictive models of heat losses in runner systems and thin-walled castings.

One such model [23a] solves the two-dimensional energy equation for flow in a parallel sided runner by means of the finite element method. Temperature dependent viscosity was not included. Few general conclusions can be drawn from this study as a specific dimensioned runner was modelled. However, the model did confirm that all significant heat transfer occurs at the leading edge of the metal which validates an assumption implicit in the dimensional analysis discussed earlier. The model also inferred that solidification may occur along the wall near the leading edge of the metal and that some or all of this freezing zone may be melted by the subsequent flow of hot metal through the runner.

Modelling the transient flow of molten metal having a free surface into a large mould cavity poses a considerable computational problem [73, 81]. However, Stoehr and Hwang [90] have made considerable progress in overcoming these difficulties using the marker-and-cell technique [2, 99] in two-dimensional simulations. This technique allows the velocity and pressure fields within the fluid to be calculated together with the free surface position from the time the fluid starts to enter the cavity until after the mould is filled. At present, heat losses are not included in the calculations although this would present little difficulty at the expense of increased computational times. Comparison of the model's predictions with high speed motion pictures of liquid entering transparent moulds having the same geometry show encouraging agreement between the predicted and observed flow patterns.

The development of such models brings closer the day when the tendency for entrapment of gas, dross or inclusions and erosion of the mould can be determined prior to the pouring of any metal leading to more efficient foundry practice. Also the availability of these models will lead to greater accuracy in the modelling of post solidification defects such as residual stress generation and shrinkage cavities [49, 62].

Natural convection

Natural convection is induced in a body of fluid by buoyancy forces due to density perturbations within the fluid. These density variations can result from either temperature or compositional inhomogeneities set up during filling of the mould cavity or during solidification itself. The existence of thermal convection in a casting is significant because of its effect on the transition between columnar and equiaxed dendrite growth. This is of primary importance in determining the grain structure within the casting and hence its mechanical properties.

Ruddle [76] appears to have been the first to predict the occurrence of convection in metal castings arising from his observations of a flat temperature profile in the centre of a test casting. However, at that time (1957) there was no confirmatory evidence. The effect of natural convection was first demonstrated in experiments with superheated lead and tin in moulds which were filled with a fine wire grid [18, 19]. It was first shown that the mesh did not contribute to solidification by increasing nucleation. Further results then demonstrated that for a particular superheat the columnar zone was significantly increased with the grid present compared to no grid. Hence, it was concluded that by reducing convection, the transition between columnar and equiaxed growth is suppressed. Therefore, the ability to predict when and for how long natural convection occurs and its effect on the temperature profile is important in determining first whether the casting will be sound and second the actual grain structure of the product. There have been a large number of studies of natural convection in enclosed cavities, too numerous to be adequately reviewed here. Consequently, the following discussion concerns only those studies which also included solidification.

Thermal convection

There have been a number of studies of solidification in two-dimensional or axisymetric cavities subjected to a suitable heat flux driving laminar thermal convection [1, 4, 11, 12, 22–24, 26, 30–36, 41, 67, 68, 72, 86, 91, 98, 104]. Many of these studies employ finite difference solutions of the vorticity and energy equations although finite element solutions have also been used [30, 68]. Significant among the many results obtained is the demonstration that there is a substantial time delay before solidification begins. During this time convection dramatically changes the thermal profile such that the thermal centre shifts away from the geometrical centre. The magnitude of this shift is a function of both the aspect ratio of the container and the Grashof number (effectively a free convective Reynolds number) although no functional relationships have been derived [23].

An alternative approach was considered by Voller et al. [98] who solved the primitive Navier–Stokes and energy equations by a finite difference method. The energy equation was modified to account for latent heat release which was modelled as a source term distributed uniformly over each control volume containing the solidification front. This does not allow the position of the interface to be located exactly. As this is required for the specification of the flow boundary conditions, an estimate of its position must be obtained. For a pure melt or eutectic alloy it was recommended that the velocity field be set to zero in those control volumes for which more than 50% of the latent heat has been released. For non-eutectic alloys a model based on Darcy's law was suggested for the mushy zone. No comparisons with experimental results are given for either of the models discussed.

Any model which attempts to solve the Navier–Stokes or vorticity equations as part of a solidification simulation is impractical for regular foundry applications due to excessive computational requirements even if adaptive computational schemes [83] are used. One way in which this difficulty can be overcome is to increase the magnitude of the thermal diffusivity to simulate the additional convective flux if the convection is laminar or the convective and Reynolds fluxes if the flow is turbulent. For laminar flow it is straightforward to show that the effective thermal diffusivity is $O(\alpha Ra^{1/4})$ in which Ra is the Rayleigh number. This approach was first suggested by Heitz and Westwater [37] who modelled convection using a conduction based solidification program by enhancing the liquid conductivity by a factor which was a function solely of the melt width. This technique averages the effects of the thermal boundary layer and quiescent core across the whole of the melt. In an attempt to verify this approach, Harrison and Weinberg [36] carried out experiments on the solidification of tin. The effective diffusivity was then obtained from measurements of the temperature gradient within the melt and the propagation speed of the solidification front. The effective diffusivity was found to be constant when the temperature difference across the melt, ΔT, exceeded 2°C. The result predicted by dimensional analysis quoted above suggests that the effective diffusivity should vary as $\Delta T^{1/4}$ in the fully convective regime. In view of the difficulties associated with the experimental determinations of the effective diffusivity, the difference between the analytical and observed results is not thought to be significant. This is encouraging as it suggests that a functional relationship relating the effective diffusivity to the problem invariants could be derived by a series of numerical experiments. However, it should be remembered that this model does not give the true temperature profile in the melt, either for laminar or turbulent flows nor does it take into account the orientation of the casting with respect to gravity. In spite of these deficiencies the simplicity of the model makes it attractive for use in practical solidification simulations in the foundry. Hence further development and testing is desirable.

There have also been a few studies of solidification in a vertical cavity subjected to an imposed vertical heat flux. If the applied heat flux vector is anti-parallel to the gravity vector then the melt becomes unstable at Rayleigh numbers greater than some critical value, the magnitude of which depends on the Prandtl number of the melt. The subsequent motion arising from this instability is known as Benard convection. Chiesa and Gutherie [11, 12] studied the effect of Benard convection on the freezing of lead and a lead/tin mixture, using temperature measurements to locate the position of the solidification front and the time at which convection ceases. A model for this process was developed in which interfacial heat transfer was modified by the use of a steady-state turbulent heat transfer coefficient. This led to good agreement when the ratio of heat transfer to freezing rate was high, that is with strongly turbulent convection. At lower values of this ratio there was a substantial discrepancy between the observed and predicted results. The authors also found for the alloy system that best agreement was obtained by assuming that convection ceased at the liquidus temperature, that is at the dendrite tip. This adds further weight to the argument that convection does not penetrate significantly into the mushy zone. More recent experiments with non-metallic compounds have been modelled by a similar technique [34] and good agreement obtained. These results are important in demonstrating the impact of Benard convection on solidification history.

In all experiments on superheated liquid metals temperatures measured by stationary probes have been observed to exhibit high frequency oscillations. This is taken as evidence that in all cases studied the convection has been turbulent. This would be expected for low Prandtl number fluids at moderate Rayleigh numbers typical of these experiments [92]. Indeed there is still some doubt as to whether laminar convection can occur at any Rayleigh number in low Prandtl number fluids. Certainly, no steady flows have ever been observed in mercury during experiments carried out above the critical Rayleigh number for the onset of convection [92]. Also experiments using other metals have indicated that turbulence was present until there was only a very small superheat in the melt. In view of these results, there is some doubt as to whether the general predictions of the various convective models based on laminar flow have any relevance to metallic systems.

There is one simplified model of turbulent convection due to Clyne [13] which appears to hold most promise as the way in which convection controlled solidification can be included in pratical simulators. In this model, the increased resistance to heat transfer which occurs over the thermal boundary layer of thickness δ adjacent to solid boundaries is parameterised by a suitable heat transfer coefficient. Outside the boundary layer, turbulence is assumed to be sufficiently strong to eliminate all thermal gradients. In its present form the model relies on an empirical estimate of δ and does not explicitly include the effects of buoyancy. Also neglected is the considerable difference in thickness between the thin momentum boundary layer and the much thicker thermal boundary layer common in low Prandtl number fluids. Recent work [21, 43, 56, 86] suggests that these deficiencies can be overcome and models suitable for practical applications developed.

Finally, it is worth mentioning that most of the convection controlled solidification experiments discussed herein are not good analogues of industrial solidification processes. In the experiments, convection is generated by holding one boundary at either a fixed temperature above the melting point of the material or imposing a constant heat flux to maintain such a temperature. As solidification proceeds the ratio of heat in to heat removed increases until a steady state is reached. In real castings there is only a finite amount of energy contained within the melt and hence convection decreases with solidification time. Thus, in all castings there is a transition between convection-controlled and conduction-controlled solidification. This transition can exert a major influence on solidification development and grain structure such that it is important to know when this transition occurs and where the solidification front is at that time. Also in real castings other processes such as initial agitation due to mould filling and heat transfer to an initially cold mould interact with convection during the initial stages of solidification. Therefore, there is a definite need for a model which describes time dependent, free convective flow and turbulence intensity within a solidifying cavity. A further constraint is that any such model should be sufficiently simple to allow its incorporation in practical solidification simulations for foundry use.

Compositional convection

When solidifying a non-eutectic alloy, the composition of the solidified material will be different from that of the melt from which it is solidified. Mass conservation requires the release of light or heavy fluid during this process depending on whether the melt composition is above or below that of the eutectic. Depending on the orientation of the solidification front to the gravity vector these compositional changes can drive free convective flows. Also, in general, the density of a melt is much more sensitive to composition than temperature. Hence in most systems, compositional convection should not be ignored, particularly in melts containing little superheat.

Despite its significance, little work has been carried out on modelling this phenomena, particularly for application to metal systems. Most of the work has been developed within the context of geophysical systems [46, 47, 93] or crystal growth [10, 75]. One notable exception is the study by Hills et al. [38] which covers the case for a solidifying melt which is both thermally and compositionally unstable. A thermodynamically consistent theory is developed capable of modelling a mushy zone within a binary alloy. On freezing the solid phase forms as a matrix of crystals which resembles a porous medium of spatially varying permeability. This is diminished in time as further solid freezes out of the liquid phase as it percolates through the matrix. These processes are incorporated into a general model for a mushy zone. However, no attempt is made to model compositional convection within the body of the melt.

One of the earliest attempts at modelling the fully coupled problem used a finite element analysis to solve the free boundary problem describing the interaction of the velocity, temperature and solute fields in binary systems [10a]. Two systems are considered. One consists of a vertical cylinder containing dilute silicon-doped germanium in a thermally unstable environment. This system is compositionally stable although cellular flows driven by radial temperature gradients can reach the interface and cause substantial radial segregation of silicon. The effects of compositional instability were modelled simply by reversing the sign of the concentration expansion coefficient thus retaining the same thermophysical properties as the stable system. For the dual unstable system, cellular flows were intensified leading to a vertically unstable density gradient existing in a thin zone adjacent to the solidification interface. Thus the effect of the solutal field is to enhance radial density gradients near the interface which can lead to small scale instabilities in the interface reminiscent of the morphological instabilities observed in quiescent melts.

Finally, it should be noted that the models discussed here all assume the convection to be laminar. While this is valid for aqueous solutions of the type modelled by Thompson [see Ref. 45] it is unlikely to be true for most metal systems of practical concern. Under turbulent conditions, the solute flux is significantly affected by the turbulence structure [79] and geometry [43] such that these effects must be included in every practical simulation.

Concluding remarks

The objective of this paper has been to present some processes involving the flow and solidification of metals which are important to the metals industry and to describe the ways in which these processes can be modelled. The discussion has concentrated on those aspects of metal flow and solidification not covered elsewhere in this volume, although occasional overlaps have inevitably occurred. A further constraint imposed upon the discussion has been to consider the models from the point of view of practical application in the metals industry. The development of mathematical models can enhance our understanding of natural phenomena, but their greatest benefit to industry is realised when they are used for their predictive capabilities during component design and manufacture [71, 87]. It is hoped that this, together with the preceding papers in this volume, provides enough information to enable the reader to gain sufficient knowledge of current developments in metal flow and solidification modelling and hence to appreciate the benefits that such modelling may bring.

Acknowledgement

The authors are grateful to Crane Limited for partial support of this work and to their Director, Mr M.J. Austin, for permission to publish some of the results contained herein.

AFAH is grateful to St John's College, Cambridge, for the award of an Overseas Research Studentship.

References

1. M.R. Albert and K. O'Neill: Transient two-dimensional phase change with convection using deforming finite elements. In: R.W. Lewis, K. Morgan, J.A. Johnson and W.R. Smith (eds) *Computational Techniques in Heat Transfer, Volume 1.* Pineridge Press, Swansea (1985) pp 229–243.

2. A.A. Amsden and F.H. Harlow: The SMAC method, a numerical technique for calculating incompressible fluid flows. *Los Alamos Scientific Laboratory Technical Report* No: LA-4370 (1970).
3. T.R. Auton, J.C.R. Hunt, K.J. Sene and N.H. Thomas: Distribution of a disperse phase in vertical or near-vertical turbulent shear flows. In: *Extended Abstracts of the International Symposium on Two-Phase Annular and Dispersed Flows*, ETF, Pisa (1984) 209–214.
4. V. Baskaran, J. Ghias and W.R. Wilcox: Modelling the influence of convection on eutectic microstructures. In: *Modelling of Casting and Welding Processes II.* The Metallurgical Society of AIME, Warrendale, Pennsylvania (1984) 115–118.
5. G.K. Batchelor: Mass transfer from small particles suspended into turbulent fluid. *J Fluid Mech* 98 (1980) 609–623.
6. G.E. Bell: On the performance of the enthalpy method. *Int J Heat Mass Transfer* 25 (1982) 587–589.
7. G.E. Bell: Inverse formulations as a method for solving certain melting and freezing problems. In: R.W. Lewis and K. Morgan (eds) *Numerical Methods in Heat Transfer Volume III.* John Wiley and Sons, Chichester (1985) pp 59–78.
8. G.E. Bell and A.S. Wood: On the performance of the enthalpy method in the region of a singularity. *Int J Numer Meth Engng* 19 (1983) 1583–1592.
9. W.J. Boettinger and S.R. Correll: Microstructure formation in rapidly solidified alloys. In: P.R. Sahm, H. Jones and C.M. Adam (eds) *Science and Technology of the Undercooled Melt.* Martinus Nijhoff Publishers, Dordrecht (1986) pp 87–108.
10. J.C. Brice: *The Growth of Crystals from Liquid.* North-Holland, Amsterdam (1973).
10a. R.W. Brown, C.J. Chang and P.M. Adornato: Finite element analysis of directional solidification of dilute and concentrated binary alloys. In: *Modelling of Casting and Welding Processes II.* The Metallurgical Society of AIME, Warrendale, Pennsylvania (1984) pp 95–114.
11. F.M. Chiesa and R.I.L. Gutherie: An experimental study of natural convection and wall effect in liquid metals contained in vertical cylinders. *Metall Trans* 2 (1971) 2833–2838.
12. F.M. Chiesa and R.I.L. Gutherie: Natural convective heat transfer rates during the solidification and melting of metals and alloy systems. *Trans ASME J Heat Transf* 96 (1974) 377–384.
13. T.W. Clyne: The use of heat flow modelling to explore solidification phenomena. *Metall Trans B* 13B (1982) 471–478.
14. T.W. Clyne: Numerical treatment of rapid solidification. *Metall Trans B* 15B (1984) 369–381.
15. T.W. Clyne: Heat flow, solidification and energy aspects of DC and strip casting of aluminium alloys. *Metals Tech* 11 (1984) 350–357.
16. T.W. Clyne, R.A. Ricks and P.J. Goodhew: The production of rapidly-solidified aluminium powder by ultrasonic gas atomisation. Part I: heat and fluid flow. *Int J Rapid Solid* 1 (1984–85) 59–80.
17. T.W. Clyne, R.A. Ricks and P.J. Goodhew. The production of rapidly-solidified aluminium powder by ultrasonic gas atomisation. Part II: solidification structure *Int J Rapid Solid* 1 (1984–85) 85–101.
18. G.S. Cole: Temperature measurements and fluid flow distributions ahead of solid–liquid interfaces. *Trans Metall Soc AIME* 239 (1967) 1287–1295.
19. G.S. Cole and G.F. Bolling: The importance of natural convection in casting. *Trans Metall Soc AIME* 233 (1965) 1568–1572.
20. A.B. Crowley and J.R. Ockendon: Modelling mushy regions. *App Sci Res* 44 (1987) 1–7.
21. L.A. Crivelli and S.R. Idelsohn: A temperature based finite element solution for phase-change problems. *Int J Num Meth Engng* 23 (1986) 99–119.

22. P.V. Desai, J.T. Berry and C. Kim: Computer simulation of forced and natural convection during filling of a casting. *Trans Am Foundrymen's Soc* 92 (1984) 519–528.
23. P.V. Desai and C. Kim: On convection in liquid metal moulds. In: R.W. Lewis, K. Morgan and B.A. Schrefler (eds) *Numerical Methods in Thermal Problems, Volume II*, Pineridge Press, Swansea (1981) pp 119–129.
23a. P.V. Desai and C. Kim: Heat losses in runner channels. In: *Modelling Casting and Welding Processes II*. The Metallurgical Society of AIME, Warrendale, Pennsylvania (1984) pp 59–66.
24. P.V. Desai and F. Rastegar: Convection in mould cavities. In: *Modelling of Casting and Welding Processes*. The Metallurgical Society of AIME, Warrendale, Pennsylvania (1981) pp 351–360.
25. L. DeSocio and G. Gualtieri: A hyperbolic Stefan problem. *Quart Appl Math* 41 (1983) 253–259.
26. C. Dietsche and U. Müller: Influence of Benard convection on solid–liquid interfaces. *J Fluid Mech* 161 (1985) 249–268.
27. E.R. Evans: Fluidity of molten cast iron. *Foundry Trade J* 128 (1955) 757–763.
28. S.C. Flood and J.D. Hunt: A model of a casting. *Appl Sci Res* 44 (1987) 27–42.
29. R.M. Furzeland: A comparative study of numerical methods for moving boundary problems. *J Int Maths Applics* 26 (1980) 411–429.
30. D.K. Gartling: Finite element analysis of convective heat transfer problems with change of phase. In: K. Morgan, C. Taylor and C.A. Brebbia (eds) *Computer Methods in Fluids*. Pentech Press, Plymouth (1980) pp 257–269.
31. C. Gau and R. Viskanta: Melting and solidification of a metal system in a rectangular cavity. *Int J Heat Mass Transf* 27 (1984) 113–123.
32. G. Gau and R. Viskanta: Effect of natural convection on solidification from above and melting from below of a pure metal. *Int J Heat Mass Transf* 28 (1985) 573–587.
33. N.J. Grant: Engineering properties and applications of rapidly solidified materials. In: P.R. Sahm, J. Jones and C.M. Adam (eds). *Science and Technology of the Undercooled Melt*. Martinus Nijhoff Publishers, Dordrecht (1986) pp 210–228.
34. R. Guenigault and G. Poots: Effects of natural convection on the inward solidification of cylinders. *Int J Heat Mass Transf* 28 (1985) 1229–1231.
35. N.W. Hale and R. Viskanta: Solid–liquid phase-change heat transfer and interface motion in materials cooled from above or below. *Int J Heat Mass Transf* 23 (1980) 283–357.
36. C. Harrison and E. Weinberg: The influence of convection on heat transfer in liquid tin. *Metall Trans B* 16B (1985) 355–357.
37. W.L. Heitz and J.W. Westwater: Extension of the numerical method for melting and freezing problems. *Int J Heat Mass Transf* 13 (1970) 1371–1375.
38. R.N. Hills, D.E. Loper and P.H. Roberts: A thermodynamically consistent model of a mushy zone. *Q J Mech Appl Math* 36 (1983) 505–539.
39. K. Ho and R.D. Pehlke: Mechanisms of heat transfer at a metal–mould interface. *Trans Am Foundrymen's Soc* 92 (1984) 587–598.
40. C.P. Hong, T. Umeda and Y. Kimura: Solidification simulation of shaped castings by the boundary element method and prediction of shrinkage cavity. *Imono* 56 (1984) 758–764.
41. S.C. Huang: Analytical solution for the buoyancy flow during the melting of a vertical semi-infinite region. *Int J Heat Mass Transf* 28 (1985) 1231–1233.
42. J.C.R. Hunt: Turbulence structure in thermal convection and shear-free boundary layers. *J Fluid Mech* 138 (1984) 161–184.
43. J.C.R. Hunt: Turbulent diffusion from sources in complex flows. *Ann Rev Fluid Mech* 17 (1985) 447–485.

44. J.D. Hunt and D.G. McCartney: Numerical finite difference model for steady state array growth. *Appl Sci Res* 44 (1987) 9–26.
45. H.E. Huppert: From multi-branched snowflakes to precious minerals. *Nature* 323 (1986) 202–203.
46. H.E. Huppert: The intrusion of fluid mechanics into geology. *J Fluid Mech* 173 (1986) 557–594.
47. H.E. Huppert and J.S. Turner: A laboratory model of a replenished magma chamber. *Earth Planet Sci Lett* 54 (1981) 144–142.
48. H.E. Huppert and M.G. Worster: Dynamic solidification of a binary melt. *Nature* 314 (1985) 703–707.
49. I. Imafuku and K. Chijiiwa: A mathematical model for shrinkage cavity prediction in steel castings. *Trans Am Foundrymen's Soc* 91 (1985) 527–520.
50. J. Isaac, G.P. Reddy and G.K. Sharma: Variations of heat transfer coefficients during solidification of castings in metal moulds. *Brit Foundrymen* 78 (1985) 465–468.
51. S.T. Johansen, F. Boysen and W.H. Ayers: Mathematical modelling of bubble driven flows in metallurgical processes. *Appl Sci Res* 44 (1987) 197–207.
52. P.B. Johns: A simple explicit and unconditionally stable numerical routine for the solution of the diffusion equation. *Int J Num Meth Engng* 11 (1979) 1307–1328.
53. P.B. Johns and G. Butler: The consistency and accuracy of the TLM method for diffusion and its relationship to existing methods. *Int J Num Meth Engng* 19 (1983) 1549–1554.
54. E.W. Jones, W.M. Sleigelmann and G.P. Wachtell: Heat transfer from molten metals to sand mould runners. *Trans Am Foundrymen's Soc* 71 (1963) 817–825.
55. H. Jones: *Rapid Solidification of Metals and Alloys*. Institution of Metallurgists, London (1982).
56. E.G. Josberger and S. Martin: A laboratory and theoretical study of the boundary layer adjacent to a vertical melting ice wall in salt water. *J Fluid Mech* 11 (1981) 439–473.
57. L. Katgerman: Computer simulation of cellular growth during rapid solidification of binary aluminium alloys. In: *Modelling of Casting and Welding Processes II*. The Metallurgical Society of AIME, Warrendale, Pennsylvania (1984) pp 135–143.
58. L. Kategerman: Effect of process conditions during melt spinning on solidification morphology of aluminium alloys. In: S. Steeb and H. Warliment (eds). *Rapidly Quenched Metals*. Elsevier Science Publishers BV, The Hague (1985) pp 819–822.
59. B.H. Kear, B.C. Giessen and M. Cohen (eds): *Rapidly-solidified Amorphous and Crystalline Alloys*. North-Holland, New York (1982).
60. M. Kölling and U. Grigull: A mathematical model to calculate the fluidity of pure metals. In: D.B. Spalding and N.H. Afgan (eds). *Heat and Mass Transfer in Metallurgical Systems*. McGraw-Hill, Washington (1981) pp 329–340.
61. W. Kunda and G. Poots: A mathematical model of ladle stirring by bath agitation in steel making. *Appli Sci Res* 44 (1987) 209–224.
62. R.W. Lewis, K. Morgan and P.M. Roberts: Determination of thermal stresses in solidification problems. In: J.F.T. Pittman, O.C. Zienkiewicz, R.D. Wood and J.M. Alexander (eds). *Numerical Analysis of Forming Processes*. John Wiley and Sons Limited, Chichester (1984) pp 405–431.
63. R.W. Lewis and P.M. Roberts: Finite element simulation of solidification problems. *Appl Sci Res* 44 (1987) 61–92.
64. M. Matsuda and M. Ohmi: Heat transfer analysis and fluidity of flowing metal in a cylindrical mould cavity. *AFS Int Cast Metals J* 6 (1981) 18–27.
65. T. Matsumiya and M.C. Flemings: Modelling of continuous strip production by rheocasting. *Metall Trans B* 12B (1981) 17–31.

66. G. Menges, V. Masberg, B. Gesenhues and C. Berry: Numerical simulation of three-dimensional non-Newtonian flow in thermoplastics extrusion dies with finite element methods. In: J.F.T. Pittman, O.C. Zienkiewicz, R.D. Wood and J.M. Alexander (eds). *Numerical Analysis of Forming Processes*. John Wiley and Sons, Chichester (1984) pp 307–350.
67. J.L. Meyer and F. Durand: Analysis of the transient effects of convection during solidification with or without electromagnetic stirring. In: *Modelling of Casting and Welding Processes II*. The Metallurgical Society of AIME, Warrendale, Pennsylvania (1984) 179–197.
68. K. Morgan: A numerical analysis of freezing and melting with convection. *Comp Meth Appl Mech Engng* 28 (1981) 275–284.
69. Y. Nagasaka, J. Ohnaka and T. Fukusako: Effect of heat transfer during pouring on solidification of steel plate castings. *Imono* 56 (1984) 22–27.
70. J.F.T. Pittman and A. Nakazawa: Finite element analysis of polymer processing operations. In: J.F.T. Pittman, O.C. Zienkiewicz, R.D. Wood and J.M. Alexander (eds). *Numerical Analysis of Forming Processes*. John Wiley and Sons, Chichester (1984) pp 165–218.
71. J. Pullins and M.K. Walther: Simulation for designing metal-casting moulds. *Comp Aided Engr* (1984) 62–68.
72. N. Ramachandran, J.P. Gupta and Y. Jahuria: Thermal and fluid flow effects during solidification in a rectangular enclosure. *Int J Heat Mass Transf* 25 (1982) 187–194.
73. T. Robertson, P. Moore and R.J. Hawkins: Computational flow model as aid to solution of fluid flow problems in the steel industry. *Ironmaking and Steelmaking* 13 (1986) 195–203.
74. S. Rogers, L. Katgerman, P.G. Enright and N.A. Darby: Modelling of liquid–liquid metal mixing. *Appl Sci Res* 44 (1987) 175–196.
75. F. Rosenberger: *Fundamentals of Crystal Growth* Vol 1. Springer, Berlin (1979).
76. R.W. Ruddle: *The Solidification of Castings*. Institute of Metals Monograph and Reports Series No 7 (2nd ed) (1957).
77. M. Sadd and J. Didlake: Non-fourier melting of a semi-infinite solid. *Trans AIME J Heat Trans* 99 (1979) 25–28.
78. P.R. Sahm, H. Jones and C.M. Adam: *Science and Technology of the Undercooled Melt*. Martinus Nijhoff Publishers, Dordrecht (1986).
79. B.L. Sawford and J.C.R. Hunt: Effects of turbulence structure, molecular diffusion and source size on scalar fluctuations in homogeneous turbulence. *J Fluid Mech* 165 (1986) 373–400.
80. W. Schneider: A local analysis of solidification in horizontal continuous casting. *Arch Eisenhüttenwes* 54 (1983) 487–490.
81. A. Schröder: Fluid flow considerations in the filling of lost moulds with open feeders. *Giessereiforschung* 37 (1985) 65–79.
82. H.J. Schulze, P.M. Beckett, J.A. Howarth and G. Poots: Analytical and numerical solutions to two-dimensional moving interface problems with applications to the solidification of killed steel ingots. *Proc R Soc Lond A* 385 (1983) 313–343.
83. T.J. Smith: Adaptive mesh schemes for computational fluid dynamics. *CFD News* 3/84 (1984) 11–15.
84. T.J. Smith: A basis for the control of the casting process. *GST Report No*: PS1940.1 GST 33/1.00.
85. T.J. Smith, A.F.A. Hoadley and D.M. Scott: On the sensitivity of numerical simulations of solidification to the physical properties of the melt and the mould. *Appl Sci Res* 44 (1987) 93–110.

86. T.J. Smith, A.F.A. Hoadley and D.M. Scott: The incorporation of natural convection effects in solidification simulation. In: R.W. Lewis and K. Morgan (eds) *Numerical Methods in Thermal Problems*, Vol V, Pineridge Press, Swansea (1987) in the press.
87. T.J. Smith and D.B. Welbourn: The integration of geometric modelling with finite element analysis for the computer-aided design of castings. *Appl Sci Res* 44 (1987) 139–160.
88. A.D. Solomon, V. Alexiades, D.G. Wilson and J. Drake: On the formulation of hyperbolic Stefan problems. *Quart Appl Math* XLIII (1985) 295–304.
89. N.D. Steward: Semi-solid casting (abstract). In: T.J. Smith (ed), *Mixing, Stirring and Solidification in Metallurgical Processes*. University of Cambridge, Department of Engineering (1985) p 7.
90. R.A. Stoehr and W.S. Hwang: Modelling the flow of molten metal having a free surface during entry into moulds. In: *Modelling of Casting and Welding Processes II*. The Metallurgical Society of AIME, Warrendale, Pennsylvania (1984) pp 47–58.
91. J. Szekely and A.S. Jassal: An experimental and analytical study of the solidification of a binary dendritic system. *Metall Trans B* 9B (1978) 389–398.
92. J.S. Turner: *Buoyancy Effects in Fluids*, 2nd Edition. Cambridge University Press, Cambridge (1979).
93. J.S. Turner, H.E. Huppert and R.S.J. Sparks: Komatiites II: experimental and theoretical investigations of post-emplacement cooling and crystallization. *J Petrol* 27 (1986) 397–437.
94. V.R. Voller: Implicit finite-difference solutions of the enthalpy formulation of Stefan problems. *I M A J Numer Anal* 5 (1985) 201–214.
95. V.R. Voller and M. Cross: Accurate solutions of moving boundary problems using the enthalpy method. *Int J Heat Mass Transf* 24 (1981) 545–556.
96. V.R. Voller and M. Cross: An explicit method to track a moving phase change front. *Int J Heat Mass Transf* 26 (1983) 147–150.
97. V.R. Voller and M. Cross: Applications of control volume enthalpy methods in the solution of Stefan problems. In: R.W. Lewis, K. Morgan, J.A. Johnson and W.R. Smith. *Computational Techniques in Heat Transfer*, Vol 1, Pineridge Press, Swansea (1985) pp 245–275.
98. V.R. Voller, N. Markatos and M. Cross: Techniques for accounting for the moving interface in convection-diffusion phase change. In: R.W. Lewis and K. Morgan (eds). *Numerical Methods in Thermal Problems*, Vol IV, Pineridge Press, Swansea (1985) pp 595–609.
99. J.E. Welsh, F.H. Harlow, J.P. Shannon and B.J. Bally: The MAC method, a computing technique for solving viscous, incompressible, transcient fluid flow problems involving free surfaces. *Techn Rept LA-3425* Los-Alamos Scientific Laboratory (1966).
100. R.E. White: An enthalpy formulation of the Stefan problem. *SIAM J Numer Anal* 19 (1982) 1129–1157.
101. K. Wollhover, Ch. Körber, M.W. Scheiwe and U. Hartmann: Unidirectional freezing of binary aqueous solutions: an analysis of transient diffusion of heat and mass. *Int J Heat Mass Transf* 28 (1985) 761–770.
102. K. Wollhover, M.W. Scheine, U. Hartmann and Ch. Körber: On morphological stability of planar phase boundaries during unidirectional transient solidification of binary aqueous solutions. *Int J Heat Mass Transf* 28 (1985) 897–902.
103. M.G. Worster: Solidification of an alloy from a cooled boundary. *J Fluid Mech* 167 (1986) 481–501.
104. J. Yoo and B. Rubinsky: A finite element method for the study of solidification processes in the presence of natural convection. *Int J Numer Meth Engng* 23 (1986) 1785–1805.
105. O.C. Zienkiewicz: Flow formulation for numerical solution of forming processes. In: J.F.T. Pittman, O.C. Zienkiewicz, R.D. Wood and J.M. Alexander (eds). *Numerical Analysis of Forming Processes*. John Wiley and Sons, Chichester (1984) pp 1–44.

Subject index

Author index